Hydraulic Conductivity - Issues, Determination and Applications

Hydraulic Conductivity - Issues, Determination and Applications

Editor

Stefanos Palmer

scitus
academics

Hydraulic Conductivity - Issues, Determination and Applications
Edited by **Stefanos Palmer**

ISBN: 978-1-68117-321-4
Library of Congress Control Number: 2016937606

www.scitusacademics.com
Box No. 4766, 616 Corporate Way,
Suite 2, Valley Cottage,
NY 10989

Printed in United States of America on Acid Free Paper

Preface

Due to the rapid development of science and technology and the increasing extraction and contamination of groundwater, many new theories, methodologies and applications have emerged in recent decades. Water can move through the soil like saturated flow, unsaturated flow, or vapor flow. Saturated flow occurs when the soil pores are completely water-filled. It depends on the permeability, level of saturation and density and viscosity of the material. Unsaturated flow takes place when the larger pores in the ground are filled with air, leaving only the smaller pores to hold and transmit water. The vapor flow occurs when vapor pressure differences develop in relatively dry soils. The steam migrates from a high vapor pressure area to a low vapor pressure zone. Hydraulic conductivity is a property of the soil that describes how easily the soil pores allow water (not steam) to move. It depends on the type of soil, porosity and configuration of soil pores. Hydraulic conductivity is probably the most important hydrogeological parameter, as it governs the flow of fluids and the migration of contaminants under the ground, in particular in soils and aquifers, together with other parameter. This book presents a comprehensive review of new measurements and numerical techniques for hydraulic conductivity.

Table of Contents

1
CHAPTER

HYDRAULIC CONDUCTIVITY AND CONTRIBUTION OF AQUAPORINS TO WATER UPTAKE IN ROOTS OF FOUR SUNFLOWER GENOTYPES

Afifuddin Latif Adiredjo1,2, Olivier Navaud3, Philippe Grieu1 and Thierry Lamaze3*

(1)Université de Toulouse, INP - ENSAT, UMR 1248 AGIR (INPT-INRA)

(2)Brawijaya University, Faculty of Agriculture, Department of Agronomy, Plant Breeding Laboratory

(3)Université de Toulouse, UPS - Toulouse III, UMR 5126 CESBIO

ABSTRACT

Background

This article evaluates the potential of intraspecific variation for whole-root hydraulic properties in sunflower. We investigated genotypic differences related to root water transport in four genotypes selected because of their differing water use efficiency (JAC doi: 10.1111/jac.12079. 2014). We used a pressure-flux approach to characterize hydraulic conductance (L_0) which reflects the overall water uptake capacity of the roots and hydraulic conductivity (Lp_r) which represents the root intrinsic water permeability on an area basis. The contribution of aquaporins (AQPs) to water uptake was explored using mercuric chloride ($HgCl_2$), a general AQP blocker.

Results

There were considerable variations in root morphology between genotypes. Mean values of L_0 and Lp_r showed significant variation (above 60% in both cases) between recombinant inbred lines in control plants. Pressure-induced sap flow was strongly inhibited by $HgCl_2$ treatment in all genotypes (more than 50%) and contribution of AQPs to hydraulic conductivity varied between genotypes. Treated root systems displayed markedly different L_0 values between genotypes whereas Lp_r values were similar.

Conclusions

Our analysis points to marked differences between genotypes in the intrinsic aquaporin-dependent path (Lp_r in control plants) but not in the intrinsic AQP-independent paths (Lp_r in $HgCl_2$ treated plants). Overall, root anatomy was a major determinant of water transport properties of the whole organ and can compensate for a low AQP contribution. Hydraulic properties of root tissues and organs might have to be taken into account for plant breeding since they appear to play a key role in sunflower water balance and water use efficiency.

Keywords: Sunflower Aquaporins Root Hydraulic conductivity

BACKGROUND

Terrestrial plants are dependent on essential leaf physiological processes such as a continuous supply with water since photosynthesis cannot be dissociated from transpiration. Water balance at the whole plant level should be regulated by coupled responses between the above-ground and below-ground parts (Shimizu et al. 2005). In the soil-plant-atmosphere continuum, the root offers the second largest resistance to water transport after the stomata (Steudle et al. 1987). Thus, gaining information on the hydraulic properties of roots might be a key step for understanding whole-plant water relations and selection of water stress-resistant species or genotypes and therefore plant breeding (Sutka et al. 2011).

In roots in which the xylem vessels are fully developed, the resistance to water transport occurs radially (Steudle and Peterson 1998; Ruggiero et al. 2003). Radial transport in roots crosses both the cell-to-cell pathway and the apoplastic pathway. The cell-to-cell route is composed of the symplastic (through the plasmodesmata) and the transcellular (involving crossing of membranes) paths. The apoplastic pathway is usually considered to have the least hydraulic resistance and is often considered to be the main route (Heinen et al. 2009). However, the presence of lignified or suberized cell walls (casparian strips in root endodermis) which constitute apoplastic barriers forces water to cross cell membranes (Shimizu et al. 2005). Several studies have attributed an important role to the cell-to-cell path. Water movement through cell membranes is facilitated by water channels, called "aquaporins" (AQPs) (Maurel 2007). AQPs are integral membrane proteins that increase the permeability of membranes to water as well as other small molecules such as CO_2, glycerol and boron (Shimizu et al. 2005; Chaumont et al. 2005). AQP proteins contain thiol groups that are sensitive to $HgCl_2$ (Savage and Stroud 2007). Assuming that mercurial inhibition of water transport occurs via the inhibition of AQPs, the strength of inhibition may indicate the extent to which the cell-to-cell (transcellular) water movement (involving water passing through membranes) is involved in the radial transport of water across the root. Therefore, to divide radial water transport in roots into cell-to-cell and apoplastic pathways, $HgCl_2$ has often been used as a specific AQP inhibitor in crops, herbs and trees (Maggio and Joly 1995; Carvajal et al. 1996; Tazawa et al. 1997; Zhang and Tyerman 1999; Wan and Zwiazek 1999; North et al. 2004; Kamaluddin and Zwiazek 2001; Shimizu et al. 2005; Sutka et al. 2011).

Intraspecific root water transport has so far been compared in only a small number of species: rice, maize, grapevine and *Arabidopsis* (Sutka et al. 2011). Sunflower is an economically important crop consumed worldwide. Although it is considered to be relatively tolerant to water stress, sunflower production can be greatly affected by drought (Pasda and Diepenbrock 1990; Grieu et al. 2008). Indeed, although sunflower extracts water efficiently and then conducts it within the plant, rates of leaf transpiration can reach very high values: up to 22 mmol H_2O m^{-2} s^{-1} (Rawson et al. 1980). It has been shown that AQPs play a role in the sunflower response to drought in both the leaf and the root (Ouvrard et al. 1996; Sarda et al. 1997, 1999). Information on the hydraulic properties of roots might be important for plant breeding. Thus, the aim of the present work was to evaluate the variations in the root hydraulic properties of four sunflower genotypes selected because of their differing whole-plant water relations under well-watered conditions (Adiredjo et al. 2014). We used pressure-induced flow through root systems since the method has been widely employed to measure the hydraulic properties of roots in various plant species. The contribution of the AQP-dependent pathway (cell-to-cell route) to water transport was characterized using mercuric chloride (mercury) as an inhibitor.

METHODS

Plant source

Four recombinant inbred lines (RILs) of sunflower (*Helianthus annuus* L.) from the collection of the Laboratory of Plant-Microbe Interactions (LIPM), INRA of Toulouse, France, were used in the experiments, namely: RIL 043, RIL 127, RIL 149 and RIL 200. The four RILs are lines from the INEDI RIL population. This population was obtained by self pollination to at least F8 from a cross between XRQ and PSC8 (Vincourt et al. 2012). These parental lines have different drought tolerance behavior (Rengel et al. 2012). The four RILs were chosen on the basis of their differing water use efficiency (WUE) response under well-watered conditions (Adiredjo et al. 2014).

Plant culture, experimental design and root analysis

The plants were grown in a growth chamber (25°C/20°C in day/night) under 14 h of light (200 μmol m^{-2} s^{-1} photosynthetically active radiation at leaf level, Fluora, L 58 W/77, Germany) and 50 ± 5% RH in 250 mL pots. They were arranged in a randomized complete design with four RILs. To minimize the effects of heterogeneity within the growth chamber, the pots were rotated every week. Root hydraulics parameters (conductance, conductivity and contribution of AQPs) were measured by determining pressure-induced sap rates in six-week-old sunflower seedlings, when above-ground parts were 15–20 cm high. The experiment was repeated three times.

Seedlings were grown in 250 mL glass pots filled with sand which could be easily washed and saturated with solution and then introduced into the pressure chamber. Thus the root system did not have to be excavated before the pressure-induced flow experiment and thereafter, excavation of the roots from the cultivated sand substrate could be gently achieved under water for determination of root characteristics. Upon completion of the exudation experiments, root fresh weight was determined. Then, the properties i.e. root length, root surface area and volume of fine roots of each root system were determined with an image analyzer WinRHIZO 2007d (Regent Instruments, Quebec, Canada). Fine roots are the smallest diameter class (0 – 0.5 mm). Finally, root dry weight was measured.

Measurement of root sap flow and $HgCl_2$ treatment

For pressure-flow experiments, upon harvest, pots were washed three times (3 × 50 mL) to saturation with water for the control treatment and $HgCl_2$ solution (500 μM) for the inhibited treatment. Saturation of the pots allowed us to determine root conductivity since under non-limiting soil moisture, plant resistance exceeds soil resistance (in wet soil, the bulk soil potential is close to 0 MPa, Ruggiero et al. 1999; 2003). Following this washing, the above-ground part was cut off with a razor blade just below the cotyledonary leaves (40–50 mm from the base). Pots with whole root systems were placed in a stainless steel pressure chamber (Soil Moisture Equipment Corp., Santa Barbara, CA, USA, Maggio and Joly 1995). Excised stems were sealed into the lid of the chamber through a silicone gasket so that part of the stem protruded and chamber pressure was gradually increased. Water expressed from each cut stem was collected using an Eppendorf tube containing dry cotton wool. The amount of sap was determined by weighing the tube before and after collection. The sap flow (Jv), expressed as the quantity of water exuded from the cut stems, was monitored every 5 min for at least 45 min after it had reached a constant rate (reached in less than 25 min). Aliquots of expressed sap were collected from each root system for later analysis of K^+ content.

In preliminary experiments (pressure-flux curves), five pressures were applied in sequentially increasing order (0.1, 0.2, 0.3, 0.4 and 0.5 MPa) to whole root systems. Flow values were logged for 25 min at each pressure, allowing a 5-min equilibration period between pressures. Because the regression of sap flow on applied pressure was linear for all genotypes (data not shown), sap flows in both control and $HgCl_2$-treated plants were determined at a constant 0.3 MPa pressure (Shimizu et al. 2005). Pressure was gradually increased up to 0.3 MPa in the chamber, and was then held constant during the measurements (flow reached a steady-state value in about 20 min). The sap flow (Jv) was then used to define (i) the whole root hydraulic conductance (L_0) calculated as the sap flow rate per unit pressure (μL s^{-1} MPa^{-1}) and (ii) the root hydraulic conductivity (Lp_r) calculated as the sap flow rate per unit root surface area and per unit pressure (m s^{-1} MPa^{-1}).

We performed preliminary experiments to determine the more suitable of two methods to treat the plants with the $HgCl_2$ inhibitor. First, after measurement of the pressure that induced sap flow in untreated roots, the pressure was released slowly before opening the chamber, the cut stem was removed from the gasket and the pot was flushed with $HgCl_2$ solution. The stem was once more sealed in the gasket secured to the lid and the pressure in the chamber again set to 0.3 MPa. However, this was a delicate procedure that often caused damage to the stem. Second, measurement of the pressure that induced sap flow was done on distinct root systems, untreated and treated (control and $HgCl_2$). Therefore, calculation of the depressive effect of $HgCl_2$ on Jv was finally achieved by considering the second method rather than the first since both gave similar results. Experiments were performed three times using three plants per RIL each time.

After flushing the pot with $HgCl_2$, maximal inhibition was achieved in less than 40 min. The reversibility of inhibition by $HgCl_2$ of pressure-induced sap flow was evaluated by flushing of the pot with 3×50 mL 10 mM mercaptoethanol (ME) . The decrease brought about by $HgCl_2$ was reversed by *ca* 90% on subsequent treatment for 30 min with ME. Some sap samples collected from control or $HgCl_2$-treated de-topped plants were diluted (*ca* 40 µL of sap +1 mL H_2O) and injected into a Dionex-D-100 ion chromatograph (USA). K^+ flux through the xylem was calculated as the product of the sap flux and concentration of K^+ in the sap.

Statistics

Data were analyzed with the PASW statistics 18 (IBM, New York, USA) package. We used the Least Significant Difference (LSD) test ($P < 0.05$) to make post-hoc comparisons between all means. Percentages of inhibition by $HgCl_2$ were calculated for each individual root system, and mean values and standard deviations were calculated for the three experiments. In total, the response of nine $HgCl_2$-treated plants was compared with those of untreated plants for each genotype.

Results

Size of root systems

Morphological parameters of sunflower root systems are presented in Table 1. RIL 043 and RIL 127 had the largest and the smallest root systems for every parameter (fresh or dry mass, surface, length and fine root volume), respectively. Differences reached up to 100% of the values, demonstrating considerable variation in root morphology between genotypes. RIL 149 and RIL 200 showed intermediate root characteristics with the ranking: RIL149 > RIL200.

Table 1. Root morphological parameters and K ⁺ flux into the xylem of four sunflower genotypes grown in a growth chamber under controlled and well-watered conditions

Genotype	*N*	RFW (g)	RDW (g)	RL (cm)	RS (cm^2)	RV (cm^3)^a	K^+ flux (µmol h^{-1} g^{-1})
RIL043	23	7.92 ±2.08	0.29 ± 0.06	1669 ± 331	210 ± 35	2.39 ± 0.74	5.08 ± 0.10 (92.6%)
RIL127	19	3.16±0.68	0.17 ± 0.05	1191 ± 211	134 ± 31	1.05 ± 0.43	2.11 ± 0.55 (121.6%)
RIL149	20	5.23±0.95	0.24 ± 0.07	1271 ± 264	161 ± 30	1.63 ± 0.42	1.76 ± 0.11 (103.3%)
RIL200	23	7.21±1.86	0.27 ± 0.08	1247 ± 457	152 ± 40	2.01 ± 0.69	6.72 ± 1.23 (116.2%)

N: number of roots, RFW: root fresh weight, RDW: root dry weight, RL: root length, RS: root surface, RV: fine root volume.
The data represent all the roots of the non-inhibited plants (control) and inhibited plants ($HgCl_2$). Values are means and standard deviations.
[a]Fine roots are the smallest diameter class (0 – 0.5 mm) determined by the WinRHIZO analysis.
[b]K^+ fluxes are for $HgCl_2$ treated plants but are also expressed in percentage of the controls (values in parenthesis).

Variation of root hydraulic properties

Figure 1 shows a representative time course of cumulative water movement of a root system before (control) and after treatment with $HgCl_2$ for the four sunflower RILs. Sap flow (Jv) from untreated or treated root systems remained virtually constant throughout the 40 min measurement period. The ranking of RILs for Jv in untreated root systems was RIL149 > RIL43 = RIL200 > RIL127. Pressure-induced sap flow was almost twice as high in RIL 149 as in RIL 127. The sap flow was strongly inhibited by $HgCl_2$ treatment in all genotypes (more than 50%).

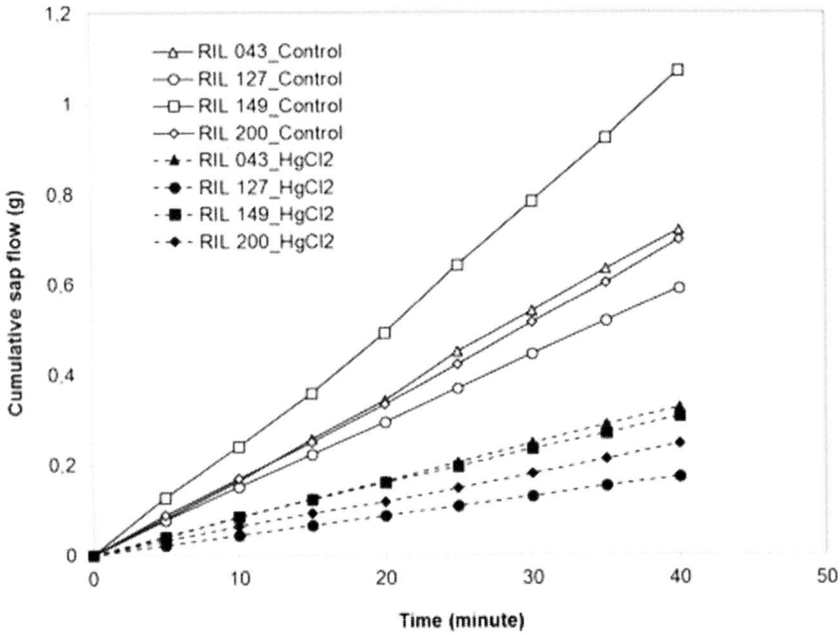

Figure 1. The figure shows an example of cumulative sap flow for one root system per RIL and per treatment (open symbols for control and solid symbols for $HgCl_2$) of RIL 043, RIL 127, RIL 149, RIL 200. The cumulative data were obtained from eight successive, five minutes apart.

Mean values of L_0 and Lp_r showed significant variation between the four RILs. The values ranged from 0.7 to 1.2 µL s^{-1} MPa^{-1} and 5.10^{-8} to 8.10^{-8} m s^{-1} MPa^{-1}, respectively. The ranking of RILs for L_0 was RIL 149 > RIL 043 > =RIL 200 > RIL 127 but the ranking was changed for Lp_r:

RIL 149 > RIL 200 > RIL 127 > RIL 043. Differences between extreme values were above 60% in both cases.

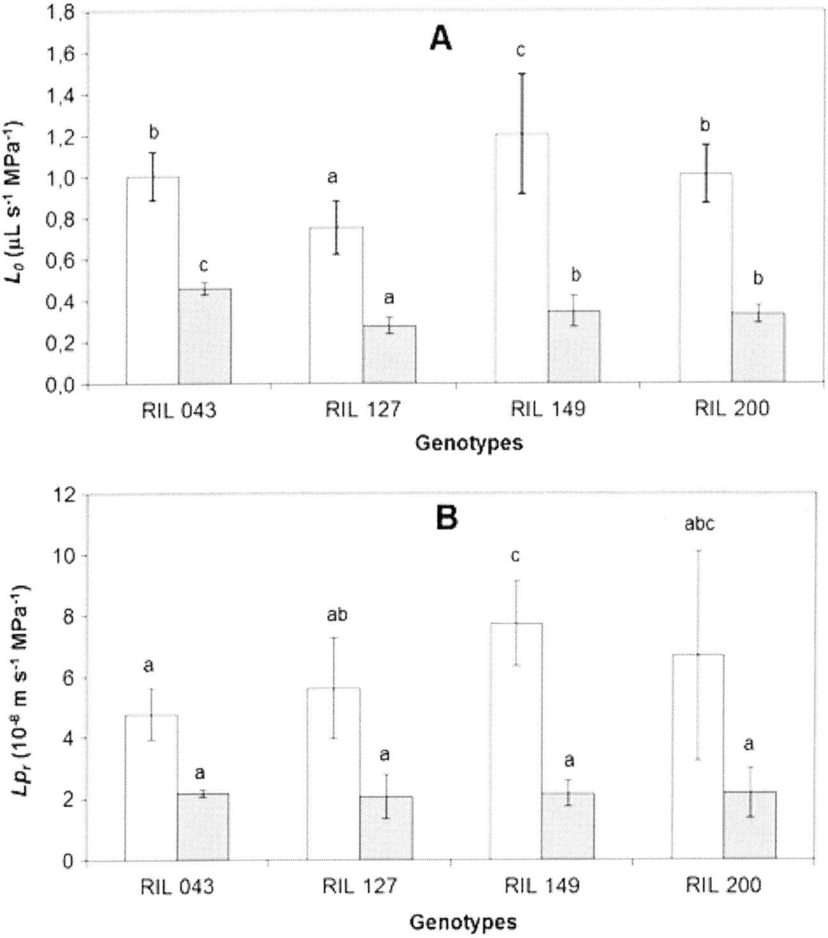

Figure 2. Means of L $_0$ (A) and Lp $_r$ (B) of RIL 043, RIL 127, RIL 149, RIL 200. The white bars are non-inhibited plants (control) and the black bars are inhibited plants ($HgCl_2$). Each value is the mean of nine plants ± standard deviation. Means within a treatment without a common letter are significantly different by $LSD_{0.05}$ test.

Contribution of AQPs to water uptake

In our experiment, the contribution of AQPs to sap flow was explored using mercuric chloride. L_0 and Lp_r fell to 30–40% of the control value and differences appeared between RILs (Figure 2A and B). $HgCl_2$-treated root systems displayed markedly different L_0 values between genotypes (Figure 2A) which ranked as follows: RIL 043 > RIL 149 > =RIL 200 > RIL 127 (L_0 for RIL 043 was about 70% higher than for RIL 127). By contrast, Lp_r values were similar for all four RILs following $HgCl_2$ treatment (Figure 2B).

The contribution of AQPs to Lp_r (AQP involvement) expressed as the relative decrease in Lp_r induced by $HgCl_2$ treatment was the highest in RIL 149 (73%) and the lowest in RIL 043 (55%) while other RILs displayed an intermediate contribution (Figure 3).

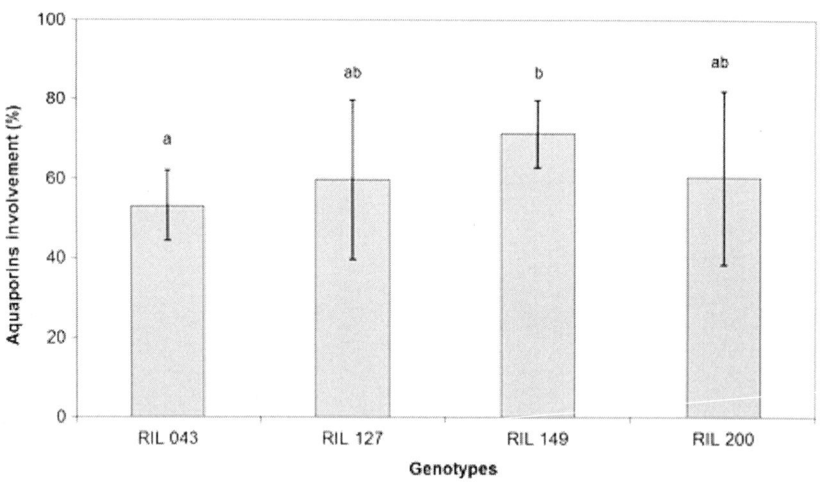

Figure 3. Means of involvement of aquaporins of RIL 043, RIL 127, RIL 149, RIL 200. Each value is the mean of nine plants ± standard deviation. Means without a common letter are significantly different by $LSD_{0.05}$ test.

The flux of K^+ into the xylem was not significantly affected by the presence of $HgCl_2$ corresponding to 92.6%, 121.6%, 103.3% and 116.2% of the controls for RIL 043, RIL 127, RIL 149 and RIL 200, respectively (Table 1).

DISCUSSION

In this study, the values of Lp_r for sunflower root systems ranged from 5.10^{-8} to 8.10^{-8} m s^{-1} MPa^{-1} and were within the range of values reported for other species (Liu et al. 2009; Sutka et al. 2011, Sakurai-Ishikawa et al. 2011), although rather at the lower end of the *Lpr* scale reported for roots of annual (crop) plants. Our values for sunflower root hydraulic conductivity were very similar to those reported for roots of 20-d-old sunflower plants (*ca* 600 µL h^{-1} g^{-1} root fresh weight MPa^{-1}, see Figure 1 and Table 1, Quintero et al. 1999) but Alfalfa had Lp_r values 10 times higher than sunflower (Li et al. 2007). The pressure that induces flow through root systems has been considered by some authors (Li et al. 2007; Liu et al. 2009) to be inappropriate to characterize "absolute" hydraulic values, because the externally applied pressure can induce flow through pathways external to the root system. Indeed, pressure chamber experiments give higher values of Lp_r than other methods using root pressure probes: resistance may vary according to the nature of the driving force for water movement (osmotic versus hydraulic) and the flow rate (Liu et al. 2009). It has been shown by Vandeleur et al. (2014) that shoot manipulation affected root hydraulic characteristics in grapevine, soybean and maize. Here, we used detopped plants to measure root hydraulics, thus a note of caution is warranted regarding the absolute values reported. However, the purpose of the present work was to compare the hydraulics of several sunflower RILs differing in their whole-plant water relationships. Root *Lp* has been repeatedly shown to change with the volume flow rate through the root system (Sakurai-Ishikawa et al. 2011, Laur and Hacke 2013). However, the flow rates induced here through pressure gradient across isolated root systems were similar to transpirational water loss rates of intact plants of the same age (Adiredjo et al. 2014).

Mercuric chloride-induced reversible inhibition of root water flow is consistent with the presence of a protein-mediated path for trans-membrane sap flow in the sunflower root. To avoid non-specific effects, concentrations have to be as low as possible and exposure time as short as possible. Coskun et al. (2012) recommended caution when using aquaporin inhibitors including Hg^{2+}. They showed membrane damage resulting from 500 µM Hg^{2+}. Here we used the same relatively high $HgCl_2$ concentration (500 µM). However, this value was similar to concentrations used in previous studies on whole root systems (Maggio and Joly 1995; Peyrano et al. 1997; Shimizu et al. 2005; Ruggiero et al. 2007). In addition, considering that we worked with sand (in which root excretion of organic compounds creates an organic matrix) and not in a hydroponics set-up, the effective concentration in the root zone was probably far below 500 µM due to immobilization of part of the Hg^{2+} by the system (Ruggiero et al. 2007). The concentration used in this study was chosen from the preliminary dose response curves performed to identify a threshold concentration that had a marked effect on sap flow (for instance, 50 μM $HgCl_2$ did not induce any depressive effect on sap flow) but that did not cause apparent irreversible toxicity effects. Indeed, the relationship between sap flow and applied pressure was highly linear suggesting that the Hg treatment did not cause broadly deleterious changes in root function during the time course of the pressure flow procedure (Maggio and Joly 1995). In addition, there was no significant difference between control and $HgCl_2$-treated roots in the amount of K^+ recovered in the xylem exudates delivered through whole-root systems demonstrating that the Hg^{2+} concentration and exposure durations used here did not poison root cells in a way that might cause them to become leaky to ions (Maggio and Joly 1995). Another convincing argument concerning the lack of general toxicity is the reversal of mercuric chloride inhibition by the scavenger 2-mercaptoethanol which is assumed to remove Hg from membranes of treated roots (Barrowclough et al. 2000). Inhibition of sap flow by $HgCl_2$ was reversed by *ca* 90% following rinsing of the root system in a mercaptoethanol solution (10 mM). In addition, sap continued to be spontaneously expressed from cut stems of excised roots several hours after mercury application demonstrating the generation of strong root pressure in the treated roots. Altogether these results indicate that $HgCl_2$ did not reduce sap flow by a general inhibition of root metabolism, but rather by a direct effect on AQPs.

Leaves need to be continuously supplied with water and carbon dioxide to fulfill their photosynthetic function. The water transport capacity of the root (L_0, root hydraulic conductance) is thus a key physiological parameter for whole-plant function since it determines the interplay between sap flow intensity and water potential gradients between soil and leaves. Differences in whole L_0 reached a high value of 60% between sunflower genotypes. L_0 was highest for RIL 149 and lowest for RIL 127 while RIL 043 and RIL 200 had similar intermediate values. L_0 reflects the overall water uptake capacity of the root and results from both the root exchange surface area and its intrinsic water transport capacity (Lp_r). Lp_r had a different ranking from L_0, with the highest value for RIL 149 (as observed for L_0) but the lowest for RIL 043 (whereas it was in second position for L_0). Lp_r in RIL 049 was found to be 70% higher than in RIL 043. Lp_r was expressed in relation to the whole root area, assuming that outer cell layers were consistently representative of the hydraulic properties of all the segments of the root system (Sutka et al. 2011). We also checked that when Lp_r was expressed in relation to root length or fine root volume, a similar difference between RIL 149 and RIL 043 was observed (see Table 1). Thus, genotype (RIL 043) can display both high whole root capacity (L_0) and small intrinsic root capacity (Lp_r). Sutka et al. (2011) described a substantial (2-fold) genetic variation in *Lpr*, establishing that *Arabidopsis* root hydraulic properties are far from uniform between natural accessions. We show here that large variations in root hydraulics also appear between sunflower genotypes.

Lp_r was sensitive to brief treatment with $HgCl_2$. This allows Lp_r to be divided into two components: cell to cell and apoplastic pathways (ignoring the dilution-diffusion process across the double layer of membrane lipids which is not sensitive to mercury). The Lp_r of apoplasmic pathways, i.e. Lp_r measured in $HgCl_2$, was identical for all genotypes. In other words, the conductance of the AQP-independent pathway (on an area basis) was similar for all RILs. Tissue mass, organization and/or cell wall structure (suberization of apoplastic barriers usually associated with root maturation which reduces water uptake capacity) may affect intrinsic root hydraulics. However, Sutka et al. (2011) reported that *Arabidopsis* accessions did not show any clear link between root suberization and the hydraulic conductivity of the AQP-independent path. In the present work, although sunflower genotypes displayed evident variations in root anatomy, no variations were observed in "intrinsic" apoplastic conductivity (Lp_r in $HgCl_2$ treated roots). This suggests that the cell-to-cell pathway (aquaporin-dependent) was the major determinant of the "intrinsic" water transport properties of the organ (Lp_r in control plants),RIL 149 showing the highest involvement of AQPs (72%) and RIL 043 the lowest (55%). The relative contribution of AQPs to root conductivity (average 60% in the present experiment with sunflower) was similar to other estimates obtained in herbaceous species (Maggio and Joly 1995; Tazawa et al. 1997; Carvajal et al. 1999; Barrowclough et al. 2000; Shimizu et al. 2005; Sutka et al. 2011; Ruggiero et al. 2007) confirming that pathways not involving AQPs can make a significant contribution to Lp_r (around 40%).

In the present study, sunflower genotypes were selected because of their contrasting behaviour with respect to water under well-watered conditions (Adiredjo et al. 2014). A variety of hydraulic profiles can be observed between the four sunflower genotypes. RIL 149 and RIL 043 had the highest L_0 but exhibited interesting and differing root properties. It appears that a "large" root anatomy (i.e. large root surface area, volume, and mass) allows RIL 043 to compensate for its lower contribution of AQPs to root hydraulics and therefore its lower Lp_r. Whole L_0 was only slightly lower in RIL 043 than in RIL 149, which had a greater L_0 and the greatest Lp_r due to high AQP involvement. By contrast, RIL 127 had the lowest whole L_0 due to small root development, despite higher intrinsic Lp_r and contribution of AQPs than RIL 043. RIL 200 exhibited intermediate values for all parameters. Interestingly, the ranking of the RILs for L_0 was the same as the ranking of the RILs for WUE determined in our previous study: RIL 149 > RIL 043 > RIL 200 > RIL 127 (Adiredjo et al. 2014). Therefore, L_0 is suggested to play a key role in sunflower water balance and WUE (Maurel 2007; Sade et al. 2010). AQPs are reported to be regulated by several stresses, particularly drought, and shoots transpiration (Martre et al. 2001, 2002; Clarkson et al. 2000; Martínez-Ballesta et al. 2003; Shimizu et al. 2005; Sakurai-Ishikawa et al. 2011; Laur and Hacke 2013; Chaumont and Tyerman 2014) often without any change in root anatomy or morphology. Under stress conditions, RIL 043, which displays the highest water transport capacity of the whole organ due to extensive root development, could be less affected than RIL 149 which presents high water transport capacities that depend on the contribution of AQPs.

CONCLUSIONS

Three main conclusions emerge from our results: (i) a large variation occurs in morphological and hydraulic profiles in sunflower, (ii) there is a varying contribution of AQPs to hydraulic conductivity but a similar root intrinsic water permeability (Lp_r, on an area basis) between genotypes and (iii) root anatomy, which appears to be a major determinant of the water transport properties of the whole organ, is able to compensate for a low AQP contribution.

ACKNOWLEDGMENTS

Afifuddin Latif Adiredjo was supported by a French Government scholarship (*Bourse du Gouvernement Français, BGF*) and a co-funding by Directorate General of Higher Education, Ministry of Education and Culture, Republic of Indonesia (*Beasiswa Luar Negeri, BLN*).

Philippe Grieu and Thierry Lamaze are Ph.D. supervisors of the first author.

REFERENCES

1. Adiredjo AL, Navaud O, Lamaze T, Grieu P (2014) Leaf carbon isotope discrimination as an accurate indicator of water use efficiency in sunflower genotypes subjected to five stable soil water contents. J Agron Crops Sci, in press

2. Barrowclough DE, Peterson CA, Steudle E (2000) Radial hydraulic conductivity along developing onion roots. J Exp Bot 51:547–557

3. Carvajal M, Cooke DT, Clarkson DT (1996) Responses of wheat plants to nutrient deprivation may involve the regulation of water-channel function. Planta 199:372–381

4. Carvajal M, Martinez V, Alcaraz CF (1999) Physiological function of water-channels, as affected by salinity in roots of paprika pepper. Physiol Planta 105:95–101

5. Chaumont F, Tyerman SD (2014) Aquaporins: highly regulated channels controlling plant water relations. Plant Physiol 164:1600–1618

6. Chaumont F, Moshelion M, Daniels MJ (2005) Regulation of plant aquaporin activity. Biol Cell 97:749–764

7. Clarkson DT, Carvajal M, Henzler T, Waterhouse RN, Smyth AJ, ooke DT, Steudle E (2000) Root hydraulic conductance: diurnal aquaporin expression and the effects of nutrient stress. J Exp Bot 51:61–70

8. Coskun D, Britto DT, Jean Y-K, Schulze LM, Becker A, Kronzucker HJ (2012) Silver ions disrupt K^+ homeostasis and cellular integrity in intact barley (Hordeum vulgare L.) roots. J Exp Bot 63:151–162

9. Grieu P, Maury P, Debaeke P, Sarrafi A (2008) Ameliorer la tolerance a la secheresse du tournesol: apports de l'ecophysiologie et de la genetique. Innovat Agronomiques 2:37–51

10. Heinen RB, Ye Q, Chaumont F (2009) Role of aquaporins in leaf physiology.J Exp Bot 60:2971–2985

11. Kamaluddin M, Zwiazek JJ (2001) Metabolic inhibition of root water flow in red-osier dogwood (Cornus stolonifera) seedlings. J Exp Bot 52:739–745

12. Laur J, Hacke UG (2013) Transpirational demand affects aquaporin expression in poplar roots. J Exp Bot 64:2283–2293

13. Li WR, Zhang SQ, Shan L (2007) Effects of water stress on characteristics of root water uptake and photosynthesis in alfalfa seedlings. Acta Agres Sin 15:206–211

14. Liu B, Steudle E, Deng XP, Zhang SQ (2009) Root pressure probe can be used to measure the hydraulic properties of whole root systems of corn (Zea mays L.). Bot Stud 50:303–310

15. Maggio A, Joly RJ (1995) Effects of mercuric chloride onthe hydraulic conductivity of tomato root systems. Evidence fora channel-mediated water pathway. Plant Physiol 109:331–335

16. Martínez-Ballesta MC, Aparicio F, Pallás V, Martínez V, Carvajal M (2003) Influence of saline stress on root hydraulic conductance and PIP expression in Arabidopsis. J Plant Physiol 160:689–697

17. Martre P, North GB, Nobel PS (2001) Hydraulic conductance and mercury-sensitive water transport for roots of Opuntia acanthocarpa in relation to soil drying and rewetting. Plant Physiol 126:352–362

18. Martre P, Morillon R, Barrieu F, North GB, Nobel PS, Chrispeels MJ (2002) Plasma membrane aquaporins play a significant role during recovery from water deficit. Plant Physiol 130:2101–2110

19. Maurel C (2007) Plant aquaporins: novel functions and regulation properties. FEBS Lett 581:2227–2236

20. North GB, Martre P, Nobel PS (2004) Aquaporins account for variations in hydraulic conductance for metabolically active root regions of Agave deserti in wet, dry, and rewetted soil. Plant Cell Environ 27:219–228

21. Ouvrard O, Cellier F, Ferrare K, Tousch D, Lamaze T, Dupuis J-M, Casse-Delbart F (1996) Identification and expression of water stress and abscisic acid-regulated genes in a drought tolerant sunflower genotype. Plant Mol Biol 31:819–829

22. Pasda G, Diepenbrock W (1990) The physiological yield analysis of sunflower (Helianthus annuus L.) Part II Climatic factors. Fett Wissenschfat Tech 93:155–168

23. Peyrano ET, Quiroga M, de Forchetti SM, Tigier H (1997) Salinity effects on hydraulic conductance, lignin content and peroxidase activity in tomato roots. Plant Physiol Biochem 35:387–393

24. Quintero JM, Fournier JM, Benlloch M (1999) Water transport in sunflower root systems: effects of ABA, Ca^{2+} status and $HgCl_2$. J Exp Bot 50:1607–1612

25. Rawson HM, Constable GA, Howe GN (1980) Carbon production of sunflower cultivars in field and controlled environment. II. Leaf Growth. Aust J Plant Physiol 7:575–586

26. Rengel D, Arribat S, Maury P, Magniette MLM, Hourlier T, Laporte M, Vares D, Carrère S, Grieu P, Balzergue S, Gouzy J, Vincourt P, Langlade NB (2012) A gene-phenotype network based on genetic variability for drought responses reveals key physiological processes in controlled and natural environments. PLoS One 7:e45249

27. Ruggiero C, De Pascale S, Fagnano M (1999) Plant and soil resistance to water flow in fababean (Vicia Faba L. major Harz.). Plant Soil 210:219–231

28. Ruggiero C, de Pascale S, Angelino G, Maggio A (2003) Developmental changes in plant resistance to water flow in Pisum sativum (L.). Plant Soil 250:121–128

29. Ruggiero C, Angelino G, Maggio A (2007) Developmental regulation of water uptake in wheat. J Plant Physiol 164:1170–1178

30. Sade N, Gebretsadik M, Seligmann R, Schwartz A, Wallach R, Moshelion M (2010) The role of tobacco aquaporin1 in improving water use efficiency, hydraulic conductivity, and yield production under salt stress. Plant Physiol 152:245–254

31. Sakurai-Ishikawa J, Murai-Hatano M, Hayashi H, Ahamed A, Fukushi K, Matsumoto T, Kitagawa Y (2011) Transpiration from shoots triggers diurnal changes in root aquaporin expression. Plant Cell Environ 34:1150–1163

32. Sarda X, Tousch D, Ferrare K, Legrand E, Dupuis JM, Casse-Delbart F, Lamaze T (1997) Two TIP-like genes encoding aquaporins are expressed in sunflower guard cells. Plant J 12:1103–1111

33. Sarda X, Tousch D, Ferrare K, Cellier F, Alcon C, Dupuis JM, Casse F, Lamaze T (1999) Characterization of closely related d-TIP genes encoding aquaporins which are differentially expressed in sunflower roots upon water deprivation through exposure to air. Plant Mol Biol 40:179–191

34. Savage DF, Stroud RM (2007) Structural basis of aquaporin inhibition by mercury. J Mol Biol 368:607–617

35. Shimizu M, Ishida A, Hogetsu T (2005) Root hydraulic conductivity and whole-plant water balance in tropical saplings following a shade-to-sun transfer. Oecologia 143:189–197

36. Steudle E, Peterson CA (1998) How does water get through roots? J Exp Bot 49:775–788

37. Steudle E, Oren R, Schulze ED (1987) Water transport in maize root. Plant Physiol 84:1220–1232

38. Sutka M, Li G, Boudet J, Boursiac Y, Doumas P, Maurel C (2011) Natural variation of root hydraulics in Arabidopsis grown in normal and salt-stressed conditions. Plant Physiol 155:1264–1276

39. Tazawa M, Ohkuma E, Shibasaka M, Nakashima S (1997) Mercurial-sensitive water transport in barley roots. J Plant Res 110:435–442

40. Vandeleur RK, Sullivan W, Athman A, Jordans C, Gilliham M, Kaiser BN, Tyerman SD (2014) Rapid shoot-to-root signalling regulates root hydraulic conductance via aquaporins. Plant Cell Environ 37:520–538

41. Vincourt P, As-sadi F, Bordat A, Langlade NB, Gouzy J, Pouilly N, Lippi Y, Serre F, Godiard L, De Labrouhe DT, Vear F (2012) Consensus mapping of major resistance genes and independent QTL for quantitative resistance to sunflower downy mildew. Theor Appl Genet 125:909–920

42. Wan X, Zwiazek JJ (1999) Mercuric chloride effects on root water transport in aspen seedlings. Plant Physiol 121:939–946

43. Zhang WH, Tyerman SD (1999) Inhibition of water channels by HgCl2 in intact wheat root cells. Plant Physiol 120:849–857

2

CHAPTER

SATURATED HYDRAULIC CONDUCTIVITY CHANGES WITH TIME AND ITS PREDICTION AT SAR AND SALINITY IN QUEVEDO REGION SOILS

Abdelmonem Mohamed Amer[1,2,3], Carmen Suarez[3], Felix Valverde[3], Roberto Carranza[3], Leonardo Matute[3], Gabriel Delfini[3]

[1]*SENESCYT, Prometeo Project, Quito, Ecuador*

[2]*Department of Soil Science, Faculty of Agriculture, Menoufia University, Shebin El-Kom, Egypt*

[3]*Faculty of Agriculture, Quevedo State Technical University-UTEQ, Quevedo, Ecuador*

ABSTRACT

Darcy's law is applied to describe the steady flow processes in which the flux remains constant with time along the conducting system. Due to the dispersion and migration of colloidal particles and lodging in the soil pores the reduction in hydraulic conductivity occurs with time in particular when the soil and the percolating solution are affected by electrolyte concentration. Hence, the aim of this study is to find empirical equations that can be used to predict the flux with time. Data for the effluent volume versus time (up to 6 hours) which was collected for three soils (located at

Quevedo-Los Rios region) treated by two salt solutions (5 and 50 meq/l) with different SAR values were used to test certain mathematical forms of equations. Only four empirical equations were found to perfectly fit the data (flux vs time) whereas, fitting the calculated and measured data of the hydraulic conductivity for all soils produced regression factors $R^2 \geq 0.99$. So, these equations can be applied to predict the hydraulic conductivity and to characterize the flow process at saturated conditions of the studied soils with great confidence. The Hoerl function model was the best of all equations for application as the fitting degrees were almost perfect for all studied soils at 5 and 50 meq/l. It was observed for all equations that one of the fitting parameters would always represent the initial hydraulic conductivity ($K^o s$) that was evaluated graphically at zero time by extrapolation.

Keywords: Saturated Hydraulic Conductivity, Empirical Models, Effluent Volume and SAR, Linear & Nonlinear Flux

1. INTRODUCTION

The basic equation for describing the steady flow of water through columns of porous materials is represented by Darcy's law:

$$q = K\frac{\Delta H}{L} \tag{1}$$

where, q is the flux or the effluent volume of the percolating solution per unit of area and time; K is the proportionality factor or hydraulic conductivity of the porous media; and $\Delta H/L$ represents the total hydraulic head gradient. Then, if Q is the volume of solution vertically flowing through a column of porous material (L cm length with a cross sectional area of A cm^2) under a constant hydraulic head (h cm), the flux in $L \cdot T^{-1}$ unit (cm/hr) can be calculated as:

$$q = \frac{Q}{At} \tag{2}$$

thus,

$$Q = A\Delta H \frac{Kt}{L} \tag{3}$$

where, $\Delta H = h + L$.

Equation (3) can be represented in a straight line whose slope is equal to $A\Delta HK/L$. It is, therefore, possible to determine the hydraulic conductivity as a parameter characterizing the flow process. Also, the slope of the linear equation is equal to K when the Equation (3) rearranged as follows:

$$Q(L/A\Delta H) = Kt \tag{4}$$

However, the validity of Darcy's law is determined by physical and/or chemical properties of either the porous media or the percolating solution and depends on the hydraulic head gradient applied [1] [2] . Also, Darcy's law may only be applied to describe steady or stationary flow processes in which the flux remains constant with time along the conducting system [3] .

In the saturated soils, K is generally considered a constant, although it may change with time due to particle rearrangement within the soil matrix. Because of changes in the physical or chemical and structural state of the porous media, changes in the hydraulic conductivity may therefore occur during the flow process. Dikinya et al., (2008) [4] measured changes in hydraulic gradients $(\Delta H/\Delta L)$ along the columns and outflow particle sizes and concentrations during pressure leaching with solutions of 100, 10 and 1 mmol/L NaCl. The lowest ionic strength has resulted in more reduced hydraulic conductivity and relatively more release of colloids associated with hydrodynamic shear and dispersion. Steady increases in hydraulic gradient $(\Delta H/\Delta L)$ and corresponding decreases in relative saturated hydraulic conductivity (K/K_o) with time were observed for the soils under investigation. The decrease in K/K_o and increase in $\Delta H/\Delta L$ were clearly influenced by the size as well as the concentration of migrating particles in the porous medium. This was attributable to swelling and dispersion within the soil matrix at these concentrations.

The effects of sodium adsorption ratio (SAR) and electrolyte concentration on hydraulic properties of soils and clays as well as the variations in the hydraulic conductivity as a function of exchangeable ion status of soils have been extensively studied [5] -[9] . Detailed information on salinity effects on soil-water properties under saturated conditions is also reported by Frenkel et al. (1978) [10] .

Dispersion and migration of colloidal particles, subsequently lodging in the soil pores, lead to reduction in hydraulic conductivity when the soil is slightly sodic and/or the percolating solution is low in electrolyte concentration. On the other hand, at high exchangeable sodium percentage (ESP) and salt concentration > 5 meq/l, the swelling of clay is the main mechanism responsible for decreases in hydraulic conductivity [5] [11] [12] . These effects may therefore, progressively reduce hydraulic conductivity according to the particular type of soil or the chemical composition of the percolating solution.

High sodium adsorption ratios and low electrolyte concentrations cause a progressive reduction in the percolating rate, resulting in a non-linear relationship between effluent volume and time, i.e., Ks becomes not constant and the Darcy's law, which characterizes steady or stationary flow processes, does not apply. So, the objectives of this study are: 1) to find some empirical equations that can correctly predict soil hydraulic conductivity, taking into consideration the changes in the physical or chemical and structural state of the porous media, which occur during the flow process; 2) to find the fitting parameters for the proposed empirical equations with respect to initial saturated hydraulic conductivity ($K^o s$) at zero time; and 3) to study the effect of SAR of two salt solutions (5 and 50 meq/l) on the saturated hydraulic conductivity of tropical soils of Quevedo-Los Rios region.

2. MATERIALS AND METHODS

To achieve the aim of this study, surface soil samples were collected at two locations to represent the Quevedo region of the Los Rios Province, Ecuador. The first location is the Felix Farm, about 30 km to Quevedo City. The second location is the INIAP Experimental Farm, Pichilingue-Quevedo. Disturbed and undisturbed soil samples were taken at a depth, 0 - 30 cm. The disturbed samples were air dried, gently crushed and sieved through a 2 mm sieve. Fractions below 2 mm were subjected to chemical and mechanical analysis. The undisturbed samples were used to determine bulk density and Hydraulic conductivity. Soil physical and chemical analyses, presented in Table 1, were done according to Black et al., (1965) [13] .

Two salt solutions (5 and 50 meq/l, representing ionic strength of 0.005 N and 0.05 N) with six SAR values; 0, 5, 10, 30, 50 and >50 or (∞) were prepared and employed in studying their effects on the saturated hydraulic conductivity of the previously mentioned soils.

Saturated hydraulic conductivity was determined by using undisturbed soil samples according to Klut's method (1972) [14] . Schematic of the head-constant device that can be used for determination of the hydraulic conductivity is shown in Figure 1, whereas the hydraulic conductivity values were calculated according to Darcy's law.

$$SAR = \frac{Na^+}{\sqrt{\frac{Ca^{++} + Mg^{++}}{2}}}$$

Table 1. Particle size distribution for the depth 0 - 30 cm of the studied soils.

Soil location	pH	EC (dS/m)	SAR*	C. sand (%)	F. sand (%)	Silt (%)	Clay (%)	Texture grade
1) Felix Farm	6.82	0.73	1.41	38.50	20.50	24.10	16.90	Sandy loam
2) INIAP-Cacao Field	7.24	1.10	2.19	2.10	16.40	36.70	44.80	Clay
3) INIAP-Corn Field	6.92	1.96	3.54	5.20	14.40	37.90	42.50	Clay

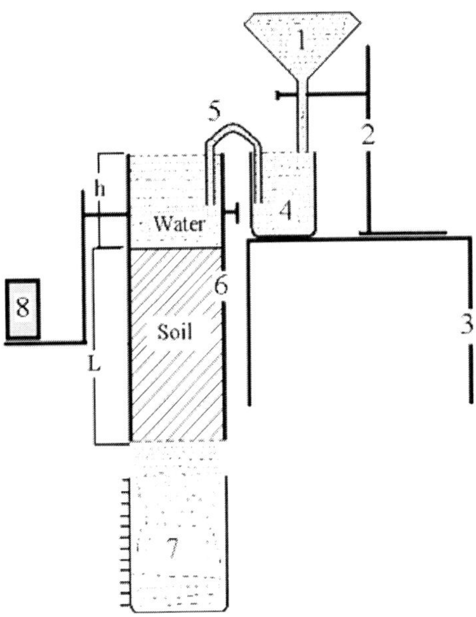

Figure 1. Classic schematic for the hydraulic conductivity method with constant head. Key: 1—standard flask; 2—car- rier; 3—table wood; 4—glass calyx; 5—glass tube on U form; 6—plastic diaphanous column; 7—glass calyx; 8— mineral avoirdupois.

The hydraulic gradient was kept constant during the experiments with a constant head device. All experiments were carried out to evaluate the effect of SAR and electrolyte (salt) concentrations on the drainage properties of studied soils. So, the effluent volume as a function of time was recorded.

Data for the effluent volume versus time (up to 6 hours) for the soils that treated with two salt solutions with six different SAR values was collected and used for testing a large number of the known forms of equations (26 models are listed in Table 2). Characteristic parameters of the proposed equations are the saturated hydraulic conductivity at zero time ($K^o s$) and coefficients a, b, c that takes into account the progressive reduction in the initial hydraulic conductivity ($K^o s$) of the soils.

The relationships between effluent volume and time were statistically analyzed to find the regression equations and then to find the initial hydraulic conductivity ($K^o s$) as well as to find the fitting parameters of the empirical equations for predicting drainage rate from soil columns under saturated conditions. The measured data were fitted to all mathematical forms and the regression coefficients were calculated using a computer program named Curvefit by Thomas S. Cox, 1984. Only equations that predicted the effluent volume versus time with a correlation factor (R^2) near unity was considered.

3. RESULTS AND DISCUSSION

3.1. Characteristics of Flow Process and Empirical Models

To characterize the flow process, the saturated hydraulic conductivity (K_s) was calculated according to Darcy's law as $Q(L/A\Delta H)/t$, where $Q(L/A\Delta H)/Kt$. This relation represents the equation of a straight line whose slope is equal to K and the term $(L/A\Delta H)$ is just a constant value. The equation of the straight line can be written as $Y = Kt$, where Y is the effluent flux. On the other hand, the other suggested equations should be used when Darcy's law could not be applied when nonlinear relationships were observed. Nonlinearity is usually due to particle rearrangement within the solid matrix, i.e. swelling or dispersion and migration of colloidal

Table 2. Tested equations for best-fit of the measured data of effluent volume changes with time.

Equations fitted to hydraulic conductivity data using Curvefit program	
1) $Y = a + b * X$ STR. LINE	14) $Y = a * X \wedge (b/X)$ MOD GEOMETRIC
2) $Y = b * X$ LINE THRU ORG	15) $Y = a * e \wedge (b * X)$ EXPONENTIAL
3) $Y = 1/(a + b * X)$ REC. STR LINE	16) $Y = a * e \wedge (b/X)$ MOD EXPONENTIAL
4) $Y = a + b * X + c/X$ LIN AND RECIP	17) $Y = a + b * \ln(X)$ LOGARITHMIC
5) $Y = a + b/X$ HYPERBOLA	18) $Y = 1/(a + b * \ln(X))$ RECIP LOG
6) $Y = X/(a * X + b)$ RECIP HYPERBOLA	19) $Y = a * b \wedge X * X \wedge c$ HOERL FUNCTION
7) $Y = a + b * X + c/X * X$ 2ND ORD HYP	20) $Y = a * b \wedge (1/X) * X \wedge c$ MOD HOERL
8) $Y = a + b * X + c * X * X$ PARABOLA	21) $Y = a * e \wedge ((X - b)/2)$ NORMAL
9) $Y = a * X + b * X * X$ PAR AT ORIGIN	22) $Y = a * e \wedge ((\ln(X) - b) \wedge 2/c)$ LOG NORMAL
10) $Y = a * X \wedge b$ POW ER	23) $Y = a * X \wedge b * (1 - X) \wedge c$ BETA
11) $Y = a * b \wedge X$ MOD. POWER	24) $Y = a * (X/b) \wedge c * e \wedge (X/b)$ GAMMA
12) $Y = b \wedge (1/X)$ ROOT	25) $Y = 1/(a * (X + b) \wedge 2 + c)$ CAUCHY
13) $Y = a * X^{(b * X)}$ SUPER GEOMET	26) $Y = K_s/\alpha(1 - e^{-\alpha t})$ ARINGHIRRI & CAPURRO

Table taken from a curve fitting computer program called Curvefit by Thomas S. Cox. 1984, based on equations listed in Curve fitting for programmable calculations by William M. Kolb. Published by IMTEC, P.O. Box 1402 Bowie MD 20716.

particles along the soil columns with subsequent plugging of conducting pores [5] [11] . For instance, data illustrated in Figure 2 and Figure 3 for the surface depth of the Felix Farm sandy loam soil show Linear relationships between effluent volume of percolating solution and time which were obtained at high electrolyte concentration (50 meq/l) with all SAR values except for low SAR values (0 and 5) which were nonlinear.

Also, nonlinear relationships were observed at low electrolyte concentration (5 meq/l) at SAR values 0, 5, 10 and 30 in the Felix Farm sandy loam soil as well as at SAR 5 and 10 for other heavier soils (INIAP farm). Thus indicate a variation or decrease in the initial hydraulic conductivity of the soil during the flow process.

The empirical equations (26 equations) were applied to fit the data for effluent volume vs time. These data were used for characterizing the flow process when hydraulic conductivity is not a constant factor under saturated conditions. However, this case is occurred when the Darcy's equation does not produce a straight line but a curve with decreasing slope with time, following changes in the physical or chemical state of the soil during the percolating process.

The following empirical equations were observed to almost perfectly fit the experimental data:

$$Y = X/(a * X + b) \quad \text{(5) (Reciprocalhypérbola)}$$

$$Y = a + b * X + c * X^2 \quad \text{(6) (Parábola)}$$

$$Y = a * b^X * X^C \quad \text{(7) (Hoerl function)}$$

$$Y = Ks/\alpha\left(1 - e^{-\alpha t}\right) \quad \text{(8) (Aringhirri\& Capurro model)}$$

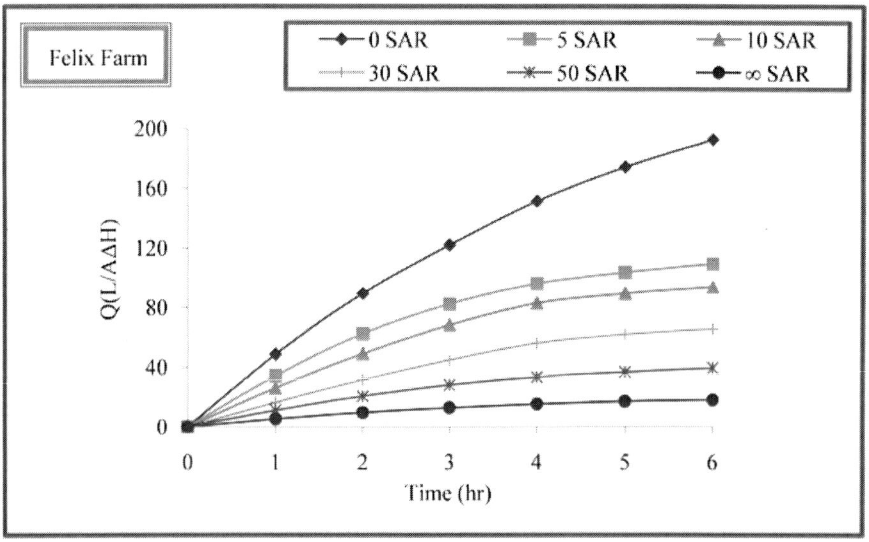

Figure 2. Relationships between effluent volume {Q(L/AΔH)} and time for Felix Farm sandy loam soil (0 - 30 cm) as a function of SAR at 5 meq/l.

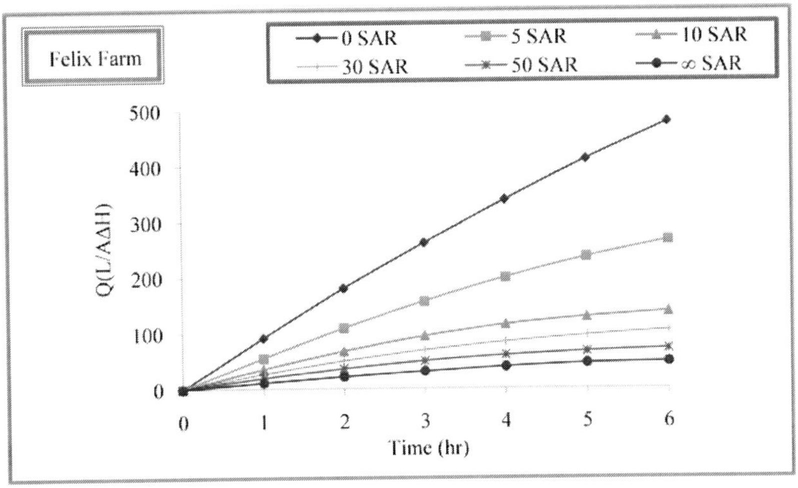

Figure 3. Relationships between effluent volume {Q(L/AΔH)} and time for Felix Farm sandy loam soil (0 - 30 cm) as a function of SAR at 50 meq/l.

where, Y is equal to $Q(L/A\Delta H)$, and the X indicates elapsed time (t), Q is the effluent volume (cm^3), A is the cross sectional area of the soil column (cm^2). L is the length of soil column (cm), ΔH is the sum of the soil length and the constant water head above soil. Ks is the hydraulic conductivity of the porous media (cm/hr), and a, b & c are empirical soil parameters to be fitted to each individual soil and SAR value.

It should be noted that Y was not taken as the effluent volume Q but rather $Q(L/A\Delta H)$ because adding the constant value $(L/A\Delta H)$ was observed to produce a straight line with a slope equal to Ks value on plotting Y vs t. However, dividing Y as $Q(L/A\Delta H)$ by the elapsed time t will produce the value of the saturated hydraulic conductivity (Ks) at time t.

3.2. Initial Hydraulic Conductivity

The initial hydraulic conductivity ($K^o s$) was observed to be presented in all the proposed equations as one of the constants of each equation. The presence of $K^o s$ as a part of all the used equations indicates that the initial hydraulic conductivity is assumed to stay constant if only no changes occur to the soil structure with time. Changes in the physical or chemical state of the soil during the percolation process will result in decreasing the value of Ks with time. Each of the four equations accounted for the decrease in Ks with time, and predicted the hydraulic conductivity change that reflects the current state of the soil.

Evaluation of the initial hydraulic conductivity ($K^o s$) at time zero involves dividing by zero, so, it can be done graphically for the relation of $Q(L/A\Delta H)/t$ vs time t for the studied soils. Graphing Ks versus time will produce a line or a curve that can be extrapolated to the Y axis and obtain the $K^o s$ value at zero time such as in Figure 4 and Figure 5. Also a regression equation can be obtained for the relationship Ks vs t and substitution of t equal to zero will give Ks at zero time. The regression equations between $Y = Ks$ and $x = t$ were obtained. The equations for Felix Farm sandy loam soil (0 - 30 cm) with electrolyte solution of 5 meq/l and SAR values of 0, 5, 10, 30, 50, and ∞ respectively, were as follows:

$$Y = -1.4804x + 13.184, \;\; R^2 = 0.9879$$

$$Y = -1.1468x + 7.9304, \;\; R^2 = 0.9851$$

$$Y = -0.6707x + 5.0721, \;\; R^2 = 0.9696$$

$$Y = -0.4514x + 3.2214, \;\; R^2 = 0.989$$

$$Y = -0.2093x + 1.485, \;\; R^2 = 0.991$$

And for electrolyte solution of 50 meq/l, they were:

$$Y = -1.2739x + 24.282, \;\; R^2 = 0.9916$$

$$Y = -1.317x + 15.631, \;\; R^2 = 0.9919$$

$$Y = -1.3846x + 10.563, \;\; R^2 = 0.9953$$

$$Y = -0.9418x + 7.6211, \;\; R^2 = 0.9935$$

$$Y = -0.7711x + 5.6546, \;\; R^2 = 0.9954$$

$$Y = -0.4121x + 3.4221, \;\; R^2 = 0.951$$

As for the representation of the initial hydraulic conductivity (K°s) through the fitted parameters a, b, and c in the suggested four models, it was observed that the first model (Reciprocal Hyperbola) contained two constants a, and b, while K°s was noticed to equal the reciprocal of the constant b $\left(K^{o}s = 1/b\right)$. The second model (parabola) contained three constants a, b and c with the constant b representing K°s. The third model (Hoerl function) included three constants a, b and c, where the constant a represented the K°s.

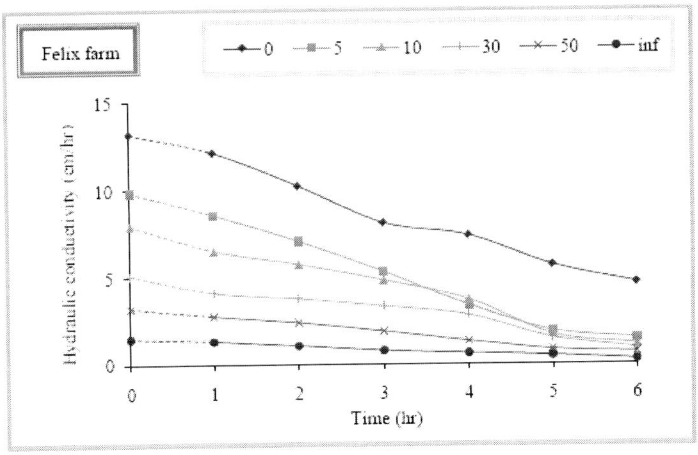

Figure 4. Initial hydraulic conductivity (K°s) as obtained by extrapolation of Ks vs t curve at 5 meq/l with different SAR values for Felix Farm sandy loam soil.

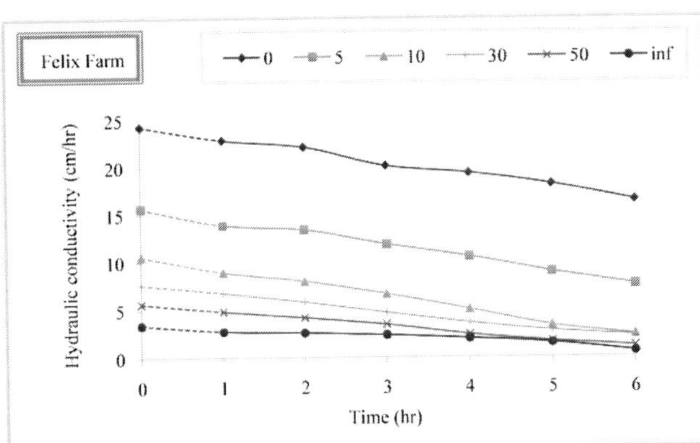

Figure 5. Initial hydraulic conductivity (K°s) as obtained by extrapolation of Ks vs t curve at 50 meq/l with different SAR values for Felix Farm soil.

Table 3. Calculated fitting parameters of the Reciprocal Hyperbola equation at different SAR values for Felix Farm sandy loam soil.

Parameters	Reciprocal Hyperbola					
	$K's$	a	$b = 1/Ks$	c	R^2	$Ks = 1/b$
SAR	5 meq salts/l					
0	11.6	0.009	0.081	0	1.0	12.345
5	8.5	0.022	0.098	0	0.998	10.204
10	6.5	0.021	0.137	0	0.996	7.299
30	4.08	0.024	0.226	0	0.996	4.424
50	2.75	0.054	0.316	0	0.999	3.164
∞	1.3	0.119	0.683	0	0.999	1.464
	Regression for the relation $K's$ vs $1/b$ is $Y = 1.0938x + 0.1522$; $R^2 = 0.9903$					
SAR	50 meq salts/l					
0	21.43	0.001	0.046	0	1.0	21.739
5	13.36	0.004	0.072	0	1.0	13.888
10	8.852	0.013	0.102	0	0.998	9.803
30	6.617	0.018	0.137	0	1.0	7.299
50	4.799	0.028	0.185	0	0.999	5.405
∞	2.747	0.027	0.349	0	0.997	2.865
	Regression for the relation $K's$ vs $1/b$ is $Y = 0.996x + 0.5713$; $R^2 = 0.9982$					

Table 4. Parameters of the Parabola equation at different SAR values for Felix Farm soil.

Parameters	Parabola					
	$K's$	a	b	c	R^2	$Ks = b$
SAR	5 meq salts/l					
0	11.6	0.424	10.35	−0.488	0.999	10.35
5	8.5	0.329	7.416	−0.521	0.996	7.416
10	6.5	0.135	5.888	−0.341	0.995	5.888
30	4.08	0.05	3.78	−0.178	0.995	3.78
50	2.75	0.082	2.459	−0.152	0.997	2.459
∞	1.3	0.059	1.119	−0.068	0.997	1.119
	Regression for the relation $K's$ vs b is $Y = 0.886x - 0.0404$; $R^2 = 0.9991$					
SAR	50 meq salts/l					
0	21.43	0.206	20.629	−0.390	1.0	20.629
5	13.36	0.125	12.717	−0.394	0.999	12.717
10	8.852	0.196	8.054	−0.429	0.997	8.054
30	6.617	0.212	5.922	−0.315	0.998	5.922
50	4.799	0.138	4.302	−0.249	0.997	4.302
∞	2.747	0.034	2.554	−0.095	0.996	2.554
	Regression for the relation $K's$ vs b is $Y = 0.9769x - 0.3821$; $R^2 = 0.9994$					

Table 5. Calculated fitting parameters of the Hoerl function with SAR for Felix Farm soil.

Parameters	Hoerl function					
	K'_s	a	b	c	R^2	$Ks = a$
SAR	5 meq salts/l					
0	11.6	11.91	0.928	0.976	1.0	11.915
5	8.5	9.209	0.838	1.137	1.0	9.209
10	6.5	6.864	0.842	1.214	1.0	6.864
30	4.08	4.224	0.864	1.208	1.0	4.224
50	2.75	2.936	0.853	1.145	1.0	2.936
∞	1.3	1.355	0.900	0.983	1.0	1.355
Regression for the relation $K's$ vs a is $Y = 1.04x - 0.0642$; $R^2 = 0.9981$						
SAR	50 meq salts/l					
0	21.43	21.542	0.970	1.003	1.0	21.542
5	13.36	13.611	0.929	1.076	1.0	13.611
10	8.852	9.275	0.870	1.144	1.0	9.275
30	6.617	6.900	0.894	1.056	1.0	6.900
50	4.799	5.065	0.866	1.125	1.0	5.065
∞	2.747	2.784	0.906	1.117	0.999	2.784
Regression for the relation $K's$ vs a is $Y = 0.998x + 0.248$; $R^2 = 0.9996$						

Table 6. Parameters of the Aringhirri & Capurro equation at different SAR for Felix Farm soil.

Parameters	Aringhirri and Capurro					
	$K's$	Ks	a	c	R^2	Ks
SAR	5 meq salts/l					
0	11.6	12.07	0.184	0	1.0	12.077
5	8.5	9.575	0.315	0	0.999	9.575
10	6.5	7.069	0.235	0	0.997	7.069
30	4.08	4.328	0.177	0	0.997	4.328
50	2.75	3.027	0.260	0	0.999	3.027
∞	1.3	1.393	0.260	0	1.0	1.393
Regression for the relation $K's$ vs Ks is $Y = 1.0596x + 0.1113$; $R^2 = 0.9959$						
SAR	50 meq salts/l					
0	21.43	21.585	0.06	0	1.0	21.585
5	13.36	13.755	0.105	0	1.0	13.755
10	8.852	9.500	0.21	0	0.999	9.500
30	6.617	7.046	0.213	0	1.0	7.046
50	4.799	5.202	0.237	0	0.999	5.202
∞	2.747	2.827	0.132	0	0.997	2.827
Regression for the relation $K's$ vs Ks is $Y = 0.9957x - 0.3929$; $R^2 = 0.9991$						

The fourth model by Aringhirri and Capurro (1994) [9] included only two constants Ks and a, where the Ks constant represented the initial hydraulic conductivity (K^os).

3.3. Calculated and Measured Hydraulic Conductivity

Hydraulic conductivity (Ks) values were calculated for the surface depth of the studied soils, using the four previous equations (models; 5, 6, 7 and 8). The calculated values were compared to the measured K^os values (Y vs t) and the R^2 was recorded for all SAR values. The equations parameters for the sandy loam soil (Felix Farm) and regression equations are presented in Tables 3-6. However, only the regression equations for both INIAP soils were calculated for 0 - 30 cm depth at salt concentration treatments with concerned models. For instance, at 5 millieq/l treatment in Cacao farm-INIAP, the regression equations using the models of Reciprocal Hyperbola, Parpola, Hoerl function and Aringhirri & Capurro were as follows: Y = 1.003x + 0.2245, Y = 0.919x – 0.0289, Y = 0.999x + 0.1039, Y = 0.997x + 0.1623, respectively, while with treatment 50 millieq/l they were: Y = 0.954x + 0.6785, Y = 0.97x + 0.1507, Y = 0.969x + 0. 401, Y = 0.960x + 0.5354, respectively, whereas, Y represents calculated hydraulic conductivity (Ks) and x is the measured one (K^os). In the corn farm-INIAP, the regression equations at 5 millieq/l were: Y = 1.008x +0.0238, Y = 0.986x – 0.0051, Y = 1.003x + 0.0102, Y = 1.006x + 0.0175, and with 50 millieq/l they were; Y = 0.988x + 0.1494, Y = 0.966x – 0.0025, Y = 0.988x + 0.1073, Y = 0.988x + 0.1299, respectively.

However, the fitting R^2 between calculated and measured values were greater than 0.99, which prove that the four equations can predict the actual data with great confidence. The Hoerl function model was the best of all models, where the fitting degree was almost perfect for all studied soils at 5 and 50 meq/l salt concentration.

3.4. Predicted Hydraulic Conductivity (Ks) as Influenced by SAR and Salt Concentration

The effect of SAR on the hydraulic conductivity and drainage properties of the studied soil samples at 5 and 50 meq/l electrolyte (salt) concentrations (equivalent to solutions of 0.005 N and 0.05 N ionic strength) is shown in Tables 7-9. The saturated hydraulic conductivity (Ks) was calculated according to Darcy's equation as $Q(L/A\Delta H)/t$ where it was applicable. On the other hand, one of the four suggested equations was used when Darcy's law was not applicable when nonlinear relationships were observed. Values of measured hydraulic conductivity (Ks) at 5 and 50 meq/l followed the same trend for all studied soils and take the following order: Felix Farm (sandy loam) > ANIAP-Corn Field > ANIAP-Cacao Field clay textured or generally decreased with increasing the clay content.

These results are in agreement with those findings by Santiwong et al., (2008) [15] , Levy and Mamedov (2002) [16] , Wissmeier and Barry (2009) [17] . Also, Arienzo et al., (2012) [18] , reported that solutions with sodium adsorption ratio (SAR) of 5 - 40, where the monovalent cation was Na^+ or K^+ and the divalent cation was Ca^{2+} or Mg^{2+} were used to leach the soil columns, at electrolyte concentrations ranging from 2.5 to 640 meq·L^{-1}. Percolating solutions with SAR caused a decrease in hydraulic conductivity as electrolyte concentrations reduced. They added that, when the Na^+ solutions flowed through the soil column a greater cloudiness was observed in the percolate indicating that clay deflocculating and movement are factors leading to the reduced structural stability. Generally the obtained data of Ks showed that, the hydraulic conductivity was highest at time

Table 7. Hydraulic conductivity (Ks) changes with time at different SAR values in Felix Farm soil.

SAR values	*Ks* values (cm/h)					
	SAR (0)	SAR (5)	SAR (10)	SAR (30)	SAR (50)	SAR (∞)
Time (h)			5 meq/l			
0	11.64	8.501	6.495	4.081	2.752	1.3
1	11.0214	7.7604	5.9128	3.7414	2.5208	1.2103
2	10.1631	7.0795	5.5608	3.5834	2.3474	1.0946
3	9.2243	6.3214	5.1608	3.3971	2.1379	0.9662
4	8.5872	5.5068	4.7102	3.1838	1.9041	0.8672
5	7.8977	4.7513	4.0755	2.8179	1.6708	0.7796
6	7.2838	4.1847	3.5650	2.4819	1.4938	0.6878
			50 meq/l			
0	21.43	13.36	8.852	6.617	4.799	2.747
1	20.7888	12.6429	8.1382	6.1672	4.4096	2.5465
2	20.4419	12.3988	7.7039	5.7368	4.1192	2.4643
3	19.7198	11.8557	7.1506	5.2498	3.7894	2.3555
4	19.1840	11.2848	6.5103	4.7642	3.3881	2.2182
5	18.6467	10.6524	5.8142	4.3186	3.0183	2.0521
6	18.0358	10.0278	5.2027	3.9376	2.6909	1.8069

Table 8. Hydraulic conductivity (Ks) changes with time for different SAR values in INIAP-Cacao Field soil.

Time (h)	*Ks* values (cm/h)					
	SAR (0)	SAR (5)	SAR (10)	SAR (30)	SAR (50)	SAR (∞)
				5 meq/l		
0	6.498	5.157	3.385	2.067	1.362	0.662
1	6.3857	4.8567	3.0476	1.9221	1.2385	0.6038
2	5.9629	4.5278	2.9088	1.7473	1.1974	0.5717
3	5.6790	4.2065	2.7333	1.5675	1.1246	0.5362
4	5.4631	3.8667	2.4200	1.3876	1.0118	0.4818
5	5.3038	3.5425	2.0994	1.2540	0.9091	0.4265
6	5.1385	3.3012	1.8437	1.1173	0.8047	0.3747
				50 meq/l		
0	17.13	12.51	8.007	6.147	3.832	1.431
1	16.7749	12.2831	7.7630	5.7972	3.5333	1.3542
2	16.4255	11.9131	7.4983	5.5222	3.2776	1.2681
3	16.1060	11.6450	7.1668	5.1513	2.9757	1.1752
4	15.9128	11.3863	6.7865	4.7629	2.6204	1.0734
5	15.7188	11.1165	6.3892	4.3371	2.2495	0.9461
6	15.5089	10.6890	5.9368	4.0297	1.9726	0.8355

Table 9. Hydraulic conductivity (Ks) changes with time for different SAR values in INIAP-Corn field soil.

Time (h)	Ks values (cm/h)					
	SAR (0)	SAR (5)	SAR (10)	SAR (30)	SAR (50)	SAR (∞)
				5 meq/l		
0	4.186	3.178	2.018	1.399	0.594	0.406
1	4.1212	3.0160	1.9277	1.3313	0.5653	0.3787
2	4.0647	2.9439	1.8938	1.2903	0.5483	0.3702
3	4.0101	2.8492	1.8382	1.2427	0.5267	0.3552
4	3.9686	2.7164	1.7687	1.1900	0.5024	0.3370
5	3.8962	2.5609	1.6892	1.1255	0.4760	0.3154
6	3.8178	2.4153	1.6008	1.0515	0.4475	0.2826
				50 meq/l		
0	9.239	5.925	4.046	2.351	1.667	0.593
1	9.0368	5.7748	3.9432	2.3009	1.6281	0.5596
2	8.8813	5.5982	3.8188	2.2358	1.5843	0.5285
3	8.7589	5.3989	3.6765	2.1595	1.5245	0.4927
4	8.6425	5.1826	3.5078	2.0768	1.4529	0.4529
5	8.5252	4.9364	3.3219	1.9781	1.3743	0.4172
6	8.4201	4.6682	3.1149	1.8646	1.2884	0.3792

zero (K^os) and decreased sharply with increasing values of SAR, where the measured Ks values at the highest $SAR(\infty)$ value were 7% - 13% of that at the lowest SAR(0) value. It noted that, the K^os is highest value compared with the same parameter at any elapsed time. The main reasons to that are the soil properties (i.e. soil structure, soil particles arrangement) kept in the initial state without any changes and fine fraction (silt or clay) not migrated into lower layers and the soil porosity still and effective to water flow in soil, [1] [2] [19] [20] . These wide differences in the hydraulic conductivity values at the SAR values are clearly indicate to the negative effect Na^+ on water movement in soils [21] .

Also, the hydraulic conductivity values were higher at time zero (K^os) and decreased with times up to 6 hours at 5 and 50 meq/l for all SAR values under investigation. The rate of decrease was higher for 5 meq/l than 50 meq/l, where the K^os values at 5 meq/l were low compared with the same values at 50 meq/l for SAR values.

4. CONCLUSION

Dispersion and migration of colloidal particles, subsequently lodging in the soil pores, lead to reduction in saturated hydraulic conductivity (Ks), hence, the relation between effluent volume and time will be non-linear and Darcy's law can't be applied. This work aims to find other equations that can correctly predict the changing flux with time with respect to Darcy's equation at non-linear flow. The changes in saturated hydraulic conductivity (Ks) with time were determined in three tropical soils located in Quevedo Region (Los Rios Province). The measured data of the investigated tropical soils were compared with that calculated by proposed equations. Initial hydraulic conductivity (K^os) has been involved with the equations that used for prediction of the change in K with time. Four empirical equations were observed to almost perfectly fit the experimental data: Reciprocal Hyperbola, Parabola, Hoerl function, Aringhirri and Capurro model. The Hoerl function model was the best of all models for application as the fitting degrees were almost perfect for all studied soils at 5 and 50 meq/l. Using K^os as a fitting parameter for all proposed equations indicates its stability if there are no changes in soil structure with time. Darcy's equation along with the four models can predict soil hydraulic properties in a wide range of either electrolyte concentrations or SAR values. Hydraulic conductivity, Ks, was highest at time zero and decreased gradually with time (up to 6 hours). Also, Ks decreased sharply with increasing SAR values. The measured Ks values for all soils at the highest

SAR value were only 7% - 13% of the measured Ks values at the lowest SAR value. The rate of decrease was higher for 5 meq/l than 50 meq/l electrolyte concentration of the percolation solution.

ACKNOWLEDGEMENTS

I am pleased to have the Prometheus Scholarship of the SENESCYT Prometeo Project of Ecuador during 2013/ 2014. The financial support of SENESCYT for this work is acknowledged.

REFERENCES

1. Amer, A.M. (2012) Water Flow and Conductivity into Capillary and Non-Capillary Pores of Soils. Journal of Soil Science and Plant Nutrition, 12, 99-112.
2. Amer, A.M. (2011) Prediction of Hydraulic Conductivity and Sorptivity in Soils at Steady-State Infiltration. Archives of Agronomy and Soil Science, 58, 1179-1194. www.tandfonline.com/loi/gags20
3. Hillel, D. (1980) Fundamental of Soil Physics. Academic Press, New York.
4. Dikinya, O., Christoph, H. and Graham, A. (2008) Decrease in Hydraulic Conductivity and Particle Release Associated with Self-Filtration in Saturated Soil Columns. Geoderma, 146, 192-200.
5. Pupinsky, H. and Shainberg, I. (1979) Salt Effects on the Hydraulic Conductivity of a Sandy Soil. Soil Science Society of America Journal, 43, 429-433.
6. Metzger, L., Yaron, B. and Mingelgrin, U. (1983) Soil Hydraulic Conductivity as Affected by Physical and Chemical Properties of Effluents. Agronomie, 3, 771-778.
7. Yousaf, M., Ali, O.M. and Rhoades, J.D. (1987) Clay Dispersion and Hydraulic Conductivity of Some Salt-Affected Arid Land Soils. Soil Science Society of America Journal, 51, 905-907.
8. Keren, R. and Singer, M.J. (1988) Effect of Low Electrolyte Concentration on Hydraulic Conductivity of Sodium/Cal- cium

Montmorillonite-Sand System. Soil Science Society of America Journal, 52, 368-373.

9. Aringhieri, R. and Capurro, M. (1994) Evaluating Saturated Hydraulic Conductivity of a Soil in Laboratory Investigations: An Empirical Model. Soil Science, 157, 77-83.

10. Frenkel, H., Goertzen, J.O. and Rhoades, J.D. (1978) Effects of Clay Type and Content, Exchangeable Sodium Percentage, and Electrolyte Concentration on Clay Dispersion and Soil Hydraulic Conductivity. Soil Science Society of America Journal, 42, 32-39.

11. Felhender, R., Shainberg, I. and Frenkel, H. (1974) Dispersion and Hydraulic Conductivity of Soils in Mixed Solutions. Transactions on the 10th International Congress of Soil Science, 1, 103-112.

12. Shainberg, I., Rhoades, J.D. and Prather, R.J. (1981) Effect of Low Electrolyte Concentration on Clay Dispersion and Hydraulic Conductivity of a Sodic Soil. Soil Science Society of America Journal, 45, 273-277. Black, G.A., Evans, D.D., White, J.L., Ensminger, L.E. and Clerk, F.E. (1965) Methods of Soil Analysis. Parts, 1 and 2. American Society of Agronomy, Madison.

13. Klute, A. (1986) Methods of Soil Analysis. 2nd Edition, ASA and SSSA, Madison.

14. Santiwong, S.R., Guan, J. and Waite, T.D. (2008) Effect of Ionic Strength and pH on Hydraulic Properties and Structure of Accumulating Solid Assemblages during Microfiltration of Montmorillonite Suspensions. Journal of Colloid and Interface Science, 317, 214-227.

15. Levy, G.J. and Mamedov, A.I. (2002) High-Energy-Moisture-Characteristics Aggregate Stability as a Predictor for Seal Formation. Soil Science Society of America Journal, 66, 1603-1609.

16. Wissmeier, L. and Barry, D.A. (2009) Effect of Mineral Reactions on the Hydraulic Properties of Unsaturated Soils: Model Development and Application. Advances in Water Resources, 32, 1241-1254.

17. Arienzo, M., Christen, E.W., Jayawardane, N.S. and Quayle, W.C. (2012) The Relative Effects of Sodium and Potassium on Soil Hydraulic Conductivity and Implications for Winery Wastewater Management. Geoderma, 174, 303-310.

18. Amer, A.M. (2012) Infiltration Functions for Prediction of Water Sorptivity and Hydraulic Conductivity in Soils. Proceedings of the BALWOIS Conference, Ohrid, 28 May-2 June 2012.

19. Brooks, R.H. and Corey, A.T. (1964) Hydraulic Properties of Porous Media. Hydrology Paper 3, Colorado State University, Fort Collins.

20. Jayawardane, N.S., Christen, E.W., Arienzo, M. and Quayle, W.C. (2011) Evaluation of the Effects of Cation Combinations on Soil Hydraulic Conductivity. Australian Journal of Soil Research, 49, 56-64.

3
CHAPTER

ROLE OF HYDRAULIC CONDUCTIVITY UNCERTAINTIES IN MODELING WATER FLOW THROUGH FOREST WATERSHEDS

Marie-France Jutras[1] and Paul A. Arp[1]

[1] Faculty of Forestry and Environmental Management, University of New Brunswick, Fredericton, Canada

1. INTRODUCTION

Soil hydraulic conductivities at saturation (K_{sat}) are highly variable in space and time. For example, K_{sat} varies along vertical and lateral flow paths depending on directional changes in soil texture, density, and structure [1, 2, 3]. Temporal changes are caused by changes in soil structure and bulk density (D_b) in response to, e.g., (i) gradual soil formation processes, and (ii) operationally induced soil compaction or de-compaction due to various land-uses [4]. Changes in weather and climate also affect K_{sat} through freezing and thawing [5, 6], swelling and shrinking [7], extent of rooting and related organic matter build-up [8]. This chapter explores how changes in hydraulic conductivity may affect modelled rates of water flow through forested watersheds, with flows referring to infiltration, percolation, run-off, interflow, base flow, and stream discharge. This is done by way of sensitivity analyses centered on two well-studied watershed studies, referring to Moosepit Brook, Nova Scotia [1, 9] and Turkey Lakes, Ontario [1, 10]. Also addressed are:

K_{sat} impacts on the retention of soil water and the transmittance of the same towards streams as influenced by evapotranspiration from open conditions to forests [14];

the relationship between K_{sat} and the state of organic matter decomposition, as characterized by the von Post index from fibric (H1) to fully humified or sapric (H10) [11, 12, 13].

The sensitivity analysis is based on using the forest hydrology model ForHyM2 [1, 15] to determine how scenario-set K_{sat} variations affect soil water retention and flow including stream discharge through the watersheds. The scenarios vary K_{sat} by changing organic matter (OM) and sand content from their actual values within the 0 to 100% per soil weight range.

2. QUANTITATIVE BACKGROUND

The equations used for estimating the sensitivity of K_{sat} on account of changes in soil texture, structure, density and organic matter content is given by [16], as follows:

$$\log_{10} K_{sat} = a + 7.94 \log_{10}(D_p - D_b) + 1.96\ \text{SAND} \tag{1}$$

$$D_b = \frac{1.23 + (D_p - 1.23 - 0.75\ \text{SAND})(1 - \exp(-0.0106\ \text{DEPTH})}{1 + 6.83\ \text{OM}} \tag{2}$$

$$\frac{1}{Dp} = \frac{OM}{Dp_{om}} + \frac{1 - OM}{Dp_{min}} \tag{3}$$

where, Dp_{om} is the particle density of OM (1.3 gcm^{-3}), Dp_{min} is the particle density of mineral soils (2.65gcm^{-3}), SAND and OM are dry soil weight fractions (fine earth fraction only), DEPTH is the mid depth of each soil layer (cm), "a" represents K_{sat} when D_p-D_b = 1 g cm^3 and SAND = 0%. Fig. 1 illustrates how variations in D_b, OM, and SAND affect K_{sat} in general.

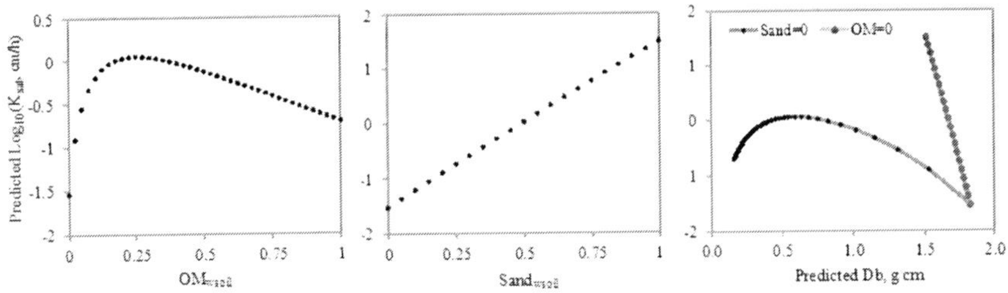

Figure 1. Left and middle: how $log_{10}K_{sat}$ varies with increasing OM, and sand fraction. Right: Changes in $log_{10}K_{sat}$ and Db when OM and Sand fraction = 0.

For organic soils, it is important to adjust a, D_b, and D_p in Eqs. 1 to 3 by extent of organic matter decomposition and humification [12, 11, 17, 13, 18, 19]. These adjustments are based on the von Post humification index ([11, 20], Table 1) as follows:

$$D_b = 0.035 + 0.0159\, vP\ (R2 = 0.93) \tag{4}$$

$$D_{pom} = \frac{D_b}{1 - \phi} \tag{5}$$

$$\phi = 100.38 - 76.7 D_b\ (R^2 = 0.99) \tag{6}$$

$$a = (2.05 \pm 0.2) - (0.046 \pm 0.004) vP^2\ (R^2 = 0.89) \tag{7}$$

where vP is the von post index (Table 2) and ϕ is the soil porosity. Fig. 2 illustrates the relationship between the von Post adjusted $log_{10}K_{sat}$ (Eqs. 1, 4-7) and $log_{10}K_{sat}$ based on literature sources. Fig. 2 shows (i) a plot of actual versus best-fitted K_{sat} values (left), and actual as well as best-fitted K_{sat} values with increasing organic matter humification in peaty soils (right).

Table 1. von Post humification index, with K_{sat}, D_b and D_p for 100% OM content according to Eqs. 1 and 4 to 7; adapted from [21] and [11]

Peat Class	von Post Index	Squeeze Test: Exudate condition	$log_{10}K_{sat}$	K_{sat} [1] $cm\ h^{-1}$	D_b [2] $g\ cm^{-1}$	D_p [2] $g\ cm^{-1}$
Fibric	H1	Water colourless	3.15	1406.48	0.05	1.44
Decomposition: none to slight; Amorphous content: low	H2	Water yellowish	2.88	756.62	0.07	1.41
	H3	Water brown, muddy; no peat	2.55	356.24	0.08	1.39
Mesic	H4	Water dark brown, muddy; no peat	2.15	141.08	0.10	1.37
Fibers still recognisable; Decomposition: moderate to strong; Amorphous content: medium	H5	Water muddy; some peat	1.66	46.20	0.11	1.36
	H6	Water dark brown; 33% peat	1.09	12.40	0.13	1.36
	H7	Any water very dark brown, 50% peat	0.43	2.72	0.15	1.35
Sapric	H8	66% peat. Water pasty	-0.32	0.48	0.16	1.34
Fibers unrecognisable; Decomposition: very strong to complete; Amorphous content: high	H9	Nearly all peat; paste uniform	-1.16	0.07	0.18	1.34
	H10	100% peat paste; no water	-2.09	0.006	0.19	1.34

[1]Eq. 1 from [16]

[2]Eqs. 4-5 from [12]

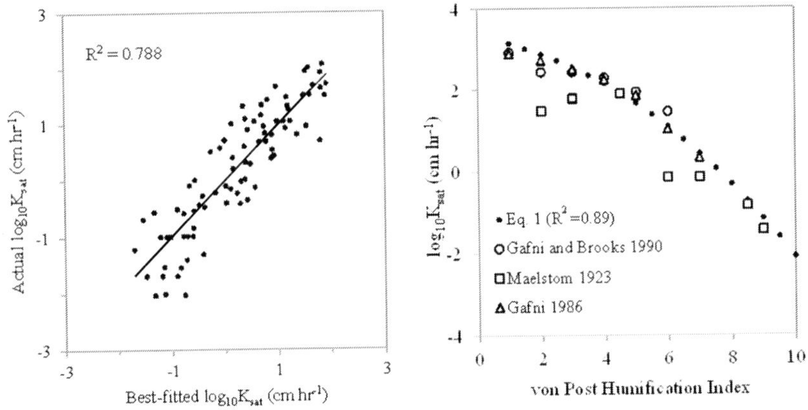

Figure 2. Best-fitted $\log_{10}K_{sat}$ versus actual data from New Brunswick and Nova Scotai, Canada, as seen in [1] and [16] (left), best-fitted $\log_{10}K_{sat}$ versus von Post humification index from literature sources (right).

3. METHODS

The two study areas, Moosepit Brook and Turkey Lakes, have contrasting terrain (generally flat versus hummocky), climate (maritime versus continental), vegetation (mostly coniferous versus deciduous), and soil parent material (ablation till versus basal till) (Table 2, Fig. 3)

Eight scenarios were adopted to examine the impacts of K_{sat} variations on water flow through these locations, as follows: the actual soil conditions in terms of soil texture and organic matter (Scenario 1, Table 3), varying the soil texture sand, silt, or clay (Scenarios 2, 3, and 4), and varying the soil organic matter content (Scenarios, 5, 6, 7, and 8).

Scenario 1:

1. Actual soil texture and OM content

Scenarios 2 to 4: Changing soil texture

1. Sand = 95% sand, 1% silt, 4% clay
2. Silt = 7% sand, 87% silt, 6% clay
3. Heavy clay = 25% sand, 25% silt, 50% clay

Scenarios 5 to 8: Changing organic matter content

1. Half actual OM
2. Double actual OM
3. No OM throughout entire soil profile
4. 100% OM throughout entire soil profile using the von Post profile of FF = 3, A&B = 5, C = 9, Subsoil = 10.

Table 2. Site description for the Moosepit Brook and Turkey Lakes watersheds.

Watershed characteristics	Moosepit Brook	Turkey Lakes
	Nova Scotia (NS)	Ontario (ON)
Abbreviation	MP	TL
Latitude (N)	44°28'	47°03'
Longitude (W)	65°03'	84°25'
Area (ha)	1670	1050
Elevation (m)	100-150	350-400
Slope (%)	1	8
Deciduous:coniferous	50:50	100:0
Rooting habit	Medium	Deep
Forest floor depth (cm)	5	7
Mineral soil: depth (cm): texture	50; SL	60; SIL
Subsoil: depth (cm): texture	70; LS	100; LS
Bedrock	Metamorphic greenschist slate	Metavolcanic basalt
Land Formation	Glacial till	Ablation till on basal till
Topography	Rolling	Undulating to rolling
Mean yearly temperature (°C)	7.02	4.52
Mean yearly snow depth (cm)	5	23
Mean yearly rainfall (mm)	1140	790
Model Run Years	1999-2004	1997-2004

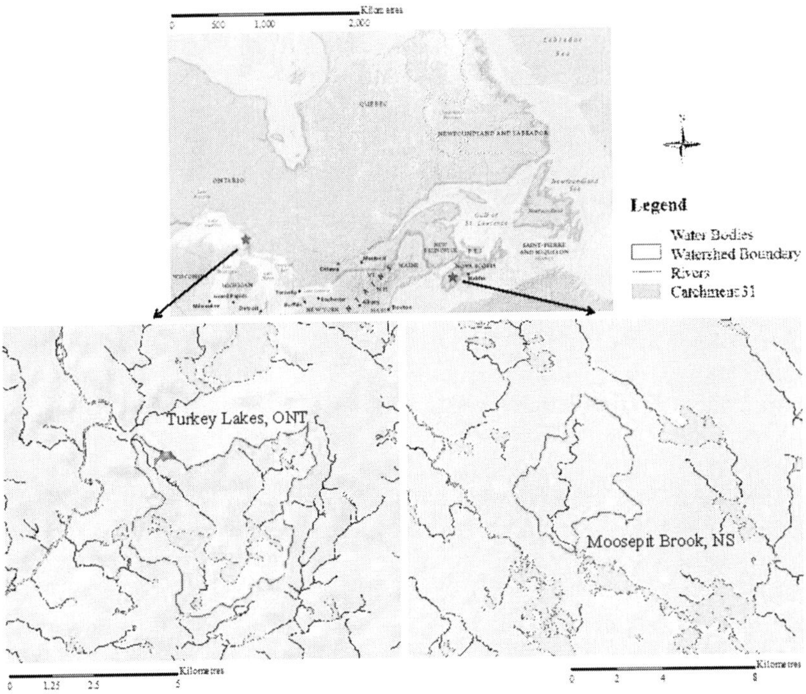

Figure 3. Locator maps for the Turkey Lakes (left) and Moosepit Brook (right) study areas.

Actual watershed inputs: Scenario 1

Table 3. Actual scenario soil input for Moosepit Brook and Turkey Lakes.

Horizon	Moosepit Brook, NS					Turkey Lakes, ONT				
	Depth (cm)	Sand (%)	Silt (%)	Clay (%)	OM (%)	Depth (cm)	Sand (%)	Silt (%)	Clay (%)	OM (%)
A	21	66	24	10	4.5	15	66	24	10	5
B	21	66	24	10	9	30	66	24	10	3
C	50+	82	12	6	4.5	50+	22	65	13	2

The sand texture percentages for scenarios 2-4 demonstrate the effects of varying texture on K_{sat} from sandy and sandy loam soils to silty and clayey soils (Fig. 4). The organic matter levels for scenarios 5-8 were chosen to demonstrate the effects of changing the organic matter on content from very small in mineral soils to fully organic soils. For the 100% organic soil condition (Scenario 8), three sub-scenarios were chosen to account for variations in forest cover from 100 % (fully forested), 50% (varying from forested to boggy) and 0% (open moss and shrub-covered bogs with no trees). This is to demonstrate how varying K_{sat} levels from high to low increase the amount of water available for evapotranspiration

Each scenario was used for initializing the ForHyM2 requirements for soil texture and organic matter by soil layer, with the A and B layers representing the top soil conditions, and the C layer representing the subsoil conditions Table 2). Layer-specific values for D_p, D_b and K_{sat} were then generated automatically via Eqs. 1 to 3. All other site-specific input requirements for daily weather (rain, snow air temperature), slope, aspect, elevation and soil layer depths were kept the same. Scenario 1 was used to refine the Eq. 1 estimates for K_{sat}, by adjusting the K_{sat} adjustment multipliers for surface run-off, interflow (forest floor, A&B layers combined), baseflow (C layers combined), infiltration, and soil percolation from the forest floor to the topsoil, and from the topsoil to the subsoil. The calibrations were done by matching modeled with actual stream discharge a the daily level, using local weather records for daily rain, snow and air temperature as model input. Modelled snowpack depth was also calibrated using daily snowpack data. The ForHyM2 model runs were done for 1999 – 2004 for Moosepit Brook, and for 1997-2002 for Turkey Lakes.

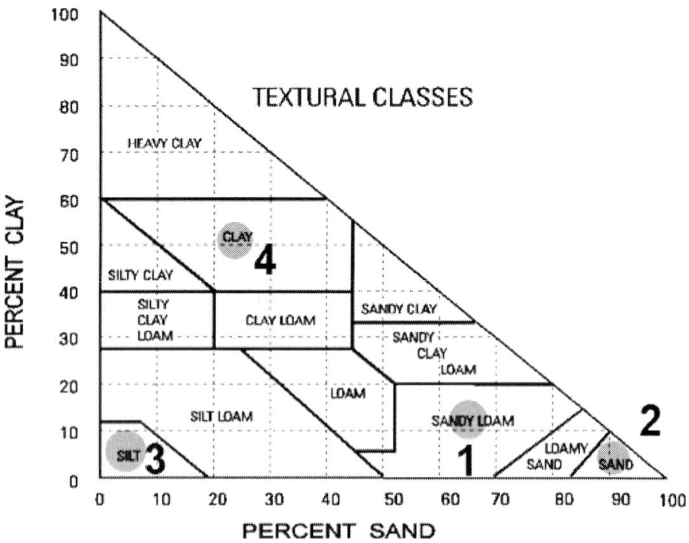

Figure 4. Mineral texture class triangle for fine soil showing texture classes for scenarios 1 - 4 (adapted from CANSIS 2000).

4. RESULTS

The results of this analysis are shown in Tables 3 to 8 and in Figs. 5 to 12 for the Moosepit Brook and Turkey Lakes study areas. Tables 3 and 4 inform about the Scenario-based changes on Dp, Db and K_{sat} for each of the two sites by topsoil and subsoil. The Db numbers indicate that the subsoil at both locations is compacted, with K_{sat} values typically 10 to 50 times lower in the subsoil than in the topsoil. Since the soil texture is sandier at Moosepit Brook than at Turkey Lakes, K_{sat} values remain higher in the subsoil at Moosepit Brook than at Turkey Lakes. Changing the topsoil texture from the actual values changes K_{sat} by about 5x upwards, and by about 10x downwards at both locations. These K_{sat} changes are similar for the somewhat coarser subsoil at Moosepit Brook. In contrast, subsoil K_{sat} is not much affected by increasing the clay and silt content, but increases with increasing sand content towards 95% by a factor of 147

Table 5 and Fig. 5 inform about the 5-year cumulative effects of the texture and OM changes on ForHyM2-modelled run-off, forest floor interflow, topsoil interflow, baseflow and stream discharge in terms of modelled mm per study period, and also in terms of modeled flow rate percentages per stream discharge. As shown, the interflow and baseflow percentage contributions to stream discharge so compiled are very sensitive to K_{sat} as well as basin slope: for intermediate K_{sat} values, interflow would dominate the base flow contributions to stream discharge within the steeper watershed at Turkey Lakes (average slope = 8%). The reverse would occur at the flatter Moosepit Brook watershed (average slope = 1%). Low subsoil permeability at Turkey Lakes would further accentuate this difference. In detail, base flow would dominate in both watersheds or at any location within the watersheds with high soil permeability and where the subsoil would not be blocked by impervious bedrock. In contrasts, locations with low overall soil permeability and low slopes would be most variable in terms of their cumulative run-off, interflow and baseflow contributions, varying from mostly baseflow to mostly interflow (Fig. 6). For example, mineral soils with high silt content (Scenario 3) would support more lateral flow in the topsoil as opposed to soils with high sand content (Scenario 2). Doubling the OM in the mineral soil (Scenario 5) would also increase baseflow, whereas reducing OM (Scenario 6) would induce the opposite. The extent of water infiltration in Scenario 4, as modeled, would be midway between Scenarios 2 and 3.

Table 4. Results for various levels of sand and OM against K_{sat}, D_b and D_p for Moosepit Brook and Turkey Lakes.

Site	Scenarios	K_{sat}, cm h^{-1}		D_b, g cm^{-3}		D_p, g cm^{-3}	
		Mineral	Subsoil	Mineral	Subsoil	Mineral	Subsoil
Moosepit Brook	1: Actual	48.40	5.95	0.95	1.61	2.48	2.59
	2: Sand	162.90	29.15	0.93	1.50	2.48	2.59
	3: Silt	3.05	0.15	1.00	1.86	2.48	2.59
	4: Heavy clay	7.15	0.50	0.99	1.80	2.48	2.59
	5: Double OM	60.60	13.30	0.72	1.48	2.33	2.54
	6: Half OM	31.35	2.70	1.14	1.70	2.56	2.62
	7: No OM	12.60	0.75	1.41	1.83	2.65	2.65
Turkey Lakes	1: Actual	39.80	0.10	1.09	1.85	2.55	2.61
	2: Sand	136.25	14.70	1.06	1.54	2.55	2.61
	3: Silt	2.50	0.05	1.15	1.92	2.55	2.61
	4: Heavy clay	5.80	0.10	1.15	1.84	2.55	2.61
	5: Double OM	56.05	0.30	0.90	1.68	2.45	2.57
	6: Half OM	30.55	0.05	1.19	1.85	2.59	2.61
	7: No OM	14.50	0.00	1.39	2.06	2.65	2.65

Figs. 6 to 9 inform about the changes in daily variations in run-off, interflow and baseflow for both locations as the soil texture changes from actual to sandy, silty and clayey (Scenarios 1 to 4, respectively, Figs. 6, 7), and soil organic matter content changes actual to 0.5 and 2 x, and 100% (Scenarios 1, and 5 to 8, Figs. 8, 9). As shown, these flows would peak faster with increasing K_{sat} (increasing sand and organic matter content), and would saturate the lower soil layers more quickly with decreasing K_{sat} and decreasing pore space, or increasing bulk density. Among the scenarios, the largest textural change on the flow regime was incurred by increasing the silt content within the already compacted subsoil at Moosepit Brook. Note that organic soils with 100% sapric organic matter would also have very low interflow and baseflow rates, and would therefore lead to relative fast soil saturation as well.

Assessing the waterflow through peatland locations within each of the two watersheds, and setting the state of decomposition of the peat equal to H1, H4, H7 and H10 produced the results compiled in Table 6. As shown, organic soils mostly composed of fibric to mesic peat (H1) would support deep percolation and baseflow, whereas organic soils mostly composed of humic peat (H10) would contain pooled water from the subsoil upwards to the surface, thereby encouraging surface run-off

Table 5. Lateral stream discharge by cumulative and percent runoff, interflow, and base flow for scenarios 1-8 for Moosepit Brook (1999-2004) and Turkey Lakes (1997-2004).

Site	Scenario	Runoff (mm)	%[2]	Interflow FF (mm)	%[2]	Interflow A&B (mm)	%[2]	Base flow (mm)	%[2]	Total Discharge (mm)
Moosepit Brook	1	2.7	0.1	202.3	4.8	682.8	16.1	3341.2	79.0	4229.0
	2	2.6	0.1	202.4	4.7	512.0	11.9	3580.4	83.3	4297.0
	3	7.0	0.2	209.5	5.1	2653.0	64.4	1251.5	30.4	4121.0
	4	2.7	0.1	202.4	4.8	1609.7	38.3	2391.4	56.9	4206.0
	5	2.6	0.1	202.3	4.8	450.0	10.6	3578.0	84.5	4233.0
	6	2.7	0.1	202.4	4.8	964.0	22.8	3054.8	72.3	4224.0
	7	2.7	0.1	202.3	4.8	1675.8	39.8	2329.9	55.3	4210.7
	8	0.0	0.0	594.8	13.2	3771.1	83.4	153.7	3.4	4519.6
	8[1]	0.0	0.0	612.4	11.6	4495.9	85.3	161.0	3.1	5269.3
	8[2]	0.0	0.0	668.6	10.6	5487.3	86.7	171.1	2.7	6327.0
Turkey Lakes	1	1.1	0.0	375.9	9.1	2990.0	72.2	772.9	18.7	4139.0
	2	3.3	0.1	376.1	8.8	301.0	7.1	3588.5	84.1	4269.0
	3	114.9	2.8	521.2	12.8	2635.0	64.8	796.5	19.6	4068.0
	4	19.4	0.5	389.1	9.4	1965.0	47.3	1782.1	42.9	4156.0
	5	2.2	0.1	376.1	9.1	2314.0	55.8	1454.4	35.1	4146.0
	6	0.5	0.0	375.8	9.1	3273.0	79.2	485.6	11.7	4135.0
	7	0.0	0.0	374.8	9.1	3457.0	83.9	290.7	7.1	4122.0
	8	0.0	0.0	69.8	1.5	4250.5	92.1	292.9	6.3	4613.1
	8[1]	0.0	0.0	70.9	1.2	5463.3	93.6	305.0	5.2	5839.2
	8[2]	0.0	0.0	72.6	1.0	6609.0	94.6	305.0	4.4	6986.5

[1] 50% coverage [2] 10% coverage

[2] % values refer to the calculated percent contributions of run-off, FF interflow A&B interflow and baseflow to stream discharge.

Note also from Table 4 and 6 that the changing K_{sat} values for decomposing peat would also have strong effects on forested peatland evapotranspiration and on stream discharge: the lower K_{sat}, the higher would be the rate of water retention and subsequent forest water uptake and evapotranspiration during the growing season (Fig. 10). In contrast, the higher K_{sat}, the faster water would be lost due to quick baseflow (Fig. 11). Outside the growing season, run-off increases, as modeled and as to be expected (Fig. 10 and 11)

Table 6. Lateral stream discharge by cumulative and percent runoff, interflow, and baseflow for scenario 8 to represent 100% peat surface deposits (von Post index set at H1, H4, H7, and H10 for the entire profile) underneath forest cover at each of the two locations.

Site	von Post	Runoff (mm)	%	Interflow FF (mm)	%	Interflow A&B (mm)	%	Base flow (mm)	%	Total Discharge (mm)
Moosepit Brook	H1	0.00	0.0	1.09	0.0	53.00	1.1	4618.27	98.8	4672.35
	H4	0.00	0.0	20.27	0.4	222.67	4.9	4291.43	94.6	4534.38
	H7	0.00	0.0	240.34	5.9	184.86	4.5	3676.06	89.6	4101.26
	H10	3275.13	97.3	74.12	2.2	13.32	0.4	4.67	0.1	3367.25
Turkey Lakes	H1	0.00	0.0	0.08	0.0	0.84	0.0	5954.36	100.0	5955.27
	H4	0.00	0.0	1.03	0.0	7.64	0.1	5641.93	99.8	5650.61
	H7	0.00	0.0	55.87	1.2	8.18	0.2	4655.45	98.6	4719.50
	H10	3057.12	97.7	50.55	1.6	4.51	0.1	16.86	0.5	3129.03

The extent water retention in terms of mm per soil layer is illustrated in Fig. 12 for the two study locations as modeled for the actual soil (Scenario 1) and for organic soil conditions (100% organic matter content, Scenario 8), starting the soil moisture content at field capacity for January 1. For the slowly draining peatland scenario (Scenario 8), subsoil moisture levels would increase from field capacity towards saturation in about one year. For the well-drained upland soil conditions (Scenario 1), K_{sat} values would be sufficiently high so that soil moisture conditions would fluctuate around the field capacity, depending on season as well as rainfall and snow melt events

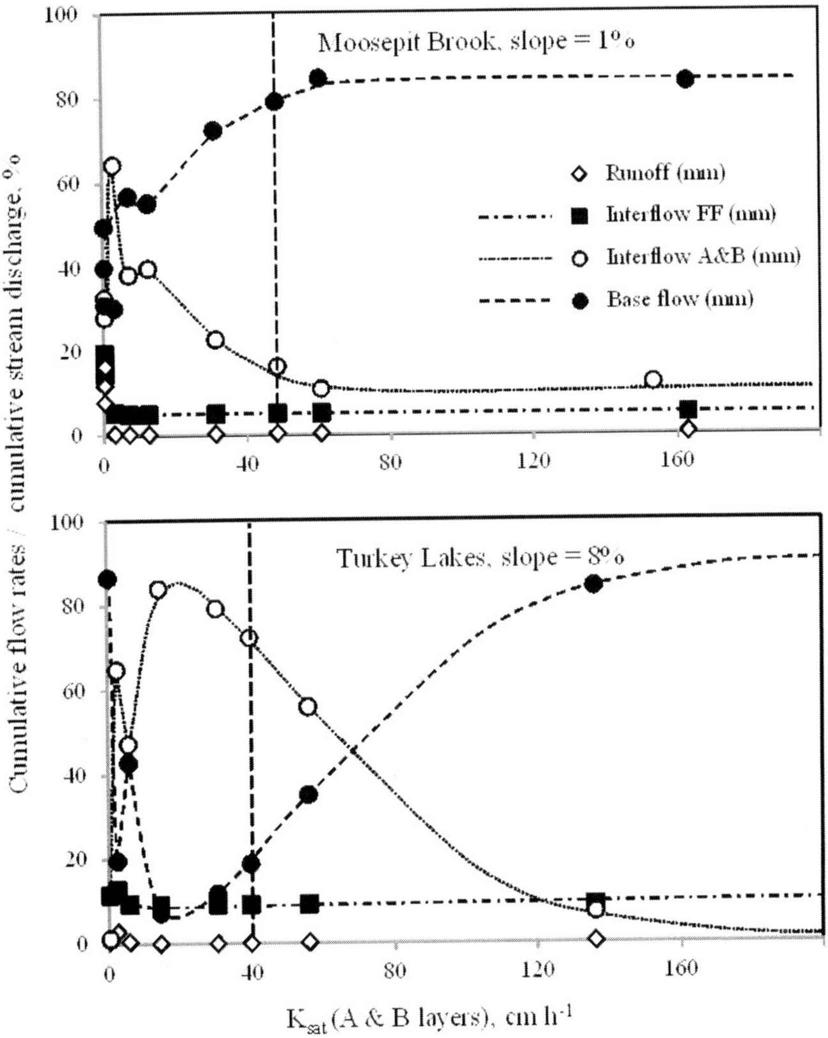

Figure 5. K_{sat} of the A&B layers by cumulative stream discharge % for Moosepit Brook (top), and Turkey Lakes (bottom) across all 8 scenarios (vertical dashed line represents the actual scenario)

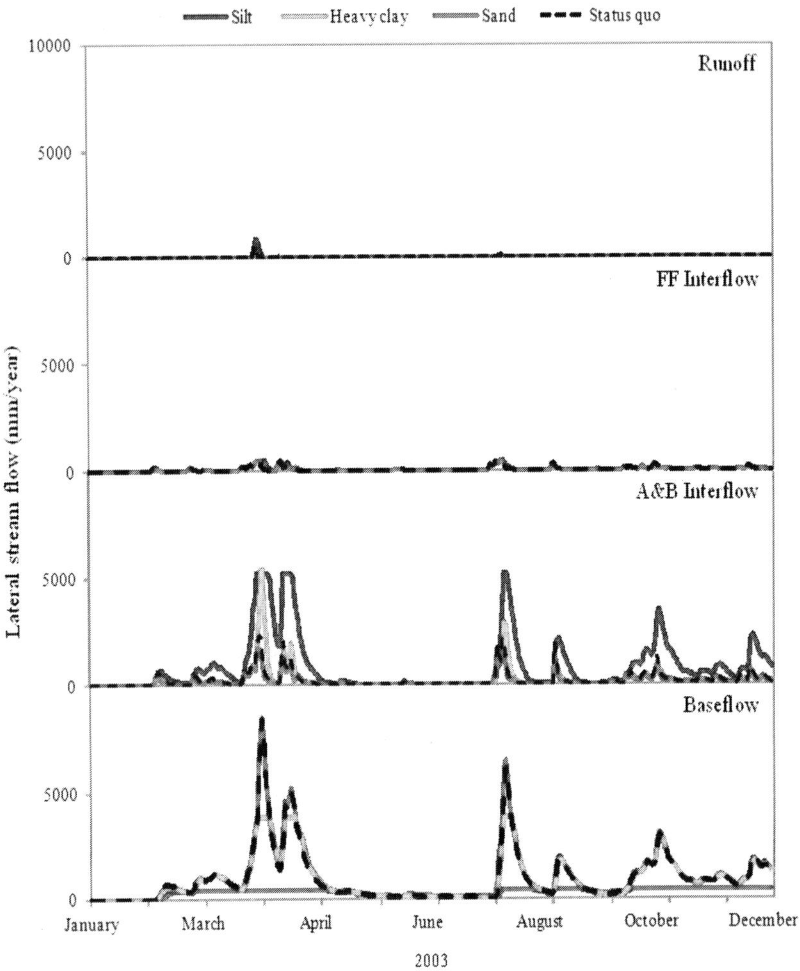

Figure 6. Run-off, forest floor and A&B interflow and base flow for Moosepit Brook, by scenario from actual to sandy, silty and clayey (Scenarios 1 to 4, respectively; 2003)

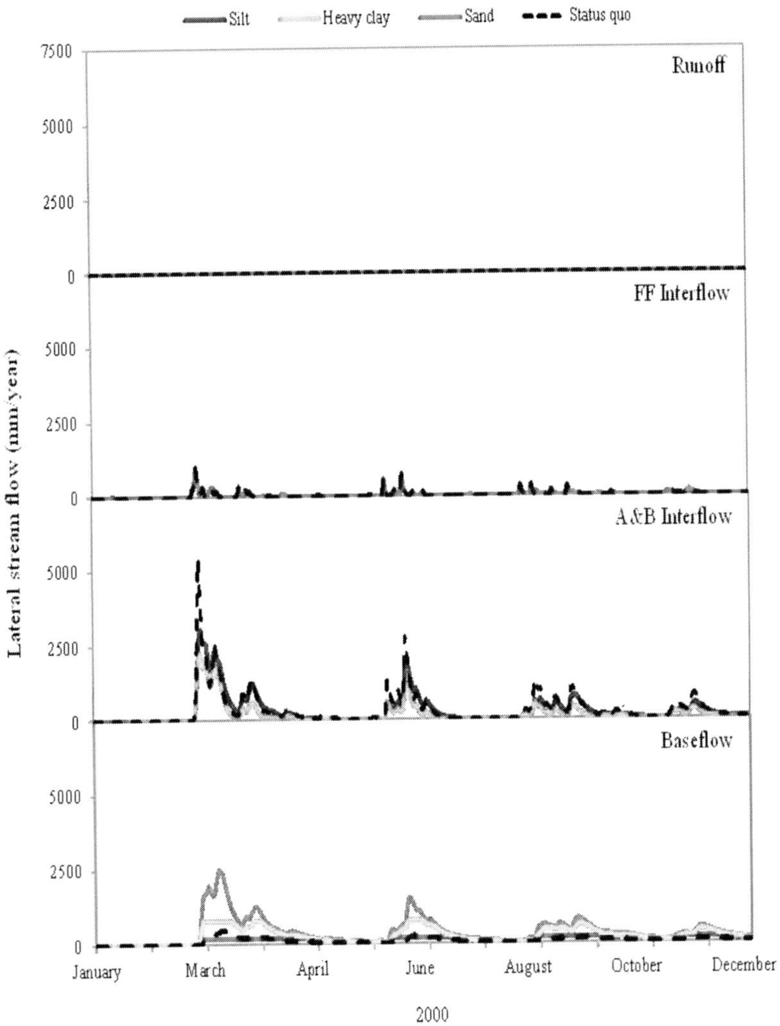

Figure 7. Run-off, forest floor and A&B interflow and base flow for Turkey lakes, by scenario from actual to sandy, silty and clayey (Scenarios 1 to 4, respectively; 2000), no runoff for any of the scenarios

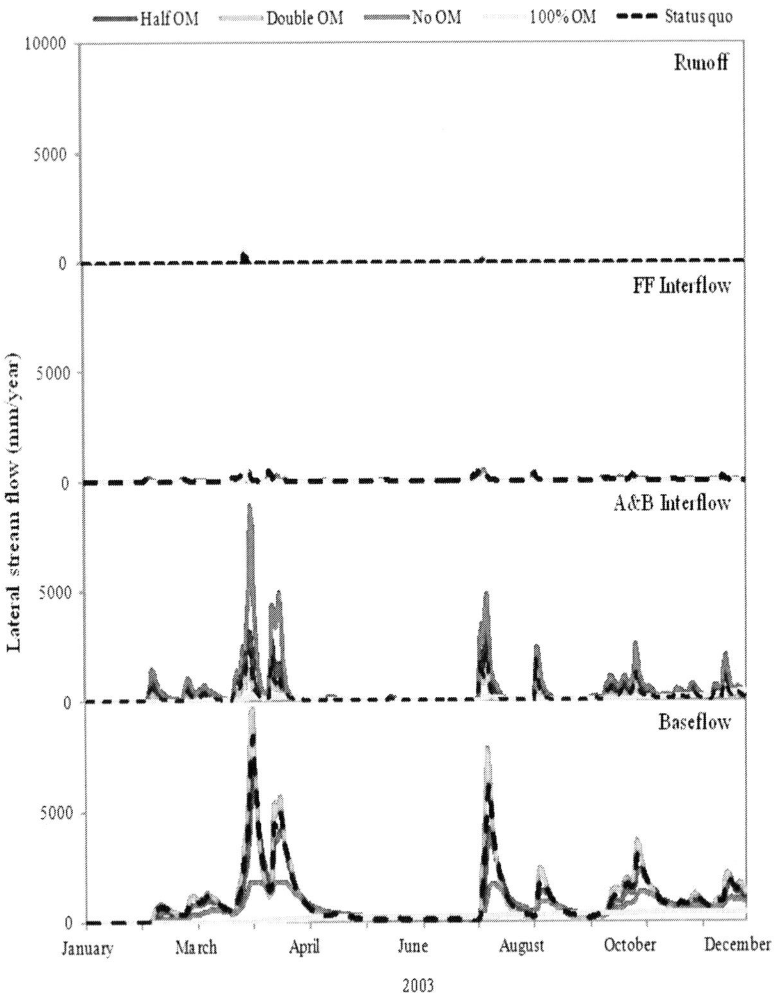

Figure 8. Run-off, forest floor and A&B interflow and base flow for Moosepit Brook, by scenario from actual to no, 0.5x and 2x actual organic matter content, and 100 % sapric organic matter (Scenarios 1 and 5 to 8, respectively; 2003)

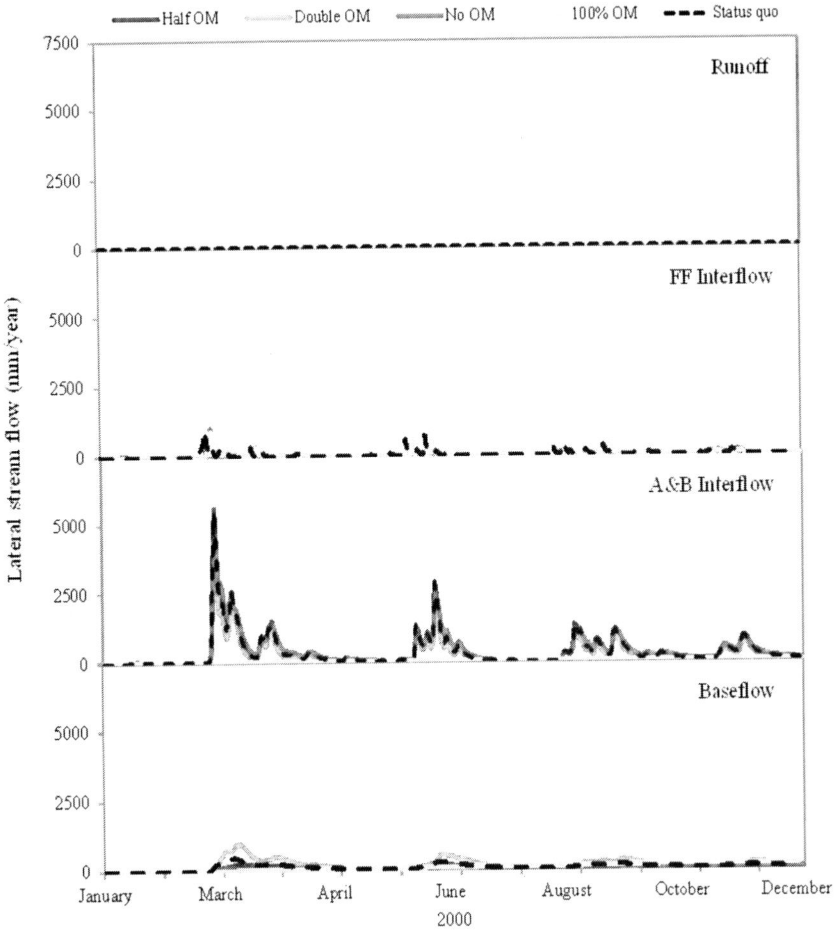

Figure 9. Run-off, forest floor and A&B interflow and base flow for Turkey Lakes, by scenario from actual to no, 0.5x and 2x actual organic matter content, and 100% sapric organic matter (Scenarios 1 and 5 to 8, respectively; 2000), no runoff for any of the scenarios.

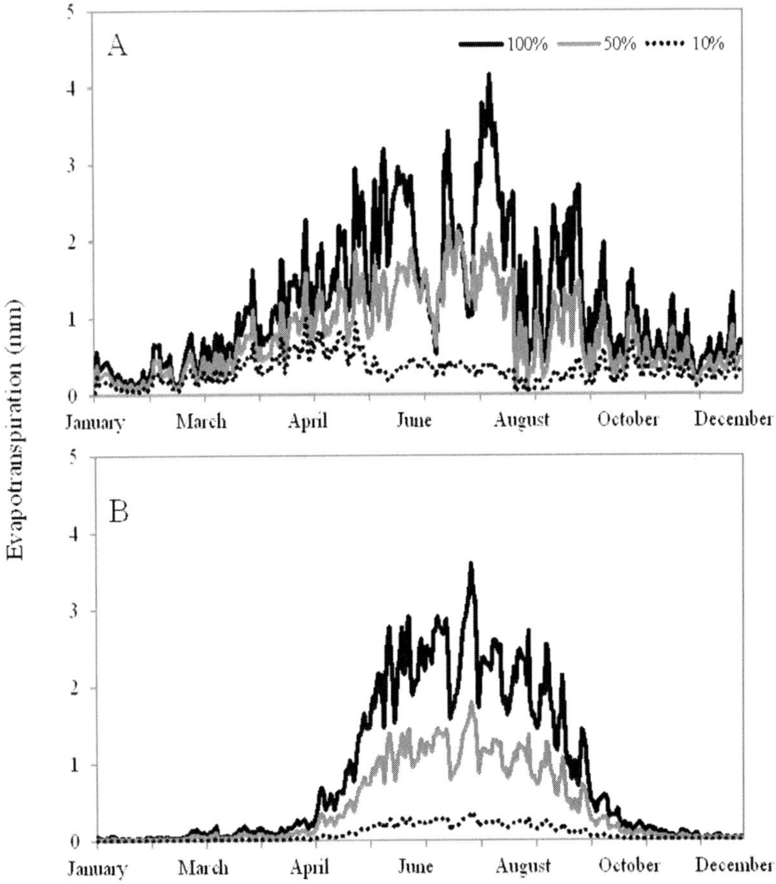

Figure 10. Evapotranspiration at Moosepit Brook (A) during 2003 and Turkey Lakes (B) during 2000, for Scenario 8 (100% OM), with actual (100% vegetation), Scenario 8[1] (50% vegetation), and Scenario 8[2] (10% vegetation). Note the difference in the extent of the growing season: wide for Moosepit Brook (maritime climate), and narrow for Turkey Lakes (continental climate).

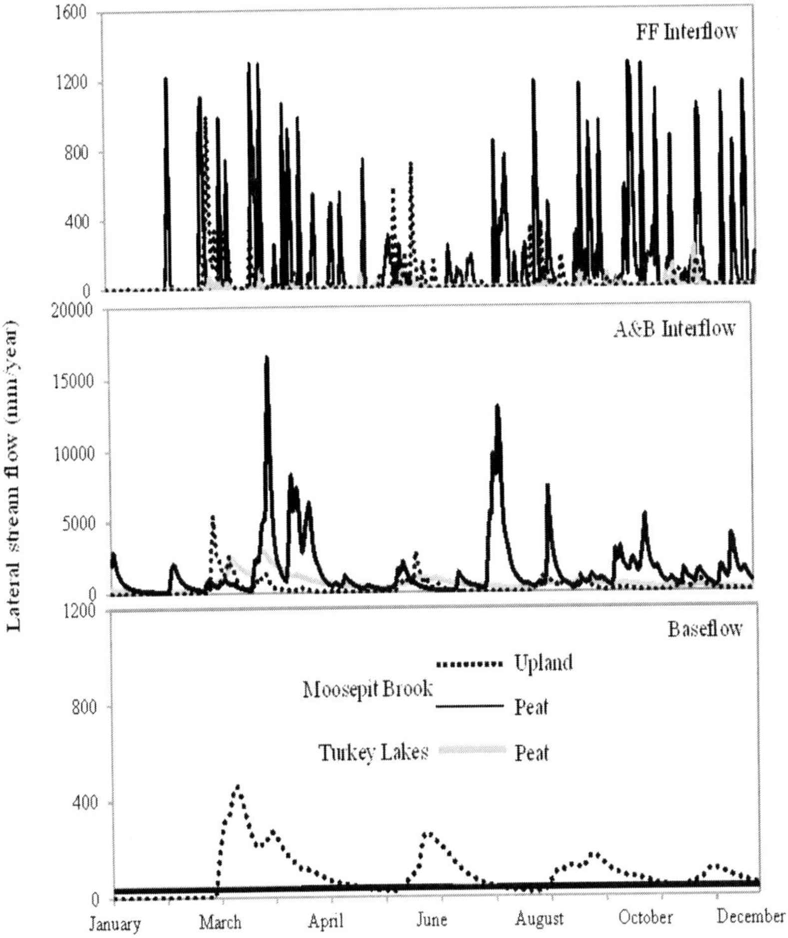

Figure 11. ForHyM2 estimated rates for daily forest floor and A&B interflow and base flow for peatland locations with a fibric - mesic – sapric layer profile at Moosepit Brook and Turkey Lakes, with 100% forest cover. Also shown: upland interflows and baseflows for the Moosepit Brook watershed (2000).

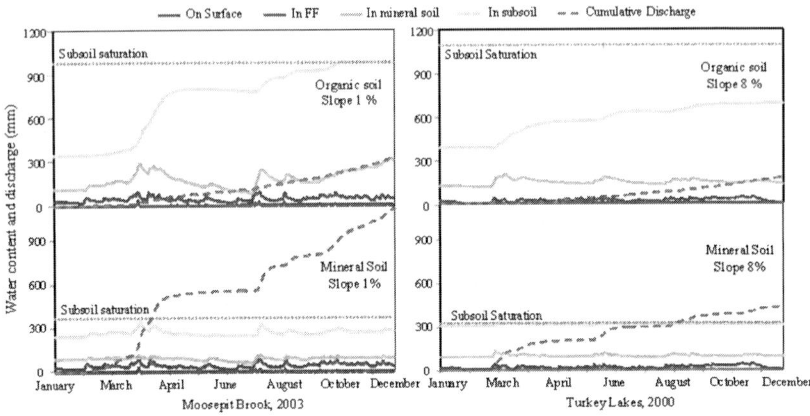

Figure 12. Soil water content on the surface, in the forest floor, in the mineral soil, and in the subsoil, as well as the cumulative discharge for Scenario 8 regarding organic soil (100% OM, top), and actual mineral soil conditions (bottom), for Moosepit Brook (2003, left) and Turkey Lakes (2000, right). Simulations start with unsaturated soil condition. Discussion

The above watershed-based K_{sat} evaluations have shown that the effective K_{sat} values for downward and lateral flow generally vary by a factor of 2 to 3 in comparison to corresponding values generated via Eqs. 1-3 [1]. As illustrated via Table 5 and subsequent figures, these variations lead to uncertainties in quantifying how water percolates through watersheds as run-off, interflow and baseflow (Fig. 5). These uncertainties also affect the flow response time, ranging generally from small delays to extended periods of flow as K_{sat} values decrease (Figs. 6 to 9). Across watersheds, however, flows tend to be well synchronized, regardless of major differences in texture, density, and organic matter content [28]. Typically, watersheds with the more compacted soils and therefore low K_{sat} values will be more peaked and will therefore be flashier than watersheds that allow deep percolation [25, 26, 27, 2]. The strongest impact of shallow to deep flow would deal with the water quality: deep water percolation during summer would lead to cooler and purer stream and seepage water with elevated pH than shallow water percolation [6]. During winter, deep percolation and persistent base flow would be warmer compared to the frost-affected surface water on poorly drained soils [6, 5]. Water flowing along the surface would also be more colored towards brown and more acidic than the more filtered and mineral-exposed water flowing at greater soil and subsoil depth [28]

While organic matter and soil density would not change drastically throughout undisturbed watersheds, such changes would occur during and after times of intense surface operations, especially under poor weather conditions. For example, forest operations during times of poor soil trafficability lead to ruts and increased soil compaction [29, 30]. In turn, soil compaction leads to lower K_{sat} values and therefore lower infiltration and hence higher surface run-off rates, thereby accelerating soil erosion and subsequent sediment transfer to streams and lakes [4]. Trails across the slopes of watersheds also affect downslope flow by compacting the soil underneath the trails, which means more water retention upslope along the trails, therefore leading to weather-effected trail destabilization, unless ditches and cross drains are installed to divert the water away from the trail beds [31]. Changes in forest cover could lead to changes in rooting space, which would – in turn – reduce the organic matter content within top and subsoils. This reduction would then alter the overall interplay between surface runoff, interflow and baseflow. Similarly, variations in climate from wet to dry (induces soil shrinking, may reduce root biomass), from frozen to non-frozen (induces collapse of frozen soil structures) would also affect K_{sat} and flow through soils by affecting the organic matter build-up, the state of soil organic matter humification, and overall changes in granular, blocky and columnar soil structures

The main advantage of the above K_{sat} formulation is that it allows for daily weather-related projections concerning downward and lateral water flow rates in forested to non-forested watersheds from times when soils are at saturation to times when soils are dry. At times of soil saturation, this quantification can then be used to estimate the effects of flow on soil stability and stream discharge. At times of drought, this quantification is can be used to estimate the effects of no flow on the remaining water reserves within soils and watersheds with and without peatland components (Fig. 12). Using ForHyM2 has the additional advantage of conducting these calculations year-round, summers through winters, based on already existing daily weather records, and extending these by way of daily, weekly, monthly or annual weather forecasts

5. ACKNOWLEDGEMENTS

Financial support for this research was received from Environment Canada, Alberta Sustainable Resource Department, and NSERC (Discovery Grant and CRD project grants).

REFERENCES

1. Jutras MF, Arp PA. Determination of hydraulic conductivity from soil characteristics and its application for modelling stream discharge in forest catchments. In (Ed.) LE, editor. Hydraulic Conductivity - Issues, Determination and Applications.: InTech; 2011. 189-202.

2. Brady NC, Weil RR. The Nature and Poperties of Soils. 13th ed. Helba S, editor. Upper Saddle River: Prentice Hall; 2001.

3. Crawford JW. The relationship between structure and the hydraulic conductivity of soil. European J Soil Sci. 1994;45: 493-502.

4. Harr RD, Fredriksen RL, Rothacher J. Changes in streamflow following timber harvest in southwestern Oregon. Forest Service Resoure Paper PNW-249. Portland, Oregon: Pacific Northwest Research Station, USDA; 1979.

5. Chi X. Hydrogeological assessment of stream water in forested watershed: temperature, dissolved oxygen, pH, and electrical conductivity. MSc. Thesis. Fredericton: University of New Brunswick, Forestry; 2008.

6. Steeves MT. Pre- and post-harvest groundwater temperatures and levels, in upland forest catchments in Northern New Brunswick. MSc. F Thesis. Frederiction, NB: University of New Brunswick, Department of Forestry; 2004.

7. Bouma J. Field measurement of soil hydraulic properties characterizing water movement through swelling clay soils. J Hydol.. 1980; 45: 149-158.

8. Löfkvist J. Modifying soil structure using plant roots. PhD Thesis. Uppsala: Swedish University of Agricultural Sciences, Department of Soil Science; 2005.

9. Yanni S, Keys K, Meng FR, Yin X, Clair T, Arp PA. Modelling hydrological conditions in the maritime forest region of south-west Nova Scotia. Hydrol. Processes. 2000b; 14: 195-214.

10. Arp PA, Yin X. Predicting water fluxes through forests from monthly precipitation and mean monthly air temperature records. Can. J. For. Research. 1992; 22: 864-877.

11. Verry ES, Boelter DH, Paivanen J, Nichols DS, Malterer T, Gafni A. Physical properties of organic soils. Chapter 5. In Kolka RK,

Sebestyenm SD, Verry ESBKN, editors. Peatland biogeochemistry and watershed hydrology at the Marcell Experimental Forest. Boca Raton: CRC Press; 2011. 135-176.

12. Päivänen J. Hydraulic conductivity and water retention in peat soils. Acta Forestalia Fennica. 1973; 129: 1-70.

13. Letts MG, Roulet NT, Comer NT, Skarupa MR, Verseghy DL. Parametrization of peatland hydraulic properties for the Canadian land surface scheme. Atmosphere-Ocean. 2000; 38: 141-160.

14. Boelter DH. Important physical properties of peat materials. In Proceedings, third international peat congress; 1968 August 18-23. Quebec, Canada: Department of Energy, Minds and Resources and National Research Council of Canada; 1968. 150-154.

15. Meng FR, Bourque PA, Jewett K, Daugharty D, Arp PA. The Nashwaak experimental watershed project: analysing effects of clearcutting on soil temperature, soil moisture, snowpack, snowmelt and streamflow. Water, Air Soil Pollute. 1995; 82: 363-374.

16. Balland V, Pollacco JAP, Arp PA. Modeling soil hydraulic properties for a wide range of soil conditions. Ecol. Model. 2008; 219: 300-316.

17. Päivänen J. The bulk density of peat and its determination. Silva Fennica. 1969; 3: 1-19.

18. Gafni A, Brooks KN. Hydraulic characteristics of four peatlands in minnesota. Can. J. Soil Sci. 1990; 70: 239-253.

19. Silc T, Stanek W. Bulk density estimation of several peats in northern Ontario using the von Post humification scale. Can. J. Soil Sci. 1977; 57: p. 75.

20. Szajdak LW, Szatylowicz J, Kolli R. Peats and peatlands, physical properties. In Gliński K, Horabik J, Lipiec J, editors. Encyclopedia of Agrophysics.: Springer; 2011. 551-555.

21. Letts MG, Roulet NT, Comer NT, Skarupa MR, Verseghy DL. Parametrization of peatland hydraulic propoerties for the Canadian land surface scheme. Atmosphere-Ocean. 2000; 1: 141-160.

22. Dube S, Plamondon A, Rothwell R. Watering up after clear-cutting on forested wetlands of the St. Lawrence lowland. Water Resource Research. 1995; 31: 1741-1750.

23. Boelter DH. Hydraulic Conductivity of Peats. Soil Sci. 1965; 100: 227-231.

24. Boelter DH, Verry E. Peatland and water in the northern lake states. General Technical Report NC-31. St. Paul, MN: Forest Service, North Central Forest Experiment Station, U.S. Dept. of Agriculture; 1977.

25. Lull HW. Soil Compaction on Forest and Range Lands: Forest Services, U.S. Department of Agriculture; 1959.

26. Kramer PJ, Boyer JS. Chapter 4: Soil and Water. In Water relations of plants and soils.: Academic Press, Inc; 1995. 84-114.

27. Kozlowski TT. Soil compaction and growth of woody plants. Scand. J. For. Research. 1999; 6: 596-619.

28. Jutras MF, Nasr M, Castonguay M, Pit C, Pomeroy J, Smith TP, et al. Dissolved organic cardon concentrations and fluxes in forest catchments and streams: DOC-3 model. Ecol. Model. 2011; doi:10.1016/j.ecolmodel.2001.03.035.

29. McNabb DH, Startsev AD, Nguyen H. Soil wetness and traffic level effects on bulk density and air-filled porosity of compacted boreal forest soils. Soil Sci. Soc. J. America. 2001; 65: 1238-1247.

30. Froehlich HA, McNabb DH. Minimizing soil compaction in Pacific Northwest forests. Proceedings Forest Soils and Treatment Impacts Conference. Knoxville, TN; 1984.

31. Jamshidi R, Jaeger D, Raafatnia N, Tabari M. Influence of two ground-based skidding systems on soil compaction under different slope and gradient conditions. Int. J. For. Eng. 2008; 19: 9-16.

4

CHAPTER

METHOD OF RELATING GRAIN SIZE DISTRIBUTION TO HYDRAULIC CONDUCTIVITY IN DUNE SANDS TO ASSIST IN ASSESSING MANAGED AQUIFER RECHARGE PROJECTS: WADI KHULAYS DUNE FIELD, WESTERN SAUDI ARABIA

Oliver M. Lopez [1], Khan Z. Jadoon [1] and Thomas M. Missimer [2,*]

[1] *Water Desalination and Reuse Center, King Abdullah University of Science and Technology, Thuwal, 23955-6900, Saudi Arabia*

[2] *U. A. Whitaker College of Engineering, Florida Gulf Coast University, 10501 FGCU Boulevard South, Fort Myers, FL 33965-6565, USA*

ABSTRACT

Planning for use of a dune field aquifer for managed aquifer recharge (MAR) requires that hydraulic properties need to be estimated over a large geographic area. Saturated hydraulic conductivity of dune sands is commonly estimated from grain size distribution data by employing some type of empirical equation. Over 50 samples from the Wadi Khulays dune field in Western Saudi Arabia were collected and the grain size distribution, porosity, and hydraulic conductivity were measured. An evaluation of 20 existing empirical equations showed a generally high degree of error in the predicted compared to the measured hydraulic conductivity values of these samples. Statistical analyses comparing estimated versus measured hydraulic conductivity demonstrated that there is a significant relationship between hydraulic conductivity and mud percentage (and skewness). The modified Beyer equation, which showed a generally low prediction error, was modified by adding a second term fitting parameter related to the mud concentration based on 25 of the 50 samples analyzed. An inverse optimization process was conducted to quantify the fitting parameter and a new empirical equation was developed. This equation was tested against the remaining 25 samples analyzed and produced an estimated saturated hydraulic conductivity with the lowest error of any empirical equation. This methodology can be used for large dune field hydraulic conductivity estimation and reduce planning costs for MAR systems.

Keywords: managed aquifer recharge; dune sediments; grain size distribution; hydraulic conductivity; porosity; empirical equations

1. INTRODUCTION

Within the context of water management, dune aquifers have been used to store, treat and recover water for more than 60 years [1,2,3,4,5,6,7]. In arid lands, rainfall events are sparse and commonly produce flash flood events that are not conductive to recharge of most aquifers (e.g., wadi aquifers) [8]. Dune aquifers are known to be effective at allowing natural recharge [9,10,11,12,13]. Increasingly, dune aquifers are being considered for use in storing captured storm water or treated wastewater for future use [14]. Therefore, there is a need to understand the hydraulic properties of dune aquifers to assess water storage and recovery potential using the concept of aquifer storage and recovery (ASR).

Dunes are an important part of the landscape in arid regions, especially within the desert areas of Australia, the Sahara, and Arabia, where dune field cover constitutes 31%, 28%, and 26%, respectively, of the total land area [15]. Investigations of dune fields have been conducted to assess the physics of wind-transported sediment mechanisms [16], the sediment grain size properties [17,18,19], and the movement of the dunes because of geohazard potential [20]. Little consideration has been given to assessment of the sediment hydraulic properties in terms of hydraulic conductivity (saturated), porosity, and specific yield [21,22,23].

Empirical equations are commonly used to estimate saturated hydraulic conductivity from grain size data in various sediment types to populate grids in models or to set limits for the maximum rate of recharge for unsaturated zone flow [24,25,26]. There are over 30 published empirical equations commonly used to perform this task with each one having a different set of assumptions for its use. Unfortunately, many of these equations were developed using very limited databases with the comparative number of measured grain size distributions analyses and hydraulic conductivity measurements being limited to a population as small as 20. Considerable improvements to the accuracy of these empirical equations for analysis of specific depositional environments have been made recently [26]. Herein, we analyze the use of the inversion mathematical method to calibrate a new equation for relating the grain size distribution to the hydraulic conductivity of interior dune sediment at a specific location and compare the developed equation to the improved method of Rosas et al. [26].

In this research, the grain size distribution, porosity, and hydraulic conductivity variations in an interior dune field are measured in an effort to (1) provide a characterization of these parameters along the cross-section of a dune, and (2) develop an improved empirical relationship between grain size distribution and hydraulic conductivity for this dune field, using a mathematical inversion method, (3) compare the developed equation to other empirical equations used to estimate interior dune field hydraulic conductivity, and (4) investigate the impact of the mud percentage (silt and clay) fraction on the hydraulic conductivity. Furthermore, a calibration procedure was proposed to estimate the fitting parameter of the developed empirical relationship by considering twenty five samples collected in the Khulays dune field, then applying the equation to prediction of another set of 25 samples for assessment of predictive error. This methodology could be applied to various other depositional environments to facilitate estimates of hydraulic conductivity without having to measure it in a large numbers of samples in the laboratory. This research applies to prediction of saturated hydraulic conductivity. It is clearly understood that the packing of the dune sediment has an impact on the saturated hydraulic conductivity and that the empirical equations do not take this into consideration, so the calculated values can be assessed to be similar to dune sands located within 10 to 20 m of the surface, which are not well compacted.

Four research hypotheses where developed for testing and are: (1) most empirical equations currently used to estimate saturated hydraulic conductivity from grain size distribution of dune sands have large errors; (2) a significant statistical correlation exists between the percentage of mud (silt and clay) in dune sand and the hydraulic conductivity; (3) a significant statistical relationship exists between saturated hydraulic conductivity and the dispersion (and other statistical moments) in interior dune sands, and (4) inversion modeling can be applied to calibrate a field specific empirical equation relating saturated hydraulic conductivity and grain size distribution in interior dune sands.

2. BACKGROUND AND METHODS

2.1. Site Description

The Wadi Khulays aquifer system, located about 50 km northeast of Jeddah, Saudi Arabia (Figure 1), consists of several valleys that drain from the western slope of the Arabian Shield into the Wadi Khulays plain near the Red Sea coast [27]. A dune field was formed by the prevailing northwesterly wind transporting alluvial deposits from the northern sheet wash and Wadi plains of Wadi Khulays to the south side, centered at about 22°6′ N, 39°14′ E [28,29]. Dune heights range from 3.5 to 8 m, depending on the location within the field. The dune sediment is predominantly quartz sand with some other siliciclastic minerals. Dune sediments lie on desert pavement that dips slightly westward and contains a significant percentage of mud and has a low hydraulic conductivity. Moisture was found at the base of the dune sediments. No significant vegetation occurs within the dune field.

2.2. Field Study and Laboratory Methods

The sediments were evaluated using standard laboratory methods for the determination of hydraulic conductivity, grain size distribution and porosity. First, samples were collected from the upper 5 to 10 cm across several dunes at five locations beginning at the toe and ending downwind of the avalanche face within the inter-dune area (Figure 2A). The dunes sampled occurred in two 800 m long transects (green markers on Figure 2B,C). A total of 50 samples were collected for full analysis. An additional 24 samples were collected from dune crests covering a surface area of about 37,500 m2 and were analyzed only for grain size distribution (Figure 2B, blue markers).

Figure 1. Map of Saudi Arabia showing the location of the dune field site.

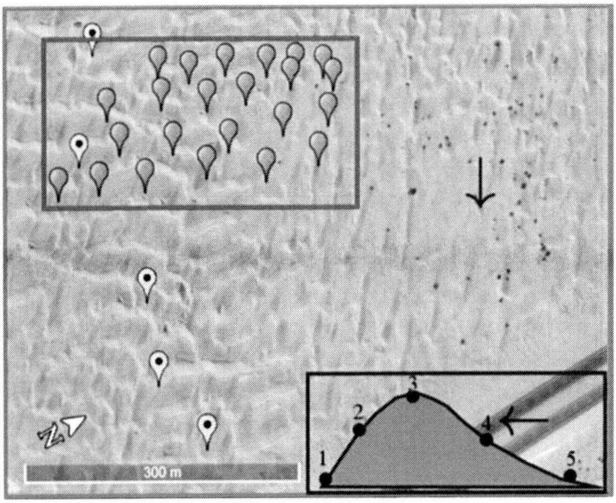

Figure 2. Wadi Khulays dune field sample locations (black arrows show the predominant wind direction). Green markers show the location of dunes selected for full sampling (analysis of grain size, porosity and hydraulic conductivity across the dune). Blue markers show a second sampling for grain size analysis only from dune crests. The blue region corresponds to the analysis shown in Figure 8. Inset: Diagram showing the sample location across individual dunes.

The grain size distribution was determined using the standard sieving technique as described by Tanner and Balsillie [29] and the American Society of Testing and Materials [30]. About 70 g of sample were analyzed using 34 standard sieves with size increments corresponding to 0.25-phi units (phi is equal to −log to the base 2 in mm). The amount of mud (silt and clay) in a sample was determined as the weight of the sample left in the pan after passing through all of the sieves with the smallest being 0.0063 mm. The sieving process involved the continuous rotation and tapping of the sieves by the use of a Ro-Tap machine for 30 min as recommended by Tanner and Balsillie [29]. The weight retained within each sieve was measured to an accuracy of 0.01g and the statistical moments were calculated following the approach described in Tanner and Balsillie [29], using custom software developed by Rosas et al. [26].

The hydraulic conductivity was measured using a standard constant head permeameter based on the methodology described by Wenzel [31] and following the American Society for Testing and Materials standard D2434-68 [32]. A constant head was applied to an approximate 10 cm column filled with saturated sediment sample and water flow was from bottom to top through the cylinder. The hydraulic conductivity was obtained by applying Darcy's Law of flow through porous media and was reported in meters per day.

In order to obtain an estimate of total porosity of each sample, a 250 mL (cc) graduated cylinder was first filled partially with a known volume of water. Sediment was carefully added to the cylinder and allowed to settle and compact. Additional water and sediment were added to the column. Care was taken to not allow air entrapment within the saturated column. The column was also compacted slightly by tapping the side of the cylinder with a rubber mallet to approximate natural system packing conditions near surface. The volume of sediment was determined along with the volume of water added. When the water level in the cylinder rose above the surface of the sediment, a correction was made to the water volume added (subtraction). Then, the estimated porosity was determined by dividing the volume of water by the volume of sediment.

2.3. Mathematical Analysis

Among the 21 empirical equations that estimate the hydraulic conductivity from the grain size distribution, most of them use a relationship in the form shown in Equation (1), where n is the porosity and β is a coefficient that may depend on the shape of the grain, type of depositional environment or other factors, d10 is the effective grain diameter (at which 10% of the sample is finer), g is the gravitational acceleration and ν is the kinematic viscosity of water.

$$K = \beta \frac{g}{\nu} \xi(n) d_{10}^2 \tag{1}$$

Rosas et al. [26] compared the effectiveness of these equations for different depositional environments, and provided depositional environment-dependent β coefficients, as well as offset values (bias) for beach sands, offshore marine sands, and dune sands (coastal and interior), as well as for river sediments. In this study, the four equations that showed the best fit for interior dune sands, as recommended in Rosas et al. [26], were selected as target equations to be modified by the inclusion of other grain size distribution parameters, based on the significance of their correlation with the hydraulic conductivity values. The inclusion was done by adding a second term to the target equations, which contains the mud percentage and a fitting parameter. The inclusion of a second term in the Beyer Equation [33] is shown below:

$$K\left[\frac{m}{day}\right] = P_1 \frac{g}{v} log\left[\frac{500}{C}\right] + P_2 exp(-mud) \tag{2}$$

where C is the coefficient of uniformity defined as C = d60/d10 (d60 is the diameter at which 60% of the sample is finer), P1 and P2 are the fitting parameters to be optimized. The inverse problem was formulated in the sense of root mean square error and the objectivity function was accordingly defined as:

$$\min \emptyset(\mathbf{b}) = \sqrt{\frac{\sum_i (K(i) - K^*(i, \mathbf{b})^2}{n}} \tag{3}$$

where K(i) and $K*$ (i,b) are the vectors containing the measured and modeled hydraulic conductivity, respectively, b = [P1,P2] is the parameter vector to be estimated, and, n is the number of samples. An inversion the first twenty-five measurements was used to estimate the fitting parameters P1 and P2. To properly deal with the nonlinearity of the objective function during minimization and, in particular, avoid being trapped in local minima, we adopted a global optimization approach. Hence, we used the shuffled complex evolution (SCE-UA) algorithm for the global optimization. This global optimization algorithm has been widely used in hydrologic modeling [34], and proved to be consistent and efficient for finding the global optimum parameter values of integrated hydrogeophysical and hydrologic models [35,36].

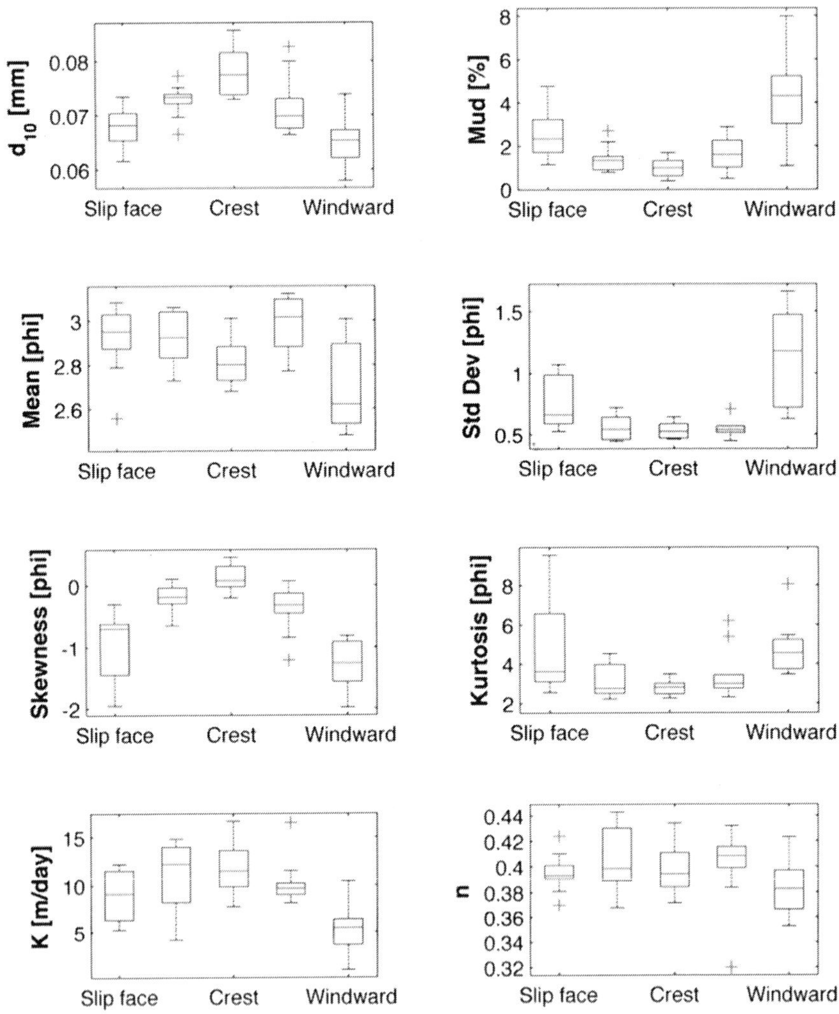

Figure 3. Summary of grain size and hydraulic properties along the cross-section of the dunes. Red crosses are outlier samples.

3. RESULTS

3.1. Grain Size, Porosity, and Hydraulic Conductivity Analyses

Measured grain size statistical parameters, d10, mud percentage, measured porosity and measured hydraulic conductivity of the 50 samples collected across the individual windward and leeward dunes are summarized in Figure 3. The grain size statistical properties d10, mud (%), standard deviation, skewness, and kurtosis, and the measured hydraulic conductivity vary in a regular pattern across the dune profile, while mean grain diameter does not vary in the same pattern (Figure 3). The dunes crests have significantly higher hydraulic conductivities and d10 values, corresponding to the lowest percentage of mud content. Porosity of the dune sands was found to be relatively uniform with a mean of 0.39 and a standard deviation of 0.02. The red-cross marks on the figure show the location of outlier samples.

3.2. Empirical Hydraulic Conductivity Estimates from Grain Size Distribution

An analysis of 21 empirical methods used to estimate hydraulic conductivity from grain size analysis shows that there is a large error between measured and estimated values using most of the methods (Figure 4A) [37,38,39,40,41,42,43,44,45,46,47,48,49,50,51]. Each of the 21 methods contains somewhat different assumptions and specific equations should be applied to specific recommended sediment types, but many of these methods have not been used properly in numerous investigations, particularly by practitioners in the engineering field. It is evident in viewing this graphic that the estimate accuracy is based on the assumptions of each equation and perhaps on the properties of the sediments at the specific location being investigated. Equations for the listed methods and assumptions for use are contained in Rosas et al. [26]. The Beyer, Harleman, Hazen (modified) and Kozeny-Carman methods showed the best fit for hydraulic conductivity of interior dune sands in this analysis. Coefficients used by these methods were modified for dune sands in Rosas et al. [26] and showed a significant improvement (Figure 4B). As shown, there is still room for additional improvement of these modified equations.

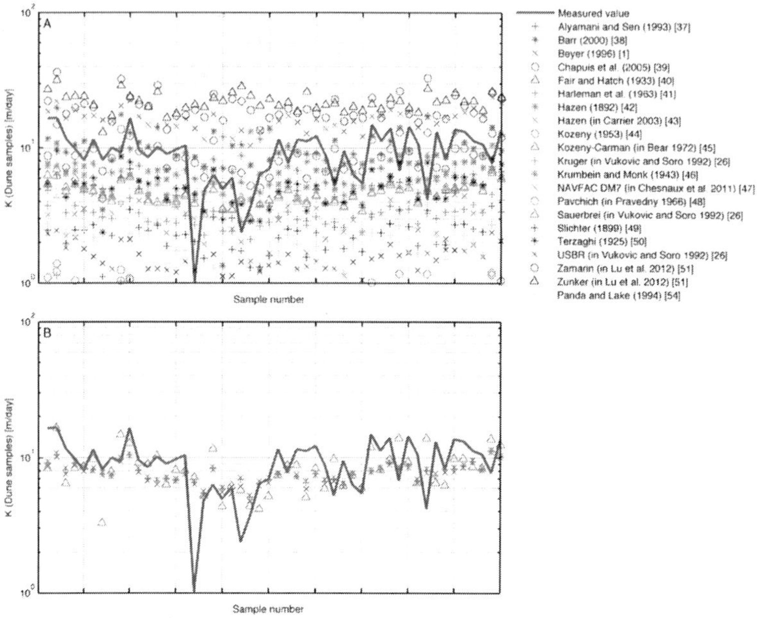

Figure 4. Estimated hydraulic conductivity using 21 empirical methods for all Wadi Khulays dune sample compared to the measured values (blue line) (A). (B) Comparison of the estimated hydraulic conductivities to the measured values using the modified methods for dune interior dune sands using the modified equations in Rosas et al. [26].

3.3. Statistical Analyses between Sediment Grain Size Properties and Hydraulic Conductivity

Linear relationships between the statistical moments and other properties of the grain size distribution vs. measured are shown as plots in Figure 5. The correlation coefficients R2 and p are shown in Table 1. The p-values are reported as −log(p) in order to compare the order of magnitude of the correlations, with a significant correlation being considered to have a value of −log(p) of 2 or more ($p < 0.01$). Even though the R2 values were very low for all the parameters considered (R2 < 0.4), significant p-values were found for the correlation between the hydraulic conductivity and d102, the standard deviation, skewness and kurtosis of the distribution in phi units (−log2 mm; [52]), as well as for the mud percentage of the sample (fraction <0.0625 mm).

Table 1. Correlation coefficients between hydraulic conductivity and grain size distribution parameters.

Correlated Parameters	R^2	$-\log(p\text{-Value})$
K, d_{10}^2	0.34	4.9
K, mean	0.04	0.8
K, standard deviation	0.30	4.4
K, skewness	0.39	6.0
K, kurtosis	0.22	3.2
K, mud	0.37	5.6
Mud, mean	0.16	2.4
Mud, standard deviation	0.72	14.2
Mud, skewness	0.45	7.1
Mud, kurtosis	0.11	1.7
K, $F = d_{10}^2 \exp(-Mud)$	0.56	4.8

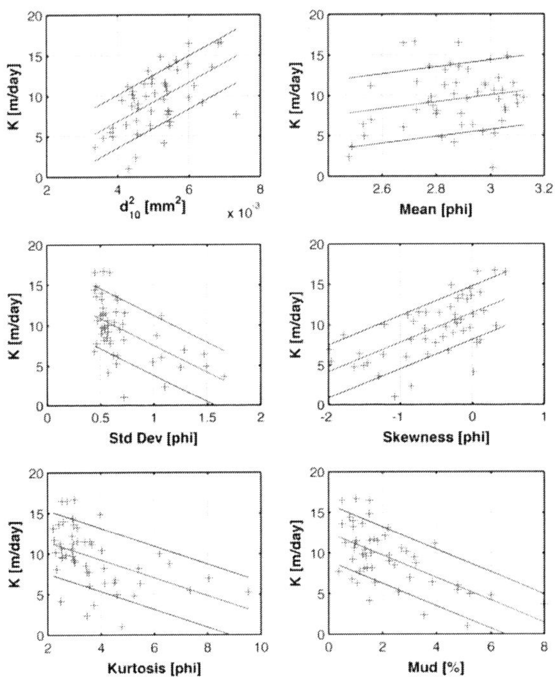

Figure 5. Relationship between mud percentage and other grain size distribution statistical parameters.

The mud percentage in the samples was positively correlated with the standard deviation (R2 = 0.72) and negatively correlated with the skewness of the distribution (R2 = 0.45) (Figure 5).

3.4. Optimization of the Target Equations by Inclusion of a Second Parameter

The percentage of mud was used for development of the fitting parameters (P1 and P2) for Equation (2). Figure 6A,B shows the convergence of the P1 and P2 parameters, respectively, to the global optimum values. For the inversion, the optimization parameter space was set to $-1 \times 10^{-2} \leq P1 \leq 1 \times 10^{-2}$ and $-10 \leq P2 \leq 10$. In the first 180 iterations the optimization algorithm searched different parameters set in the provided parameter space and later converged to the optimum values. The stopping criterion for the inversion was specified with respect to the convergence criterion in which an optimum was assumed to be reached when the objective function did not improve by more than 0.01% in 10 successive evolution loops. The estimated parameters P1 and P2 were 4.0×10^{-4} and −0.595.

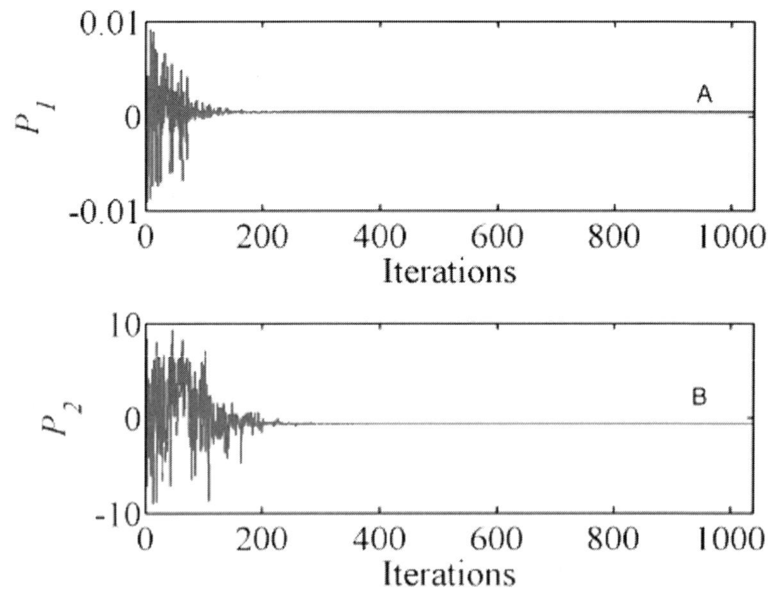

Figure 6. Convergence in the optimization of P1 and P2.

Figure 7 shows the response surfaces of the objective functions for the P1 and P2 parameter plane. The range of each parameter has been divided into 500 discrete values resulting in 250,000 objective function calculations for the plot. The white star marker corresponds to the solution found by the SCE optimization algorithm. Response surface analysis of the objective function is important as it provides valuable insights into the uniqueness of the inverse solution, the sensitivity of the model to the different parameters, and parameter correlations. The topography of the objective function shows that the parameters P1 and P2 are not correlated and a U-valley can be observed in the direction of the P2 parameter. Furthermore, in the direction of parameter P1, the V-shape valley of decreasing and increasing trend of the objective function shows that the objective function is more sensitive to the parameter P1 as compared to the P2.

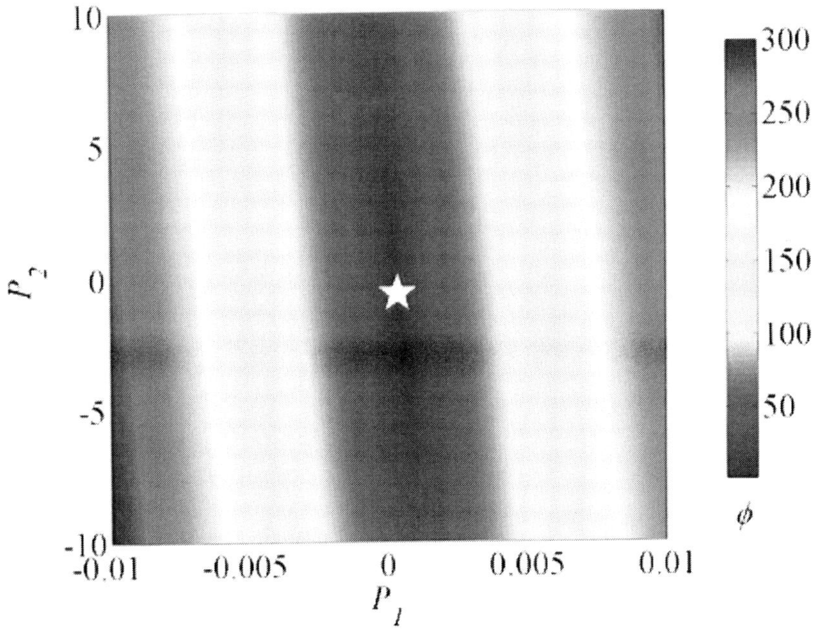

Figure 7. Response surfaces of the objective functions for P1 and P2 parameter plane.

Table 2 shows a comparison of the R-square and RMSE values obtained for the four best fitted empirical equations for interior dune sands: Beyer [33], Harleman et al. [41], Hazen [42] and Kozeny-Carmen [45], and the new equation by including mud (%) and two fitting parameters. The fitting parameters were estimated by inversion using first 25 measurements and for the rest of the measurements model hydraulic conductivity was estimated by using the estimated fitting parameters in the new equation. Values of R2 and RMSE indicate that the new equation estimates modeled hydraulic conductivity close to the measured hydraulic conductivity. The RMSE improvement is calculated with respect to Equation (2).

Table 2. Performance of empirical equations.

Equation	R^2	RMSE (m/day)	RMSE Improvement (%)
Beyer	0.60	3.04	25%
Harleman	0.59	3.01	23%
Hazen	0.59	3.03	24%
Kozeny-Carmen	0.36	3.38	38%
Equation (2)	0.61	2.44	

4. DISCUSSION

The patterns along the cross-section of the dunes were similar for the d10 value, the mud percentage and the hydraulic conductivities; dune crests showed higher hydraulic conductivities and d10 values, as well the lowest mud content, which were progressively lower (hydraulic conductivity and d10) and higher (mud percentage) towards the extremities of the dune.

Since the dunes are constantly being turned over by physical processes (wind action and gravity), the patterns reveal the vertical distribution of the properties inside the dune. Sediments within the interdune areas were poorly sorted as indicated by a higher standard deviation. Coarser gravelly grains tend to accumulate in these areas due to gravity, and these larger grains offer a protection for the finer particles against wind erosion. Most of the samples were negatively skewed except for a number of samples collected from the crests.

Most of the equations found in the literature for the determination of hydraulic conductivity from grain size distribution use the square of the d10 value and a function of the porosity [26]. The porosity of the dune sands from the Wadi Khulays dune field was determined to be relatively uniform and therefore the effect of porosity on the hydraulic conductivity could not be evaluated.

The skewness of the grain distribution, as well as the mud content showed a statistically significant correlation with the hydraulic conductivity. None of the equations used in the Rosas et al. [26] study make use of the skewness in the distribution. Panda and Lake [53] modified the Kozeny-Carman equation to include a corrective factor based on the skewness and coefficient of variation ($Cdp= \sigma/\overline{dp}$ where $\overline{dp}$ is the mean grain diameter) of the particle size distribution (psd). The equation was found to underestimate the hydraulic conductivity of the dune sands and is problematic for a number of samples with high standard deviation in the psd (Figure 4). Krumbein and Monk [46] use the standard deviation of the distribution in phi units, however, the estimated hydraulic conductivities by this method were shown to be ineffective in determining the values for dune sands (Figure 4; [26]). Moreover, evaluating the statistical moments of the grain size distribution is not a straightforward task. Because the mud percentage is significantly correlated with both the hydraulic conductivity values and the skewness of the distribution, it was selected as a proxy parameter that can be used to improve the effectiveness in determining the hydraulic conductivity from grain size distribution.

The ability to accurately predict the hydraulic conductivity of several samples from grain size distribution should significantly reduce the effort in the characterization of large areas, such as dune fields. In this study, the inclusion of the mud term significantly reduced the RMSE of the predicted hydraulic conductivity values. This allowed for the calculation of the spatial distribution of the hydraulic properties over a large area of the dune field by measuring only the grain size distribution parameters and applying Equation (2) (Figure 8). Since the porosity of the dune sands was relatively uniform, the function of the porosity was assumed to be constant. This approach may be used in large dune areas to determine the appropriate locations for water infiltration or can be correlated with the infiltration rate in dune recharge studies.

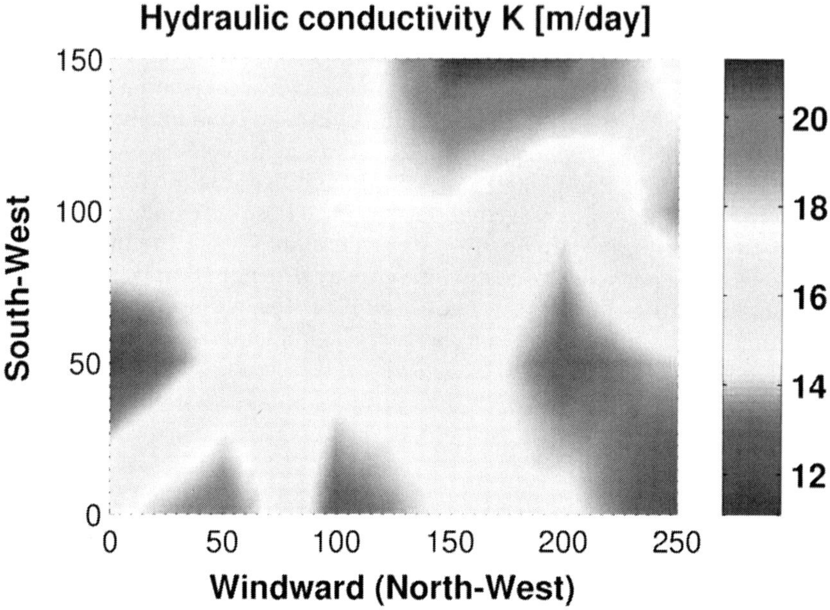

Figure 8. Spatial distribution of the hydraulic conductivity along the dune crests in a selected site within the dune field.

Since the samples used to develop the equation were relativity small in volume compared to the full dimensions of the dunes, the issue of scale needs to be addressed in that small samples commonly tend to underestimate the bulk hydraulic conductivity of an aquifer [54]. Larger scale measurement methods used to assess saturated aquifer systems, such as standard aquifer performance tests or tracer tests, almost always yield higher average hydraulic conductivity values based on the occurrence of bedding or diagenetic enhancement of hydraulic conductivity within an aquifer. While dune aquifers are not homogeneous and isotopic, they approach this condition better than almost any other type of aquifer, especially when located near or at land surface. Therefore, use of the new equation to determine the average hydraulic conductivity of a dune system will yield a reasonable estimate, perhaps on the low end of the actual average value. Within other types of aquifer systems, the error of using grain size distribution to estimate hydraulic conductivity becomes greater, especially as the heterogeneity of the sediment increases, as found by Rosas et al. [26].

5. CONCLUSIONS

Evaluating the hydraulic properties of dune sands is an important task in understanding recharge and in designing artificial storage and recovery systems in dune aquifer systems. The vertical distribution of these properties inside most dunes, except where there are internal mud layers, can be obtained by measuring the grain size distribution along the cross-section of the dune. Furthermore, there is interest in reducing the effort in estimating the hydraulic properties of dune sands over a large area. This can be achieved by the use of empirical relationships between grain size distribution parameters and the hydraulic conductivity.

Significant correlations between the hydraulic conductivity values of dune sands and their grain size distribution parameters were found, especially for the skewness of the distribution and the mud content, as well as with the d10 value, which is commonly used in empirical equations for determining the hydraulic conductivity. While the relationship of the hydraulic conductivity and the skewness is significant, its influence can be accounted for by the use of a proxy parameter, the mud content, which showed high correlations with both the skewness and hydraulic conductivity of the samples. Finally, the porosity of the dune sands is very uniform at a high porosity of 0.39, with a standard deviation of 0.02. Therefore, an accurate estimation of the hydraulic conductivity can be obtained by employing an empirical equation using the d10 value, the mud percentage of the sample and two optimized coefficients. This approach can be used to study the spatial distribution of dune sands in larger areas. Application of the new empirical equation developed from this research will improve estimates of hydraulic conductivity for interior dune fields, but may not be applied to coastal dune fields that tend to contain coarser sand, have a higher d10 value, and contain little or no mud.

The data and analysis confirms the first research hypothesis that large errors occur between estimated and measured saturated hydraulic conductivity using the published empirical equations. The second hypothesis concerning the statically significant relationship between the measured and estimated hydraulic conductivity and mud percentage is confirmed. The hypothesis concerning a significant statistical relationship between the hydraulic conductivity and the dispersion is not supported, but a statistically significant relationship with skewness was confirmed. A confirmation of the last hypothesis shows that the use of inversion modeling using the mud percentage with one of the existing empirical equations leads to improved estimation of hydraulic conductivity. The technique did make a slight reduction in error of prediction compared to the modified equations developed by Rosas et al. [26].

Detailed design of aquifer recharge systems in dune field settings does require that a number of field and laboratory tests should be performed to assess the saturated and unsaturated hydraulic conductivity (vertical and horizontal) using more sophisticated methods, such as double-ring infiltrometer tests and measurements made using undisturbed cores (if they can be collected) and aquifer performance tests where the sand is saturated. However, the relationship between grain size distribution and hydraulic conductivity can be used as a rapid scanning tool, to add data between detailed tests (populating a model grid), and for estimation in locations where large amounts of grain size data have been published (data mining and reanalysis). The estimation of saturated hydraulic conductivity from grain size analysis is therefore useful particularly in the planning of large-scale managed aquifer recharge projects, but cannot replace more detailed analytical methods used in design.

ACKNOWLEDGMENTS

Funding for this research was provided by the Water Desalination and Reuse Center, King Abdullah University of Science and Technology and from discretionary faculty funding provided by the university.

AUTHORS CONTRIBUTIONS

Oliver Lopez collected and analyzed the samples and worked on the manuscript. Khan Jadoon performed the mathematical analysis of the samples using inverse modeling. Thomas Missimer coordinated the research and wrote most of the text.

REFERENCES

1. Roosma, E.; Stakelbeek, A. Deep-well infiltration in north-Holland dune area. Groundwater 1990, 28, 778–782.
2. Martin, R.R.; Gerges, N.Z. Replenishment of an urban dune system by stormwater recharge. In Water Down Under 94: Groundwater Papers, Proceedings of the 25th Congress of the International

Association of Hydrogeologists, Adelaide, Australia, 21 November 1994; Schonfeldt, C., Ed.; Institution of Engineers: Barton, Australia, 1994; pp. 21–25.

3. Van Breukelen, B.M.; Appleo, C.A.J.; Olsthhoorn, T.N. Hydrogeochemical transport modeling of 24 years of Rhine water infiltration in the dunes of the Amsterdam water supply. J. Hydrol. 1998, 209, 281–296.

4. Van Houtte, E.; Verbauwhede, J. Reuse of wastewater effluent for indirect production of drinking water in Flanders, Belgium. In Proceedings of the AWWA Membrane Technology Conference, Atlanta, GA, USA, 2–5 March 2003.

5. Dillion, P. Aquifer storage and recovery with stormwater and reclaimed water in Australia. In Proceedings of the UNESCO-VIETNAM Workshop on Augmenting Groundwater Resources by Artificial Recharge in South East Asia, Ho Chin Minh City, Vietnam, 15–17 December 2005; pp. 286–312.

6. Thoa, N.T.K.; Arduino, G.; Giang, N.V.; Van, P.T.K.; Vuong, B.T. The model for management of aquifer recharge in sand dune area of Bac Binh, Binh Thuan. J. Earth Sci. 2006, 28, 389–410.

7. Maliva, R.G.; Missimer, T.M. Aquifer Storage and Recovery and Managed Aquifer Recharge: Planning, Hydrogeology, Design, and Operation; Schlumberger Corporation: Houston, TX, USA, 2010.

8. Missimer, T.M.; Drewes, J.; Amy, G.; Maliva, R.G.; Keller, S. Restoration of wadi aquifers by artificial recharge with treated wastewater. Groundwater 2012, 50, 514–527.

9. Dincer, T.; Al-Muhgrin, A.; Zimmerman, U. Study of infiltration and recharge through sand dunes in arid zones with special reference to stable isotopes and thermonuclear tritium. J. Hydrol. 1974, 23, 79–109.

10. Allison, G.B.; Stone, W.J.; Hughes, M.W. Recharge in karst and dune elements of a semi-arid landscape as indicated by natural isotopes and chloride. J. Hydrol. 1985, 76, 1–25.

11. Winter, T.C. Effect of groundwater recharge on configuration of the water table beneath sand dunes and on seepage in lakes in the Sandhills of Nebraska. J. Hydrol. 1986, 86, 221–237.

12. Al-Hassoun, S.A.; Al-Turbak, A.S. Infiltration in two sand dune areas in Saudi Arabia. In Proceedings of the National Conference on Hydraulic Engineering and International Symposium on

Engineering Hydrology, San Francisco, CA, USA, 25–30 July 1993; pp. 42–48.

13. Al-Hassoun, S.A. Simulation of moisture and recharge in sand dune using "SWIM" model. In Proceedings of the 20th IASTED International Conference on Modelling and Simulation (MS2009), Banff, AB, Canada, 6–8 July 2009; pp. 13–16.

14. Maliva, R.G.; Missimer, T.M. Arid Lands Water Evaluation and Management; Springer: Heidelberg, Germany, 2012.

15. Breed, C.S.; Fryberger, S.G.; Andrews, S.; McCauley, C.; Lennartz, F.; Gebel, D.; Horstman, K. Regional studies of sand seas using Lansat (ERTS) imagery. A Study of Global Sand Sea; U.S. Geological Survey Professional Paper 1052; McKee, E.D., Ed.; United States Government Printing Office: Washington, DC, USA, 1979; pp. 305–398.

16. Bagnold, R.A. The Physics of Wind Blown Sand and Desert Dunes; Dover Publications, Inc.: New York, NY, USA, 1954.

17. Ahlbrandt, T.S. Textural parameters of eolian deposits. A Study of Global Sand Seas; United States Geological Survey Professional Paper 1120-A; McKee, E.D., Ed.; United States Government Printing Office: Washington, DC, USA, 1979; pp. 21–51.

18. Lancaster, N. Grain size characteristics of linear dunes in the southwestern Kalahari. J. Sediment. Pet. 1986, 56, 395–400.

19. Lancaster, N. Geomorphology of Desert Dunes; Routledge: London, UK, 1995.

20. Al-Harthi, A.A. Geohazard assessment of sand dunes between Jeddah and Al-Lith, western Saudi Arabia. Environ. Geol. 2002, 42, 360–369.

21. Haruyuki, F.; Mitsuhiro, I.; Tahei, Y.; Takashi, T. Determining unsaturated hydraulic conductivity of dune sand a low pressure head from a steady-state evaporation experiment. Sand Dune Res. 1999, 46, 15–26.

22. Van den Berg, E.H.; de Vries, J.J. Influence of grain fabric and lamination on the anisotropy of hydraulic conductivity in unconsolidated dune sands. J. Hydrol. 2003, 283, 244–266.

23. Wang, T.; Zlotnik, V.A.; Wedin, D.; Wally, K.D. Spatial trends in saturated hydraulic conductivity of vegetated dunes in the

Nebraska Sand Hills: Effects of depth and topography. J. Hydrol. 2008, 349, 88–97.

24. Kasenow, M. Determination of Hydraulic Conductivity from Grain Size Analysis; Water Resources Publications LLC: Littleton, CO, USA, 2002.

25. Vukovic, M.; Soro, A. Determination of hydraulic conductivity of porous media from grain-size composition. In Determination of Hydraulic Conductivity from Grain Size Analysis; Kasenow, M., Ed.; Water Resources Publications: Littleton, CO, USA, 1992.

26. Rosas, J.; Lopez, O.; Missimer, T.M.; Dehwah, A.H.A.; Sesler, K.; Coulibaly, K. Hydraulic conductivity and grain size distribution: Comparison of methods for different sedimentary depositional environments. Groundwater 2014, 52, 399–413.

27. Sagga, A.M.S. Barchan dunes of Wadi Khulays, western region of Saudi Arabia: Geomorphology and sedimentology relationships. J. King Abdulaziz Univ. Earth Sci. 1998, 10, 105–114.

28. Roobol, M.J.; Kadi, K.K. Cenozoic Faulting and Traces of Tertiary Gabbro Dikes in the Rabigh Area; Saudi Geological Survey Technical Report SGS-TR-2008-6 (geological map); Saudi Geological Survey: Jeddah, Saudi Arabia, 2007.

29. Tanner, W.F.; Balsillie, J.H. Environmental Clastic Granulometry; Florida Geological Survey Special Publication 40; Florida Geological Survey: Tallahassee, FL, USA, 1995.

30. American Society for Testing and Materials (ASTM). Standard Test Method for Particle Size Analysis of Soils; ASTM Standard D422-63; ASTM: West Conshohocken, PA, USA, 1995.

31. Wenzel, L.K. Methods for Determining the Permeability of Water-Bearing Materials with Special Reference to Discharging-Well Methods; U.S. Geological Survey Water-Supply Paper 887; United States Government Printing Office: Washington, DC, USA, 1942.

32. American Society for Testing and Materials (ASTM). Standard Test Method for Permeability of Granular Soils; ASTM Standard D2434-68; ASTM: West Conshohocken, PA, USA, 2006.

33. Beyer, W. Zur Bestimmung der Wasserdurchlassigkeit von Kieson und Sanduen aus derKornverteilung. Wasserwirtsch.-Wassertech. 1964, 14, 165–169. (In German).

34. Lambot, S.; Slob, E.C.; Rhebergen, J.; Lopera, O.; Jadoon, K.Z.; Vereecken, H. Remote estimation of the hydraulic properties of sand using full-waveform integrated hydrophysical inversion of time-lapse, off-ground GPR data. Vad. Zone J. 2009, 8, 743–754.

35. Jadoon, K.Z.; Lambot, S.; Slob, E.C.; Vereecken, H. Analysis of horn antenna transfer functions and phase-center position for modeling off-ground GPR. IEEE Trans. Geosci. Remote Sens. 2011, 49, 1649–1662.

36. Jadoon, K.Z.; Weihermuller, L.; Scharnagl, B.; Kowalsky, M.B.; Bechtold, M.; Hubbar, S.S.; Vereecken, H.; Lambot, S. Estimation of soil hydraulic conductivity parameters in the filed by integrated hydrogeophysical inversion of time-lapse ground-penetrating radar data. Vad. Zone 2012, 11.

37. Alyamani, M.S.; Sen, Z. Determination of hydraulic conductivity from complete grain-size distribution. Groundwater 1993, 31, 551–555.

38. Barr, D.W. Coefficient of permeability determined by measurable parameters. Groundwater 2001, 39, 356–361.

39. Chapuis, R.P.; Dallaire, V.; Marcotte, D.; Chouteau, M.; Acevedo, N.; Gagnon, F. Evaluating the hydraulic conductivity at three different scales within an unconfined sand aquifer at Lachenaie, Quebec. Can. Geotech. J. 2005, 42, 1212–1220.

40. Fair, G.M.; Hatch, L.P. Fundamental factors governing the stream-line flow of water through sand. J. Am. Water Work. Assoc. 1933, 25, 1551–1565.

41. Harleman, D.R.F.; Melhorn, P.F.; Rumer, R.R. Dispersion-permeability correlation in porous media. J. Hydraul. Div. Am. Soc. Civ. Eng. 1963, 89, 67–85.

42. Hazen, A. Some Physical Properties of Sands and Gravels, with Special Reference to Their Use in Filtration; 24th Annual Report; Massachusetts State Board of Health: Boston, MA, USA, 1892; pp. 539–556.

43. Carrier, W.D. Goodbye Hazen; Hello, Kozeny-Carman. J. Geotech. Eng. 2003, 129, 1054–1056.

44. Kozeny, J. Das wasser in boden, grundwasserbewegung. In Hydraulik; Springer: Vienna, Austria, 1953; pp. 280–445. (In German)

45. Bear, J. Dynamics of Fluids in Porous Media; Dover Publications, Inc.: New York, NY, USA, 1972.

46. Krumbein, W.C.; Monk, G.D. Permeability as a function of the size parameters of unconsolidated sands. Trans. Am. Inst. Min. Metall. Pet. Eng. 1943, 151, 153–163.

47. Chesnaux, R.; Baudement, C.; Hay, M. Assessing and comparing the hydraulic properties of granular aquifers on three different scales. In Proceedings of Geohydro 2011, Quebec City, QC, Canada, 28–31 August 2011; p. 9.

48. Pravedny, G.H. Design and Selection of Grain Size Composition of Filter Beds for the Transition Zones of Large Dams; Energiia: Moscow, Russia, 1966.

49. Slichter, C.S. Theoretical Investigation of the Motion of Ground Waters; U.S. Geological Survey 19th Annual Report, Part 2; United States Government Printing Office: Washington, DC, USA, 1899; p. 322.

50. Terzaghi, C. Principles of soil mechanics. Eng. News-Rec. 1925, 95, 832.

51. Lu, C.; Chen, X.; Cheng, C.; Ou, G.; Shu, L. Horizontal hydraulic conductivity of shallow streambed sediments and comparison with grain-size analysis results. Hydrol. Process. 2012, 26, 454–466.

52. Krumbein, W.C. Size frequency distribution of sediments. J. Sed. Petrol. 1934, 4, 65–77.

53. Panda, M.N.; Lake, L.W. Estimation of single-phase permeability from parameters of particle-size distributions. Am. Assoc. Pet. Geol. Bull. 1994, 78, 1028–1039.

54. Schulze-Makuch, D.; Carlson, D.A.; Cherkauer, D.S.; Malik, P. Scale dependency of hydraulic conductivity in heterogeneous media. Groundwater 1999, 37, 904–919.

5
CHAPTER

EVELUATION OF THE QUALITY OF THE SOIL FROM SOIL PHYSICAL-HYDRICAL INDICATORS

Paulo Leonel Libardi[1] and Flávia Carvalho Silva[2]

[1] Exact Sciences Department– ESALQ-USP, Piracicaba, SP, Brazil

[2] Agronomics Sciences Department – UEM, Umuarama, PR, Brazil

1. INTRODUCTION

Due to the high demand for food, farmers have intensified agriculture seeking high productions. Hence, agriculture, amongst other activities, has been considered a highly potentially polluting activity of all the system water-soil-plant-environment. As a consequence, the intensive use of the soil causes its degradation. The accelerated waste will always exist if the farmer does not take proper measures to combat the causes related to various processes such as: chemical depleting and leaching, erosion, physical and biological degradation. The growing concern about the environment raised the concern about the quality of the soil. Ever since, several concepts of soil quality have been proposed, however, currently "soil quality" is defined as the capacity of the soil to keep biological productivity, environment quality and the vegetal and animal lives healthy on earth [1]. There has currently been a wide discussion towards environmental patterns and indicators, specially in Brazil, where there are very few studies and even a lack of systematization over the subject, since there is plenty of data that could provide support for an extensive discussion over the topic. The evaluation of the quality of the soil could be carried out by the monitoring of its features or physical, chemical and biological characteristics. In this approach, the expression "soil quality indicators" is being used, since it is the parameter or reference that best translate the conditions of a specific environment compartment. Among them, some attributes or physical indicators that might go through a few medium term changes have been recommended, such as density, porosity, aggregation and compression state. Hydrical conductivity, water retention, storage and density of water flow in the soil may also be indicators of great importance to assess the quality of the soil. Although such parameters are not frequently studied in Brazil, in foreign literature they are reported to vary according to different soil preparations and management. Detailed familiarity of the water dynamics during the development of a culture provides essential elements for the stablishment or improvement of agricultural management practices that aim to optimize the productivity. Water is a fundamental factor in the development of a culture, interfering mainly in the development of the root system and the absortion and tranference of nutrients for the plants. When a certain culture is irrigated, or when it rains, the water penetrates the soil and redistributes in its interior. If the quantity added by these processes exceeds the soil infiltration capacity, the excess moves sideways over the soil surface. The theoretical-practical knowledge of the runoff water, infiltration and evapotranspiration, the movement of the water in the interior of the soil profile, the absortion of the water by the plants, the draining, etc, is extremely relevant to obtain an increase in the productivity of the cultivars and to prevent the environmental degradation.

2. BODY

When hydrical conductivity of a soil in mentioned, it is implicit that a sample of soil is being considered a porous media and as a fluid that moves in it an aqueous solution (hydrical). The methods of direct determination of hydrical conductivity of the soil may be classified in laboratory methods and field methods. In the laboratory methods, samples may be used with deformed structure or samples with undeformed structure and they are subdivided for saturated and unsaturated conditions. For the measurement of the hydrical conductivity of saturated samples a simple equipament is used, denominated permeameter, and for unsaturated samples the hydrical conductivity may be measured under the steady-state condition, in which big or small columns are used and also under transient conditions. The measurement of hydrical conductivity in the field may also be carried out under the saturation condition (below the groundwater) and under unsaturated conditions (above the groundwater). The Instantaneous Profile Method [2]; [3]; [4] is the most commonly used in the field and it is applied for unsaturated conditions in situations in which the groundwater is nonexistent or very deep; it also has the advantage of being possibly used in heterogenous soil [3]. To apply this method in the field, a portion of sufficiently large area of soil must be chosen, so that the processes in the center are not affected by the boundaries. This area must be conveniently instrumented for the measurement of the amount of water in the soil and the matric potencial, along the profile until the depth of interest. The surface of the soil is, then, mantained under water depth, in a way that the profile, until de depth of interest is as humid as possible. Reaching this condition, which may be perceived by the virtually no variation of the content of water and the matric potencial with the infiltration time along the profile, infiltration is interrupted and the surface of the soil is covered with a plastic cover to avoid evaporation and the water inlet through the surface. The water in the profile redistributes, then, by the process of internal drainage, and as this process occurs, periodical measurements of water and matric potencial are carried out, data to determine the hydric conductivity K according to the water content θ, that is, the function $K(\theta)$ [5]. The measurements of the water content for this purpose have normally been carried out through a) techinque of neutron moderation or, simply, neutron probe (for example [6], b) tensiometers, installed in several depths, together with their respective retention curves [7] and, more recently, c) time-domain reflectometry technique or, simply TDR [8]. Other techniques are or will be able to be utilized, however, for the measurement of the water content, since the thermogavimetric method, taken as a pattern, the gamma radiation attenuation techinque and new techniques that have been developed as the computerized tomography (CT), Magnetic Resonance Imaging (MRI), electromagnetic induction, etc.[9]. In Brazil, there are few studies about hydric conductivity for different preparations and soil management and, besides that, the experimental results and contradicting, implicating in detailed analysis of the physical-hydrical properties of the place where the research is being carried out, as this indicator depends on the type of soil, management systems, vegetation, etc. Empirical parameters of K adjustment function θ, allowing function $K(\theta)$ of Oxisol, under different management systems, were obtained by [10] in laboratory. These authors verified discrepancies in the conductivity behavior for different managements and depths, with the same tendency observed in the field determination (Table 1). More recently, [11] observed, in Latosol, that the average hydrical conductivity of saturated soil until 1m deep is larger in bare soil than in cultivated soil (Table 2).

Table 1. Empirical parameters to calculate unsaturated soil conductivity as a function of adjusted managements from data obtained in the field (1); taken from [10].

Depth (m)	Woods			Growing upland rice			Irrigated lowland rice		
	a	b	r^2	a	B	r^2	a	b	r^2
0,2	53,47	-22,22	0,85	30,95	-13,78	0,79	59,56	-27,10	0,94
0,3	54,61	-21,61	0,97	48,84	-20,02	0,75	57,76	-27,47	0,95
0,4	54,30	-21,33	0,91	47,26	-18,60	0,94	46,98	-22,05	0,93
0,5	50,86	-20,98	0,92	57,32	-22,42	0,93	52,08	-22,56	0,97
0,6	35,32	-14,91	0,88	37,92	-15,71	0,91	59,33	-24,96	0,93
0,7	39,08	-19,52	0,95	40,86	-19,16	0,91	41,17	-16,96	0,97
0,8	33,38	-14,47	0,92	40,87	-17,08	0,96	43,16	-18,48	0,89
0,9	35,27	-15,65	0,98	38,35	-15,23	0,93	35,58	-15,00	0,94

Empirical parameters a and b of the equation $\ln K = a\theta + b$ e r^2 = determination coefficient.

Table 2. Average hydrical conductivity of saturated soil (mh^{-1}) as a function of management systems and depth taken from Angelotti Neto & Fernandes (2005).

Depth (m)	Management	
	Bare	Cultivated
0,2	$3{,}25.10^{-3}$	$3{,}55.10^{-4}$
0,4	$2{,}07\ 10^{-3}$	$6{,}16.10^{-4}$
0,6	$4{,}45\ 10^{-3}$	$1{,}03.10^{-3}$
0,8	$2{,}79.10^{-3}$	$1.93.10^{-3}$
1,0	$3{,}74.10^{-3}$	$2{,}55.10^{-3}$
Average	$3{,}26\ 10^{-3}$ a	$1{,}30.10^{-3}$ b

Averages followed by the same letters differ from each other to 5% in probability by the Tukey test.

The soil, water reservoir for the plants, is affected by the management and culture practices, changing the dynamics and the retention of water in its pores. This retention of water in the soil matrix is controlled by two types of force, capillary and adsorption, which are denominated matric forces and originated the term matric potential of water in soil. The retention curve connects the amount of solution in the soil in equilibrium with the tension applied. It is also known that once the retention curve is obtained one can calculate, from it, the distribution of pores size in the soil. With the water retention curve or, simply, retention curve, the grower, during the management of water, will know exactly when and how much water he will need to add to the system of production to supply the plant properly with no loss of the root system for internal drainage. It was [12], in his work searching for an equation to quantify the movement of water in the soil under unsaturated conditions, who introduced in the soil science this relation between the content of water and the matric potencial. The methodology of retention curve determination had its beginning, apparently, as a [13] work that utilized a funnel fitted with a porous plate in the lower part with which potential matric values or the water retention in the soil were fixated and it measured the corresponding content of water after the equilibrium. For this reason, this kinf of funnel available in the market is also called Haines funnel [5] and it is used to determine of the most humid part of the retention curve (tension levels below 10 kPa). After that [14, 15, 16] developed a porous plate pressure chamber to measure higher tensions (10 to 2000 kPa) also known as Richards pressure chamber. Nowadays, both these devices are routine equipment to determine the retention curve in soil physics laboratories. Some variations of the funnel may be found (but the theoretical principle is the same) under names such as tension table, sand tank, etc. [17]. It is important to make it clear that the retention curve can be obtained through wetting or drying, resulting in the hysteresis phenomenon [18, 19]. In this context, to utilize the results practice, there is a need to specify the curve branch which is being used [20]. The complete retention curve of a soil (from 0 to 1500 kPa), besides reflecting the water behavior in the soil in terms of water availability for the plants, is also a reflex of the important physical properties of the soil such as texture, structure, pores distribution, consistence, etc. [21, 22, 23]. In this context, it is important to mention the recent work of [24] about the physical quality of a soil through the retention curve. Basically, what the author does is define a parameter S that is nothing but a slope of the retention curve (expressed as the content of water in the soil as a function of the tension logarithm of the water in the soil) in its inflection point and based on several arguments, the higher the S, in absolute values, the better the physical quality of the soil. As an illustration, in figure 1 below, the exemple introduced by Dexter himself [24] of two retention curves of the same degraded (or compressed) and non-degraded soil. Besides that, as regarded to water conduction through soils, the retention curve may be an alternative to the direct hydric conductivity of the soil K as a function of the content of water in the soil θ, that is, this function K (θ) can be theoretically calculated from water retention data more easily measured in laboratories. [25, 26, 27, 28, 29, 30].

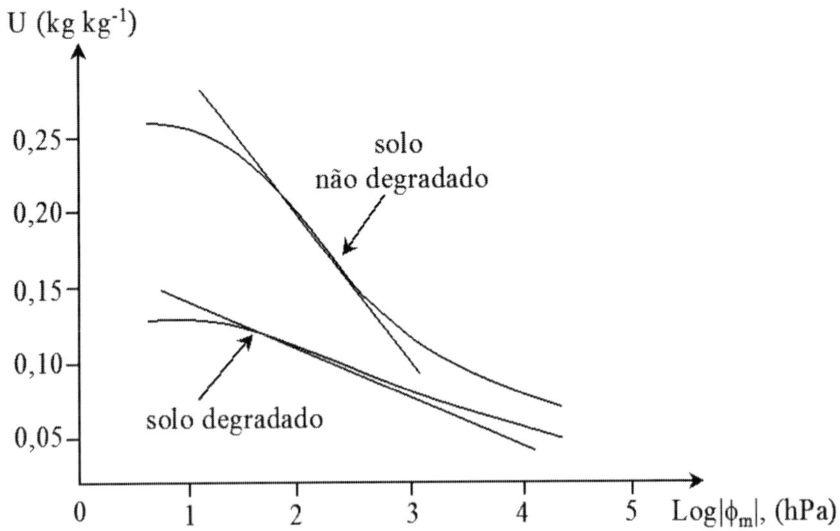

Figure 1. Retention curves of the same soil in two situations: degraded and non-degraded [24].

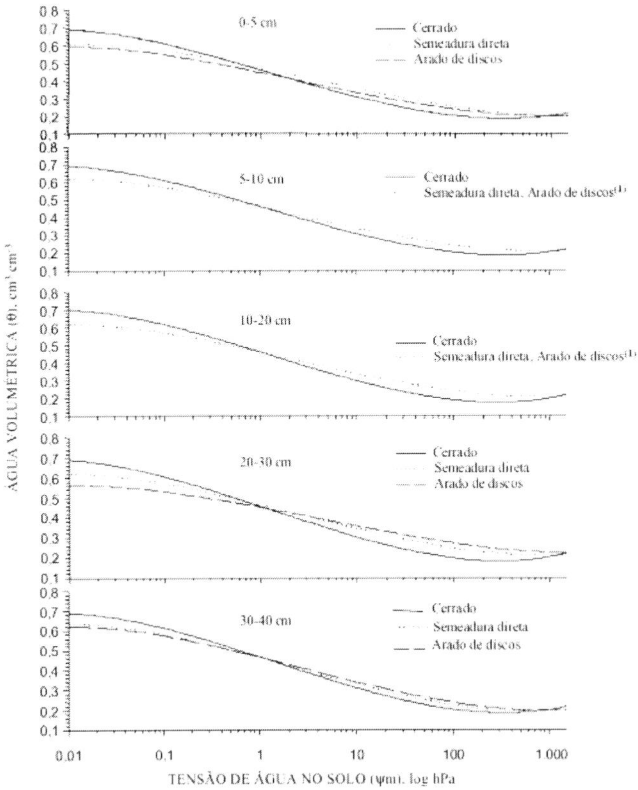

Figure 2. Retention curves of five different depths of Oxisol under cerrado, tillage and plowing discs preparation [31].

In figure 2 retention curves of five different depths of Oxisol under cerrado are presented, tillage and plowing discs preparation. Data, obtained by [31], show, according to the authors, that, in the tension belt between 100 and 1500 kPa, particularly in the systems that involve cultivation, in many depths, the curves showed an aspect nearly rectilineal, asymptotic to the abscissa axis, indicating the existence of ultra-micropores with low water-storage capacity. [32] concluded, from results obtained in figure 3, that in the layer from 0,4-0,8 m, significant increase in the water retention in the soil under the irrigated system was noticed between the matric potencial of −33 to −1.500 kPa, comparing to the woods soil, due to a more expressive reduction of macropores in this soil.

Knowing how the plants utilize the water retained in the soil and how they respond to the level or storage in the profile, may be a viable solution to establish effective management strategies aiming a better possible use of water reserves in the soil by the crops. Water storage in the soil represents the water that the soil is storing in a specific layer in a specific depth. In figure 4 the results obtained by [33] are illustrated and determine the variation of water storage in the soil for three managements, between current humidity and PWP, for eight different redistribution times and five depths in determination experiment of function K(θ) by the instantaneous profile method. They concluded that until the 0,25 m layer, the under rainfed soil always retained more water than the woods soil and irrigated, having a higher variation between the woods and two other management systems.

Nitrogen leaching is extremily important because it can decrease the quantity of ammonium and nitrate in the topsoil sensibly and, consequently, reduce the nutrient availability; when in excess, leaching constitutes a potential danger of groundwater contamination by nitrate [34]. Over the last few decades, surface and underwater sources contamination with nutrients, particularly N and P, has become a significant subject for people in general, including the farmers. High concentrations of nitrate in the water for human consumption are worrying, as they cause methemoglobinemia, commonly known as blue baby syndrome. This is a problem that occurs only in children under six months old and pregnant women. Besides that, water contamination by nitrates and P has been connected with another abnormality denominated hypoxia (low oxygen levels) in water from Gulf of Mexico, which inhibits the production of shrimp and other acquatic species in this region [35]. For the USA, the maximum concentration of nitrate for water to be considered potable and not cause any harm to human health is 10 mg L^{-1} and 1mg L^{-1}, respectively, being the same values adopted by Brazilian Legislation, but for the European Union the accepted value of nitrate reaches 50 mg L^{-1} [36]. N leaching occurs in descending order for NO_3^- > NH_4^+ > Organic-N. Most leaching in form of nitrate occurs due to its negative charge being repelled by soil colloids where the same charge prevails [37]. The amount of N that is lost in leaching varies a lot according to the N dose, the method of application of fertilizers, the speed of mineralization and immobilization by plants and microorganisms, the amount of rain and the soil properties that influence its capacity to retain water, whichever are the texture, structure and porosity [38]. To quantify the elements due to leaching, it is necessary to know the water flow density in the soil and the element concentration in the soil solution, since their product generates the flow density of the element in the soil. The water flow density in the soil represents the water flow per unit cross-sectional area of soil and may be determined by Darcy-Buckingham equation [5]. Studies about water dynamics in field conditions emphasizing water flow in the crop root zone are not frequent and, often,

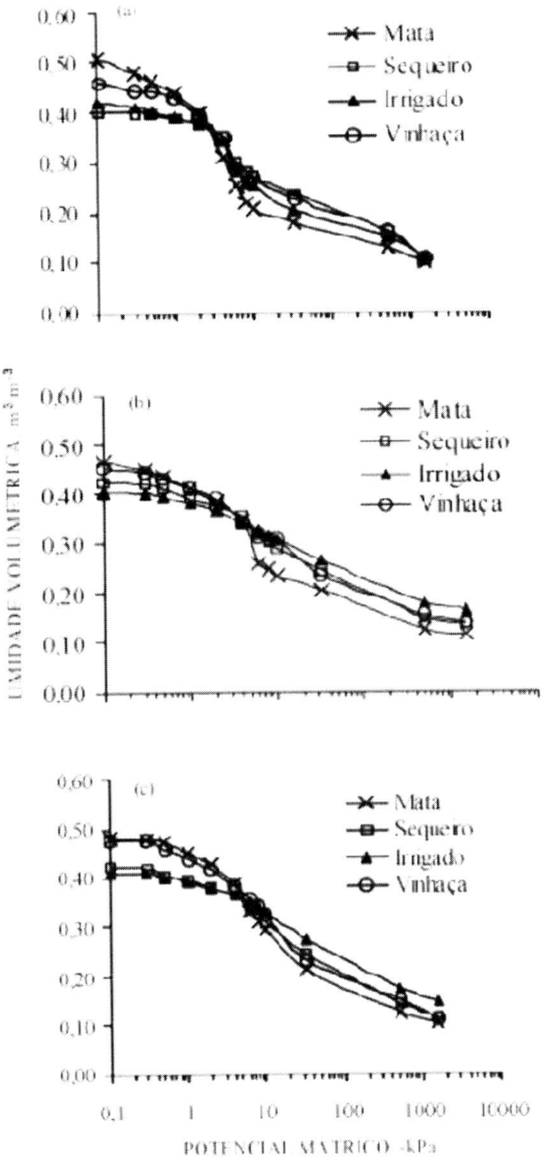

Figure 3. Retention curves in the layers (a) 0-0,2 m, (b) 0,2-0,4 m e (c) 0,4-0,6 m, for the use and management of the soils studied systems [32].

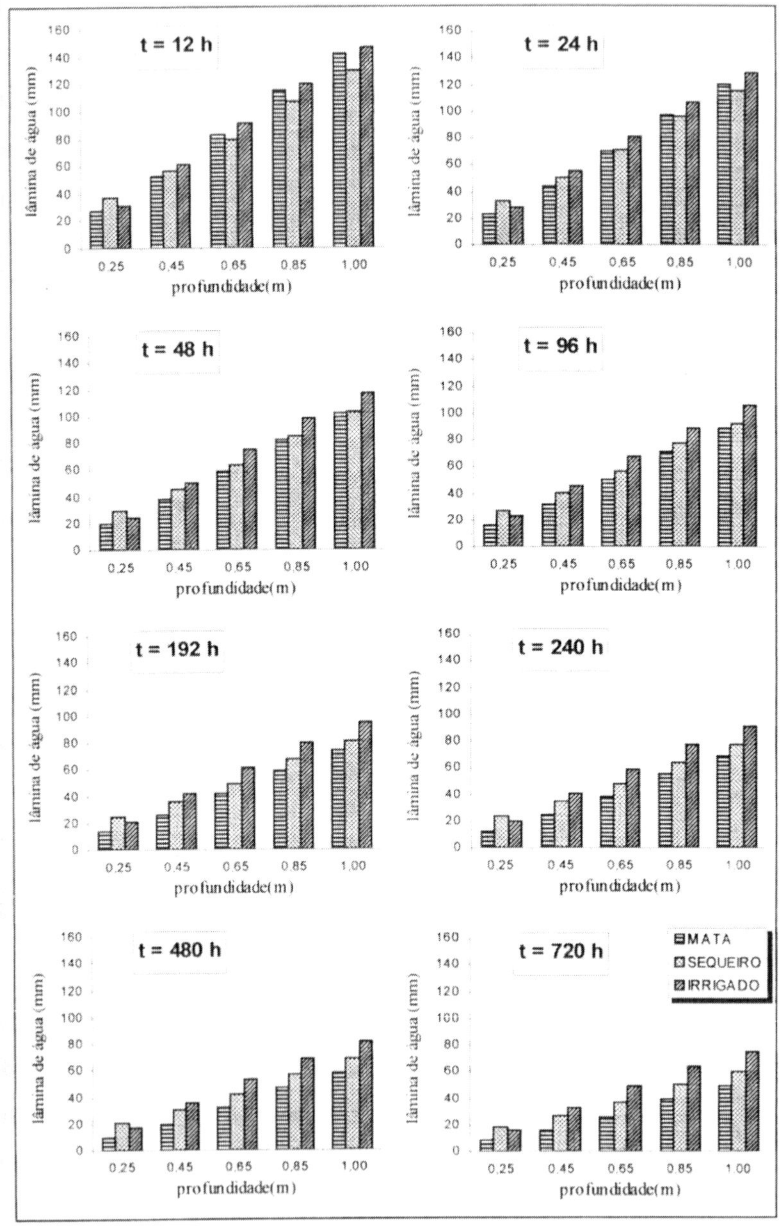

Figure 4. Variation of water storage in the soil for three managements, between the current humidity and PWP, for eight different drainage times and five depths [33]

Table 3. Water drainage and total nitrate leaching provenient from the fertilizer (PF) to 0,80 m depth, in the succession corn-black oat-corn [39]

Treatments 1	Treatments 2	DAE	Corn cultivation – harvest 2003/2004			
			Precipitation (mm)	Water drainage to 0,80 m (mm)	Leaching-NO^-_3 (kg ha^{-1}) Total	PF
30	90	0-30	110,1	103,62 a	10,27 a	0,20 a
60	60			129,91 a	13,97 a	0,18 a
30	90	30-60	200,6	165,67 a	33,14 a	1,79 a
60	60			180,11 a	20,00 b	0,28 b
30	90	60-90	239,4	155,71 b	24,93 b	0,94 a
60	60			218,63 a	61,64 a	0,86 a
30	90	90-120	64,9	1,54 b	0,01 a	0,07 a
60	60			3,65 a	0,01 a	0,07 a
30	90	Total	615,0	**426,54 b**	**68,35 b**	**3,00 a**
60	60			**532,29 a**	**95,61 a**	**1,39 b**

Treatments 1	Treatments 2		Between the first corn cultivation and black oat			
			Precipitation (mm)	Water drainage to 0,80 m (mm)	Leaching -NO^-_3 (kg ha^{-1}) total	PF
30	90	Total	296,0	151,18 b	2,81 b	-
60	60			243,24 a	4,77 a	-

Treatments 1	Treatments 2	DAE	Black oat – harvest 2004			
			Precipitation (mm)	Water drainage to 0,80 m (mm)	Leaching -NO^-_3 (kg ha^{-1}) total	PF
30	90	0-40	0,0	6,68 b	0,00 a	-
60	60			11,88 a	0,00a	-
30	90	40-80	146,5	15,02 b	0,62 b	0,01 a
60	60			61,69 a	2,64 a	0,02 a
30	90	Total	146,5	**21,70 b**	**0,62 b**	**0,01 a**
60	60			**73,57 a**	**2,64 a**	**0,02 a**

Treatments 1	Treatments 2		Between black oat and the second corn cultivation			
			Precipitation (mm)	Water drainage to 0,80 m (mm)	Leaching -NO^-_3 (kg ha^{-1}) total	PF
30	90	Total	136,3	56,78 b	0,07 a	-
60	60			142,14 a	0,03 a	-

Treatments 1	Treatments 2	DAE	Corn cultivation - harvest 2004/2005			
			Precipitation (mm)	Water drainage to 0,80 m (mm)	Leaching -NO^-_3 (kg ha^{-1}) total	PF
30	90	0-30	181,9	113,16 b	4,80 a	-

60	**60**			180,17 a	6,45 a	-
30	**90**	**30-60**	231,7	132,50 b	11,79 a	-
60	**60**			211,40 a	9,80 a	-
30	**90**	**60-90**	78,4	0,61 b	0,00 a	-
60	**60**			4,74 a	0,00 a	-
30	**90**	**90-120**	164,4	133,37 a	6,43 b	0,13 a
60	**60**			134,59 a	12,12 a	0,13 a
30	**90**	**Total**	**656,4**	**379,64 b**	**23,02 a**	**0,13 a**
60	**60**			**530,90 a**	**28,37 a**	**0,13 a**

Obs. The treatments 30-90 and/or 60-60 refer to the application of N in the sowing in the 6-8 leaf stage. DAE stands for "days after emergency". Averages followed by the same letter do not differ from each other in significance level 5% by Tukey test. 1 and 2 refer to the nitrogen fertilization in corn, respectively in the sowing and coverage.

incomplete, due to the great complexity of the necessary experimental procedures. However, as mentioned previously, knowing internal drainage is essential to estimate the leaching of chemicals down to the root system of the crop under analysis. When [39] assessed a nitrate leaching (total and provenient from fertilizers) to 0,80 m depth, in a succession of cultures corn-black oat-corn, under no-tillage implemmentation, utilizing fertilizer labeled ^{15}N, they obtained the results that are presented in Chart 3. It can be observed through the data in the chart, as concluded by the authors, that total leaching nitrate loss to 0,80 m depth, in the first corn crop, in 120 kg ha^{-1} of N dose, was approximately 96 and 68 kg ha$^{-1,}$ for parcelling with 60 kg ha^{-1} and 30 kg ha^{-1} of N in sowing, respectively, of which only 3 and 1 kg ha^{-1} of N were provenient from the nitrogen-rich fertilizer. In the second year of cultivation, nitrate leaching was far lower than compared to the first year and the leachate nitrate provenient from the fertilizer was negligible (average 0,23%).

[40] continuing studies with leaching in the crop succession consisted of: maize in 2006 followed by brachiaria plus fallow and finally maize in 2007. Internal drainage decreased with the increase of applied N levels to the crop succession, changing from 31.5 to 73.4 % of the total rainfall (97 mm) in the first maize, from 26.1 to 58.1 % of the total rainfall (695 mm) in brachiaria + fallow, and from 56.6 to 87.4 % of the 419 mm of total rainfall in the second maize crop. The leaching of total nitrate (from fertilizer and other sources) was very low in the first maize crop for all applied N levels and significant for the rates of 120 and 180 kg ha-1 in the periods of brachiaria plus fallow (26.16 for 120 kg ha-1 and 39.8 for 180 kg ha-1) and of the second maize crop (approximately 23 kg ha-1 for both levels). There was no N leaching from fertilizer in the first maize crop and N leaching was very low in the brachiaria and second maize crop.

3. CONCLUSIONS

Conservative management systems, such as tillage, present an expressive improvement effect in the hydro-physical quality of tropical and subtropical soils. The soil in function of type has diverse physical, chemical and biological composition, making it difficult to have a pattern for the amount of hydro-physical indicators to evaluate their quantity. Therefore, there is a need for more detailed studies, particularly for tropical conditions that can define more precise values for the indicators of quality. Different types of soil have a great influence in the leaching process magnitude, however, other factors such as organic matter content, dose, type and time of fertilizer application, mainly the nitrogen-rich one, and also the weather, have great influence in this process. Thus, leaching studies aiming the use of fertilizers and their interaction with different soil managements are influential to Brazilian research, since there is a great lack of results for soils in tropical and subtropical weather countries.

REFERENCES

1. Doran, J.W.; Parkin, T.B. *Defining and assessing soil quality*. In: DORAN, J. W. et al. (Eds.). Defining soil quality for a sustainable environment. Madisson: ASA/SSSA, 1994. p-3-21.
2. Watson, K.K. An instantaneous profile method for determining the hydraulic conductivity of unsaturated porous materials. *Water Resources Research*, 1996; 2 709-715.
3. Hillel, D.A; Krentos, V.K.; Stilianov, Y. Procedure and test of an internal drainage method for measuring soil hydraulic characteristics in situ. *Soil Science.*, 1972; 114, 395-400.
4. Libardi, P.L.; Reichardt K.; Nielsen, D.P.; Biggar, J.W. Simple field methods for estimating soil hydraulic conductivity. *Soil Science Society of American journal*, 1980; 44, p. 3-7.
5. Libardi, P.L. *Dinâmica da água no solo.* 2.ed. São Paulo: Editora da Universidade de São Paulo, 2005. 329p.
6. Greacen, E.L. Soil water assessment by neutron method. CSIRO. Austrália, 140p, 1981.
7. Carvalho, L.A. de. Condutividade *hidráulica do solo no campo: as simplificações do método do perfil instantâneo*, 2003. Tese

(Mestrado) – "Escola Superior de Agricultura Luiz de Queiroz", Universidade de São Paulo, Piracicaba, 2003.

8. Topp, G.C.; Davis, J.L.; Annan, A.P. Eletrotromagnetic determination of soil water content: measurement in coaxial transmission lines. *Water Resources Research*, 1980; 16(3): 574-582.

9. Smith, K.A.; Mullins, C.E. Soil analysis-physical methods. New York: s.ed., 1991. 620p.

10. Klein, V.A.; Libardi, P.L. Condutividade hidráulica de um Latossolo Roxo não saturado, sob diferentes sistemas e manejo. *Ciência Rural*: Santa Maria, 2002; 32 (6), 945-953.

11. Angelotti Neto, A.; Fernandes, E.J. Condutividade hidráulica de um Latossolo Vermelho em pousio e cultivo intensivo. *Pesquisa Agropecuária Brasileira*, 2005; 40 (8), 797-802.

12. Buckhingham, E. *Studies of the movement of soil moisture*. USDA Bur. Soil Bull. 38. US Government Printing Office. Washington D.C., 1907.

13. Haines, W.B. Studies in the physical properties of soil: V. The hysteresis effect in capillary properties and the models of misture associated therewith. *Jounal Agronomy. Science*, 1930; 20, 97-116.

14. Richards, L.A. A pressure membrane extraction apparatus for soil solution. *Soil Science*, 1941; 51, 377-386.

15. Richards, L.A. Pressure membrane apparatus – construction and use. *Agronomy. Engineer*, 1947; 28, 451-454.

16. Richards, L.A. Methods of measuring soil moisture tension. *Soil Science*. 1949; 68, 95 – 112.

17. Kutílek, M; Nielsen, D.R. *Soil Hydrology*, Catena Verlag, Germany, 1994, 370p.

18. Poulovassilis, A. The influence of the initial water content on the redistribution of soil water after infiltration. *Soil Science*, Baltimore, 1983; 135, p. 265-281.

19. Klute, A. Water retention: laboratory methods. In BLACK, C.A., ed. Methods of Soil Analysis. I Physical and mineralogical methods. Madison, Winconsin, American Society of Agronomy, Inc., *Soil Science Society of America*, Inc. Publisher, 1986, 635-662.

20. Moraes, S.O. *Heterogeneidade hidráulica de uma terra roxa estruturada*. Piracicaba, 1991, 141p. (Doutorado - Escola Superior de Agricultura "Luiz de Queiroz"/USP), 1991.

21. Sharma, M.L; Uehara, G. Influence of soil structure on water relations in low humic latosols: J. Water retention. *Soil Science Society America Journal*, Madison, 1968; 32, 765-770.

22. Reeve, M.J.; Smith, P.D.; Thomasson, A.J. The effect of density on water retention properties of field soils. *Journal of Soil Science*, Oxford, 173; 24, 355-367.

23. Rawls. W.J. e Pachepsky, Y.A. Soil Consistence and Structure as predictors of water retention. *Soil Science Society. American. Journal*, 2002, 66, 1115-1126.

24. Dexter, A.R. Soil physical quality - Part I. Theory, effects of soil texture, density and organic matter and effects on root growth. *Georderma*, 2004; 120, 201-214.

25. Childs, E.C.; Colins George, N. The permeability of porous materials. Proc. Roy. Soc., *London, Ser. A*. 1950; 201, 392-405.

26. Burdine, N.T. Relative permeability calculations from pore-size distribution data. *Petroleum Trans. Am. Inst. Mining Eng.*, 1953; 198, 71-77.

27. Marshall, T.J. A relation between permeability and size distribution of pores. *Journal Soil Science*, v. 9, p. 1-9, 1958.

28. Millington, R.J.; Quirk, J.P. Permeabilty of porous solids. Trans. *Faraday Society.*, 1961; 57, 1200-1207.

29. Mualem, Y. A new model for predicting the hydraulic conductivity of unsaturated porous media. *Water Resources Research*, 1976; 12, 513-522.

30. Van Genuchten, M.T. A closed-form equation for predicting the hydraulic conductivity of unsaturated soils. *Soil Science Society America Journal*, 1980; 44, 892-898.

31. Oliveira, G.C.; Dias Júnior, M.C.; Resck, D.V.C.; Curi, N. Caracterização química e físico-hídrica de um Latossolo Vermelho após vinte anos de manejo e cultivo do solo. *Revista Brasileira de Ciência do Solo*, 2004; 28, 327-336.

32. Silva, A.J.N.; Cabeda, M.S.V.; Lima, J.F.W.F. Efeito de sistemas de uso e manejo nas propriedades físico-hídricas de um argissolo amarelo

de tabuleiro costeiro. *Revista Brasileira de Ciência do Solo*, 2005; 29,.833-842.

33. Klein, V.A.; Libardi, P.L. *Propriedades físico-hídrico-mecânicas de um Latossolo Roxo, sob diferentes sistemas de uso e manejo*. 150 p., 1998. Tese (Doutorado em Solos e Nutrição de Plantas) – Escola Superior de Agricultura "Luiz de Queiroz", Universidade de São Paulo, Piracicaba, 1998.

34. Stark, J.C.; Jarrel, W.M.; Letey, J. Evaluation of irrigation-nitrogen management practices for celery using continous-variable irrigation. *Soil Science Society American Journal*, Madison, 1980; 47, 95-98.

35. Hoeft, R.G. *Desafios para a obtenção de altas produtividades de milho e de soja nos EUA*. Piracicaba: Potafos, p.1-4, 2003 (Informações Agronômicas, 104).

36. Tundisi, J.G. *Água no Século XXI*: Enfrentando a escassez. São Carlos: RiMa IIE, 2003. 248p.

37. Gonçalves, C.N.; Ceretta, C.A.; Basso, C.J. Sucessões de culturas com plantas de cobertura e milho em plantio direto e sua influência sobre o nitrogênio do solo. *Revista Brasileira de Ciência do Solo*: Viçosa, 2000; 24, 153-159.

38. Kiehl, J.C. Nitrogênio: Dinâmica e disponibilidade no solo. In: ___. *Curso de atualização em fertilidade do solo*. Campinas: Fundação Cargill, 1987. p.139-157.

39. Fernandes, F.C.S.; Libardi, P.L.; Carvalho, L.A. Internal Drainage and nitrate leaching in a corn-black oat-corn sucession with two split nitrogen applications. Scientia Agricola, Piracicaba, 2006; 63, 483-492.

40. Fernandes, F.C.S.; Libardi, P.L. Drenagem interna e lixiviação de nitrato em um Latossolo sob sucessão milho-braquiária-milho, com diferentes doses de nitrogênio. *Revista Brasileira de Ciência do Solo*, Viçosa, 2009, 33, 1163-1173.

6
CHAPTER

A MULTI-SCALE APPROACH IN HYDRAULIC CHARACTERIZATION OF A METAMORPHIC AQUIFER: WHAT CAN BE INFERRED ABOUT THE GROUNDWATER ABSTRACTION POSSIBILITIES

Antonella Baiocchi [1], Walter Dragoni [2], Francesca Lotti [1], Simone M. Piacentini [2] and Vincenzo Piscopo [1,*]

[1] *Department of Ecological and Biological Sciences, University of Tuscia, Viterbo 01100, Italy*

[2] *Department of Physics and Geology, University of Perugia, Perugia 06100, Italy*

ABSTRACT

Hard-rock aquifers, which constitute a water supply source in many countries, are highly heterogeneous and defining a realistic model of an aquifer can be extremely complex. The objective of this study was to hydraulically characterize a metamorphic aquifer in a representative area of Italy and to identify the most appropriate approach for tapping of groundwater in this challenging environment. The results of surface fracture surveys, injection tests, pumping tests, and a simplified numerical model were compared. From the surface fracture survey, a model of the rock mass characterized by a well-developed discontinuity network and by a high frequency of discontinuities resulted. The injection tests showed the extreme heterogeneity and the lower hydraulic conductivity of the rock mass in comparison with the results of the pumping tests. The independent estimate of the hydraulic parameter resulting from numerical model highlighted a range of values higher than those resulting from the pumping tests. The study demonstrated that the continuum medium approach can be used in the case of hard-rock aquifers with a dense network of discontinuities. The multi-scale approach is recommended for investigating hydraulic heterogeneity and significantly helps to identify the most promising areas for well locations and their characteristics in relation to the style of fracturing.

Keywords: hard-rock aquifers; hydraulic properties; aquifer heterogeneity; numerical modeling; Italy

1. INTRODUCTION

Hard-rock aquifers constitute a water supply source in many regions of Africa, America, and Asia. These aquifers are generally characterized by a low yield due to their low intrinsic primary permeability and porosity. Several factors influence the occurrence of groundwater in hard rocks, such as lithology, geomorphology, tectonics, and climate conditions.

Many studies have shown that hard-rock aquifers are comprised of weathered mantle and fissured bedrock where the groundwater movement takes place. The hydraulic conductivity and storage of the weathered mantle and underlying fissured bedrock are derived from geodynamic and geomorphic processes. The fissures have been explained by several processes, such as lithostatic decompression, tectonic activity, cooling stress, and weathering processes. These two layers make up a composite aquifer followed in depth by the fresh basement, which is only locally permeable where tectonic fractures are present [1,2,3,4,5,6,7,8].

Hard-rock aquifers, such as those constituted by plutonic and metamorphic rocks, are recognized to be highly heterogeneous, particularly in relation to the different degrees of fracturing. This implies that the bedrock transmissivity can vary over many orders of magnitude [3,7,9,10,11,12,13]. Therefore, well yields are highly variable, and a high incidence of well failure occurs. Thus, defining a sustainable pumping rate is more difficult here than in other cases because of the extremely complex aquifer behavior. In this case, more so than in other cases, considerations of the discontinuous flow system, anisotropy, and heterogeneity are required to develop a reasonable model of the aquifer.

To plan for the optimal management and sustainable groundwater development of these aquifers when the bedrock is densely fractured, the continuum approach is the characterization method frequently used. This approach is less expensive than discrete-fracture or hybrid approaches. Even if several studies have concluded that the porous medium approximation is valid in densely fractured aquifers [14,15,16], the scientific debate is still open, particularly with regard to the scale at which the approximation becomes appropriate [17,18,19,20,21,22].

In Italy, where hard rocks outcrop over large areas in the Alps, Calabria, and Sardinia, the hydrogeological properties of plutonic and metamorphic rocks are not well known. The groundwater yields of these aquifers (generally less than 5 × 10−3 m3/s per km2) are lower than those in the more common carbonate and alluvial aquifers (up to 3 × 10−2 m3/s per km2), which are widely used for the water supply. The interest in hard-rock aquifers is, thus, generally scarce; nevertheless, the large extent of these rocks and the scarcity of water resources in Sardinia (where hard-rock aquifers constitute 51% of the regional area) and Calabria regions (where hard-rock aquifers constitute 39% of the regional area) justify a thorough analysis [23]. Interest in these aquifers becomes even more relevant in the framework of the ongoing climate change: most of the scenarios regarding future climate scenarios in the Mediterranean area suggest a reduction in rainfall and, thus, a reduction in water resources [24,25].

This is the context of our study, which concerns a dam site in metamorphic rock in southern Sardinia. The dam site provides an opportunity to characterize a hydraulically fractured, metamorphic aquifer using a multi-scale approach. The results of the surface fracture surveys, injection tests, pumping tests, and simplified numerical model were compared to examine the heterogeneity of the hard-rock aquifer and to verify the degree of approximation of the porous medium approach. The objective of the study was to hydrogeologically characterize these low-permeability rocks and to examine which approach could better represent the fractured aquifer. This is a key point in the identification of the appropriate tapping method in a hydrogeological environment where water resources are generally scarce, but are fundamental to the drinking and irrigation water supply.

2. STUDY AREA

The area under examination is 25 km west of the town of Cagliari, in the Sardinia region (Figure 1), where the Paleozoic basement related to the

Hercynian orogenic evolution outcrops. The Paleozoic basement consists of an originally-sedimentary succession (Lower Cambrian to Early Carboniferous) that was deformed during the Carboniferous period under the low metamorphic to anchizonal setting [26]. The Arburese Unit outcropping at the dam site consists of the Arenarie di San Vito Formation (Middle Cambrian to Lower Ordovician), which is several hundred meters thick and is composed of decimetric to metric alternations between micaceous metasandstones, quartzites, and metasiltstones. Clastic, poorly-cemented sediments, from gravels to sands and silts, of the Upper Pleistocene to the Holocene overlay the metamorphic basement [27]. The structural framework is mainly linked to the Hercynian Orogeny and Pliocene tectonic events. The Hercynian tectonics favored the thrust of the Arburese Unit over the Upper Ordovician to Lower Carboniferous succession and its deformation in large scale folding with axes of folds E–W and N–S oriented. The Pliocene tectonics shallow modified the structural setting through normal fault systems E–W, and N–S to NNW–SSE oriented [28,29].

The hydrogeological properties of the Paleozoic basement are not well known. It is only known that metamorphic rocks form a low-permeability aquifer (until 20 years ago, these rocks were considered to be impervious) where groundwater circulation occurs in the fissured zones close to the surface (on average, the first tens of meters). In the most heavily fractured zones (i.e., near the faults), the groundwater flow is more active and can occur at a greater depth [30].

The study site corresponds to the building zone of the Medau Zirimilis Dam (Figure 1), a 50-m-high rockfill dam that was completed with an upstream impervious face and with a grout curtain 10–50 m deep. The drainage basin of the reservoir, 29.5 km2 of extension, is characterized by a typical Mediterranean climate with mean annual precipitation between 700 and 900 mm and air temperature between 15 and 16 °C.

During the construction of the dam and the initial reservoir filling, several hydrogeological investigations were conducted. One of these was the measurement of the seepage through the rock foundation and laterally to the dam, that discharged in the drainage systems located at the base of the dam embankment and downstream of the grout curtain. These and other investigations highlighted the fact that the groundwater circulation involved the most fissured layers closest to the surface, including, on average, the first 50 m of the Paleozoic rocks, up to a depth of 100 m in the faulted zones. The most evident faulted zone has been found in the valley bottom at the site of the dam [30].

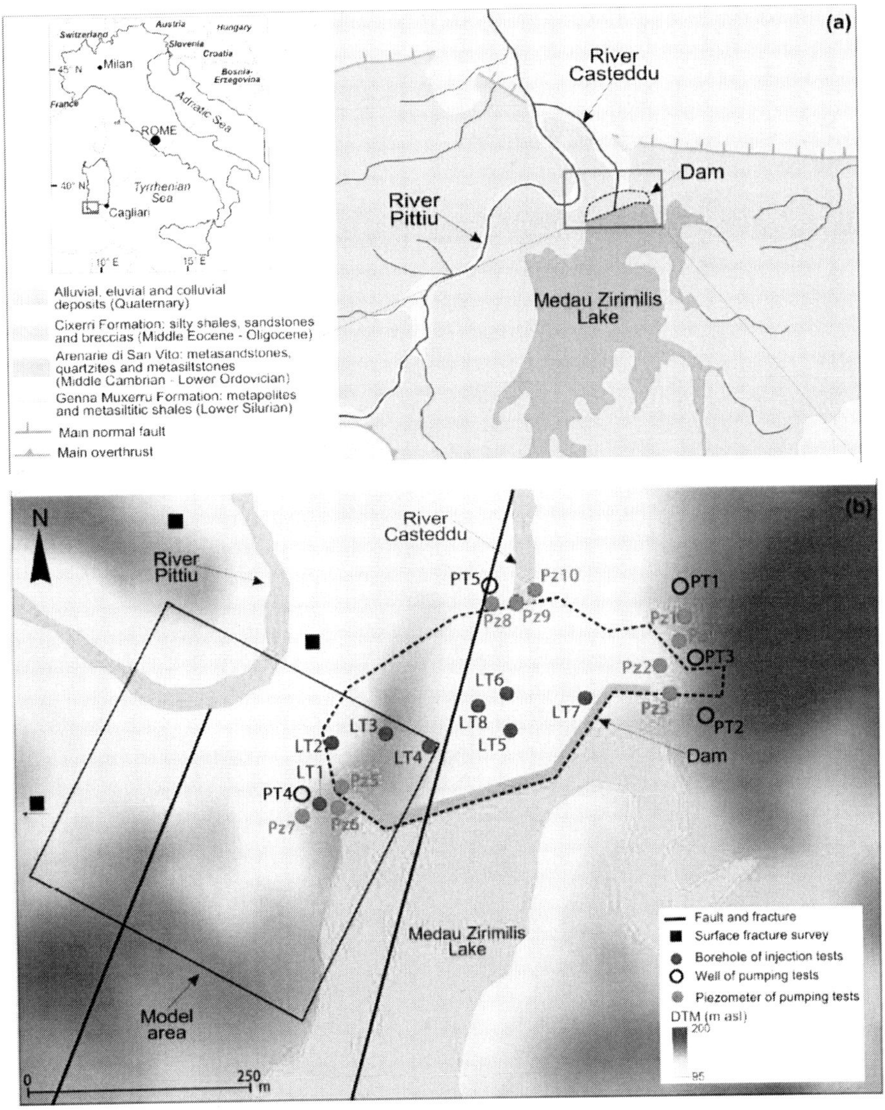

Figure 1. (a) Location of the study area; (b) Location of the surface fracture surveys, injection tests and pumping tests at the site of the Medau Zirimilis Dam.

3. INVESTIGATIONS, MATERIALS, AND METHODS

The results of the previous surveys [30] have been revised, integrated and appropriately processed to obtain the hydraulic characterization of the Palaeozoic rocks. The field investigations at the dam site included a surface fracture survey, injection tests and pumping tests (Figure 1).

The surface fracture survey concerned the Arenarie di San Vito Formation on three rock walls (ranging in size from 20 to 30 m2) with different orientations located in the western sector of the dam (Figure 1): there were two vertical planes oriented to 60° N and 350° N and one horizontal plane. For each planar surface, the discontinuity orientation, the apparent aperture (measured by a feeler gauge with minimum determinable aperture of 0.05 mm), and the spacing between the discontinuities belonging to the same set were determined. The discontinuities considered were those having a trace length greater than 80% of the size of the measured plane (from 2 × 10 to 3 × 10 m).

The injection tests (Lugeon tests) concerned single-hole tests conducted on sealed sections with packers (from 3 to 5 m in length). Five pressure steps were applied (generally 0.2–0.5–1.0–0.5–0.2 MPa at the manometer), measuring the undisturbed water level before injection and making sure that the injected discharge was stabilized. A total of 76 tests were performed in eight boreholes penetrating the Arenarie di San Vito Formation (Figure 1) to a maximum depth of 60 m; one of the boreholes was drilled with a dip angle of 60° (LT8 in Figure 1).

During the initial reservoir filling, pumping tests were performed at five wells (Figure 1), with depths between 73 and 100 m (dip angle from 10° to 30°). The pumping tests lasted between 7.5 and 45 h at a constant rate between 2.1 × 10–3 and 3.5 × 10–3 m3/s, with the observation of a drawdown in at least one piezometer nearby the tested well. The pumping data were interpreted using analytical techniques [31] and commercial software (Aquifer Test 4.1, Nova Metrix GM, Kitchener, Canada).

Based on the results of the hydraulic characterization of the site, a simplified numerical model was constructed for the western sector of the dam site (Figure 1), where one of the tested wells is located, and nearby the outcrops of the surface fracture surveys. The purpose of the numerical simulation was to examine the results of pumping tests to a further scale, different from that of the analytical solutions and of the other hydraulic tests. The model was implemented with the code MODFLOW-2000 (Groundwater Vistas 6 graphical user interface). The reconstruction of the

DTM provided the top elevation. The initial heads were obtained using the interpolation of the heads measured during the initial reservoir filling.

4. RESULTS

The results are reported with reference to different methods of the hydraulic characterization and to the numerical simulation.

4.1. Discontinuity-Network Characterization

Two discontinuity types were identified in the examined outcrops. The first type consisted of regular bedding planes that were continuous at the scale of the outcrops and delimited the various metasandstone and metasiltstone beds. The second type of discontinuity consisted of sub-vertical straight joints of varying orientations that cut the bedding planes.

The projection of the discontinuity poles on a stereographic net [32,33] made it possible to group the systematic discontinuities into three different sets (Figure 2). Table 1 lists the average values of the dip direction, dip angle and Fisher dispersion coefficient of each set.

The measured spacing approaches a log-normal distribution for the three sets. The apertures were measured on discontinuity segments when they were not altered by the weathering phenomena and ranged from <0.05 mm (instrumental measurement limit) to 1 mm. None of the different distributions tested (normal, log-normal, logistic, gamma, and Weibull) perfectly represented the measured data. The log-normal distribution best represented the majority of the data, even if it contained a greater dispersion for the minimum and maximum values of the parameter. The statistics regarding the discontinuity spacing and the apparent aperture are listed in Table 1.

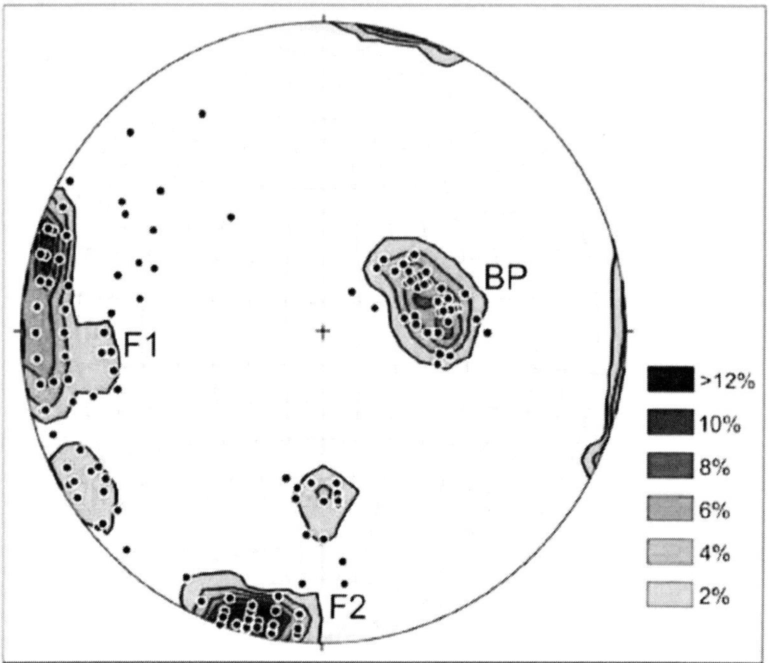

Figure 2. Equal-area projection of discontinuity poles measured in three outcrops (contour interval = 2%) showing the bedding plane (BP) and two fracture sets (F1 and F2).

Table 1. Characteristics of the three distinguished discontinuity sets, bedding plane (BP) and two sub-vertical fracture sets (F1 and F2).

Characteristic	Statistics	Set 1: BP	Set 2: F1	Set 3: F2
Orientation	Number of measurements	48	96	68
	Mean dip direction (°)	255.1	91.9	11.9
	Mean dip angle (°)	29.2	78.1	75.8
	Fisher dispersion coefficient	68.7	13.0	21.6
Spacing	Number of measurements	64	74	133
	Mean (cm)	26.20	24.45	10.47
	Standard deviation (cm)	1.77	2.01	1.38
Aperture	Number of measurements	97	187	161
	Mean aperture (mm)	0.12	0.15	0.15
	Standard deviation (mm)	0.19	0.26	0.21

A simplified model of the fracture network has been considered for a first estimate of the hydraulic properties of the rock mass.

If the fractured rock mass is assumed to have an impermeable matrix, if the discontinuities are considered as parallel plate model, with an aperture a and a spacing d, and if the groundwater moves by laminar flow in the discontinuities, the hydraulic conductivity ki of each set can be estimated according to the cubic law [34]:

$$k_i = \frac{a^3 g}{12 v d} \tag{1}$$

where g is the gravitational acceleration and υ is the kinematic viscosity of water.

The values of a and d introduced in Equation (1) are the mean values of the apparent aperture and spacing resulting from the statistical analysis (Table 1). The calculated hydraulic conductivities of the different sets ranged from 5 × 10−6 to 3 × 10−5 m/s (Table 2).

Table 2. Estimated hydraulic conductivity of each discontinuity set, bedding plane (BP) and two sub-vertical fracture sets (F1 and F2).

Set	k (m/s × 10^{-5})
Set 1: BP	0.54
Set 2: F1	1.13
Set 3: F2	2.64

If the discontinuities are well-connected and if the flows between the discontinuity sets do not interfere, the hydraulic conductivity tensor of the fractured rock mass [K] can be calculated based on the orientation of the discontinuity sets [35]:

$$[K] = \sum_{i=1}^{n} k_i \times H_{jl}^{i} \; j, l = x, y, z$$

$$H_{jl}^{i} = [I - n^T \times n] \tag{2}$$

where ki is the hydraulic conductivity of the ith-discontinuity set; I is the unit matrix; and n is the direction cosine vector whose components are expressed in terms of the dip azimuth and dip angle of the discontinuity set (with the following coordinate system: y-axis oriented north, x-axis oriented east and z-axis oriented along the vertical).

The principal axes of anisotropy and the diagonal components of [K] in these directions were determined using the eigenvalues and eigenvectors of the matrix [K]. The eigenvalues and the orientation of the principal axes of the tensor are listed in Table 3.

Table 3. Values and orientation of the major, intermediate and minor components of the hydraulic conductivity of rock mass.

Components	Value (m/s $\times 10^{-5}$)	Trend (°)	Plunge (°)
Minor	1.58	201	15
Intermediate	3.19	109	6
Major	3.84	359	74

4.2. Injection Tests

The injection tests processing first included a study of the relation between the flow rate (q) and the net injection pressure (H0). In some cases, the relations highlighted the opening and erosion processes in the discontinuities, while, in other cases, the void filling and essential laminar flow were noted.

The hydraulic conductivity kL in each test interval was calculated by applying the following equation [36]:

$$k_L = \frac{q}{lH_0}\left[\frac{1}{2\pi}\left(1 + ln\frac{l}{2r}\right)\right] \tag{3}$$

where l and r are the length of the injection section and the borehole radius, respectively; and q and H0 are the flow rate and the net injection pressure values under laminar flow, respectively.

The calculated values of kL fell within three orders of magnitude (10–8–10–6 m/s), with a geometric mean of 2.81 × 10–7 m/s. Figure 3 shows the kL values of each borehole plotted versus the average depth of the injection section. At depths greater than 25 m, there was a significant decrease in the hydraulic conductivity, with the exception of the inclined borehole (LT8 in Figure 3). Above a depth of 25 m, the boreholes that were located in the valley bottom (LT4-LT8 in Figure 3), corresponding to a faulted zone, exhibited a higher kL.

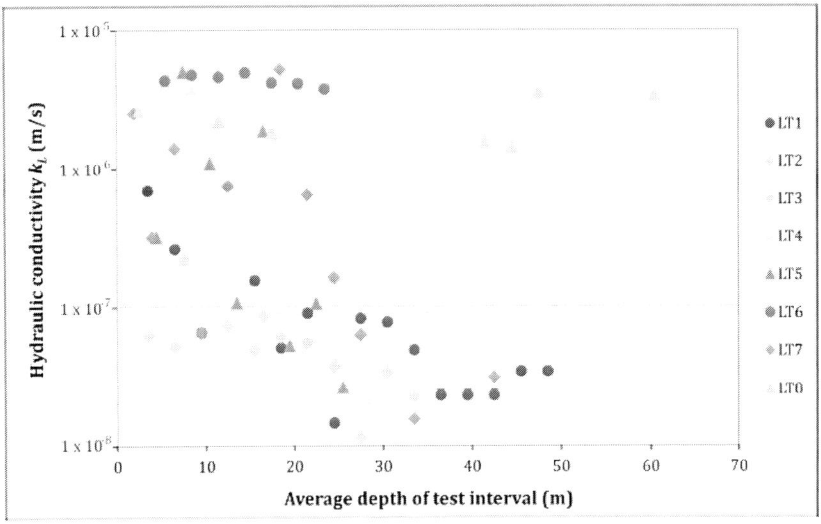

Figure 3. Hydraulic conductivity kL from the injection tests versus the average depth of the injection section.

4.3. Pumping Tests

The pumping test data diagnosis included a comparison of the drawdown plots of each piezometer with theoretical models to determine the aquifer parameters. This analysis showed that the Theis [37], leaky aquifer [38] and double porosity [39] models could explain most of the measured drawdowns; in some cases, two or three theoretical models interpreted the same drawdown-time trend without significant differences in the calculated aquifer parameters, i.e., the transmissivity and storage coefficient. For wells with more than one piezometer, there was a simultaneous response in the pumping, and generally, a similar shape for the drawdown curve for the different observation wells was reported, as observed in Figure 4.

Table 4 summarizes the results of the aquifer parameters, which were calculated as an average value when more than one model presented a good fit with the measured data and with the statistics of the curve matching. The overall transmissivity and storage coefficients varied from 2.3 × 10−4 to 3.6

× 10−3 m2/s and from 3.1 × 10−5 to 2.6 × 10−2, respectively. The transmissivities were converted into hydraulic conductivities (KP) based on the saturated aquifer thickness intercepted by the tested well. The values of KP fell in the range of 4.6 × 10−6–4.1 × 10−5 m/s, with a geometric mean of 1.68 × 10−5 m/s.

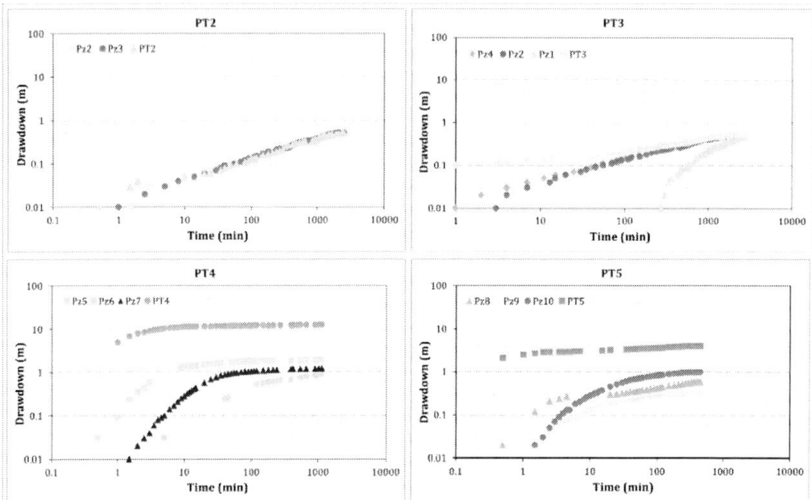

Figure 4. Drawdown-time plots resulting from the pumping tests for the wells with more observation piezometers.

Table 4. Results of the pumping tests: Q, constant rate of pumping; Hsat, saturated thickness of the aquifer intercepted by the well; T, transmissivity; KP, hydraulic conductivity; S, storage coefficient; RSS, sum of squared residuals.

Well	Q (m^3/s)	H_{sat} (m)	Piezometer	T (m^2/s)	K_P (m/s)	S	RSS
PT1	2.1×10^{-3}	50	Pz1	4.89×10^{-4}	9.78×10^{-6}	2.61×10^{-2}	0.001
PT2	2.3×10^{-3}	41	Pz2	1.44×10^{-3}	3.51×10^{-5}	1.85×10^{-3}	0.024
			Pz3	1.42×10^{-3}	3.46×10^{-5}	5.11×10^{-3}	0.024
PT3	2.4×10^{-3}	53	Pz4	1.51×10^{-3}	2.85×10^{-5}	7.29×10^{-3}	0.021
			Pz2	1.41×10^{-3}	2.66×10^{-5}	4.56×10^{-3}	0.023
			Pz1	4.46×10^{-4}	8.41×10^{-6}	1.80×10^{-2}	0.002
PT4	2.5×10^{-3}	50	Pz5	2.32×10^{-4}	4.64×10^{-6}	3.15×10^{-5}	0.045
			Pz6	9.50×10^{-4}	1.90×10^{-5}	6.26×10^{-4}	0.013
			Pz7	2.33×10^{-4}	4.66×10^{-6}	5.22×10^{-4}	0.035
PT5	3.5×10^{-3}	86	Pz8	3.57×10^{-3}	4.15×10^{-5}	9.92×10^{-4}	0.023
			Pz9	3.44×10^{-3}	4.00×10^{-5}	8.83×10^{-4}	0.001
			Pz10	8.24×10^{-4}	9.58×10^{-6}	2.64×10^{-4}	0.002

4.4. Groundwater Flow Model

The hydrogeological characterization of the site was completed by a potentiometric surface reconstruction over the entire dam site using the head data measured in 70 wells and boreholes during the initial reservoir filling when the artificial lake level was at 121 m a.s.l. [30]. The head contour line deformation due to the lake and due to the gaining river was obtained by adding head points along the boundary conditions; the dam effect (i.e., the upstream impervious face and grout curtain) was obtained by applying an impervious bound in the Kriging interpolation technique. From Figure 5, where the head distribution and the standard error ranges associated with the Kriging prediction are shown, it appears that the head contour lines generally replicate the topography. On the left of the dam, the lake feeds groundwater which discharges into River Pittiu and laterally bypasses the grout curtain flowing towards the rockfill embankment of the dam.

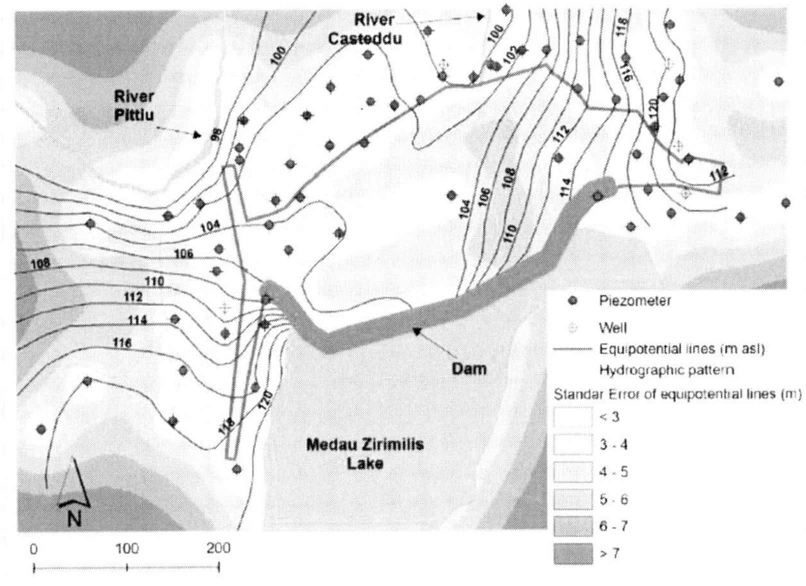

Figure 5. Potentiometric map at the site of the Medau Zirimilis Dam reconstructed by the kriging interpolation technique and the associated ranges of standard error when the artificial lake level was at 121 m a.s.l.

Thanks to the presence of physical boundary conditions (lake and river) and to the presence of both pumping tests and surface fracture survey, the western sector of the dam was selected for the numerical simulation (Figure 1).

The model involved an area of 86,750 m2 (Figure 6). The grid consisted of one layer of 8736 cells with a telescopic refinement around the pumping well (PT4) and side dimensions from 0.33 to 8 m. As the investigations revealed, the bottom of the aquifer was not well defined; the available stratigraphic logs where compared with the results of injection and pumping tests and the model bottom was set at a depth of 50 m from ground surface. The aquifer was considered continuous, unconfined and isotropic, as resulted from field investigations.

The boundary conditions included: constant head boundary for the lake (i.e., water level at 121 m a.s.l.); drain boundary condition for the gaining stream (i.e., River Pittiu) and for section of the dam embankment where flow was measured in the drainage systems; no-flow cells in sectors outside the domain of interest or along groundwater divides (western side); well boundary condition for the pumping well (i.e., PT4 well); constant flux boundary condition for the southern boundary (Figure 6). No precipitation event occurred in the days before and during the pumping tests, therefore no areal recharge was applied.

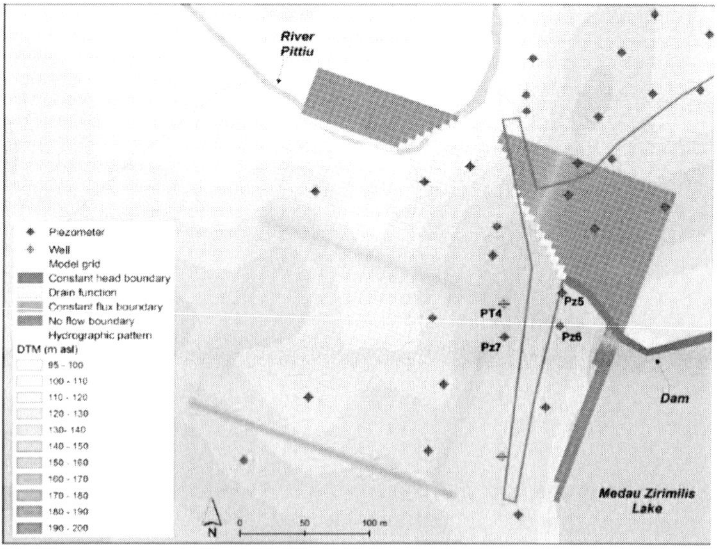

Figure 6. Grids and boundary conditions of the numerical model. DTM: digital terrain model.

The days in which the pumping test was performed, a constant pumping rate of 2.5 × 10–3 m3/s from the well PT4 was measured. The same day the lake level was 121 m a.s.l. and a flow of 6.7 × 10–3 m3/s was measured from the horizontal drain located at the base of the embankment in the left sector of the dam [30]. If we consider the portion of the drain that falls within the modeled area, a maximum outflow of 3 × 10–3 m3/s is expected from this model boundary. Outflow from River Pittiu was not measured during the days of the test.

Inflows from the southern and lake boundaries were estimated. For the southern boundary, considering the surface area of the upstream potential groundwater basin drained from this section, an area of about 2 km2 is obtained. Base flow measures are available for an instrumented basin which is morphologically and geologically comparable with the basin under investigation, resulting between 2 × 10–3 and 4 × 10–3 m3/s per km2 (i.e., River Fluminimaggiore; [40]). According to this information, an inflow from 4 × 10–3 to 8 × 10–3 m3/s is expected from the southern boundary. A second estimate considers the hydraulic gradient and transmissivity of this sector of the dam site and the length of boundary, obtaining an inflow ranging between 5 × 10–3 and 6 × 10–3 m3/s. Using this last approach for the lake boundary, an inflow ranging between 4 × 10–3 and 5 × 10–3 m3/s results.

The model underwent an initial calibration through the trial-and-error technique. The main purposes of this step were to test the numerical stability of the model and to roughly reduce the initial difference between the observed and calculated values of the observation points. This step preceded the application of the parameter estimation code PEST 13.0 [41], which was applied in two different modes: classical zone calibration and pilot point calibration [42,43].

A steady-state calibration was used as the first time period of the transient simulation. A second time period, divided into 20 steps, was introduced with pumping from PT4 at a constant rate (2.5 × 10–3 m3/s) for 19 hours. Transient data were calibrated for the pumping well (PT4) and Pz5 and Pz7 piezometers.

A sensitivity analysis of the transient model highlighted the significant influence of the bottom elevation, which was only known in coincidence of boreholes and piezometers. Therefore the results were expressed by translating the hydraulic conductivity into transmissivity.

In Figure 7, simulated contour lines at the end of 19 h and transmissivity distribution are reported. In the same figure the simulated and observed heads of all wells and piezometers are shown, as well as the trends of drawdown in the PT4 well and in the Pz5 and Pz7 piezometers. The calibrated values of the specific yield are within the order of magnitude of 10−4. The water balance is shown in Table 5 in comparison with measurements and analytical computations of the flows.

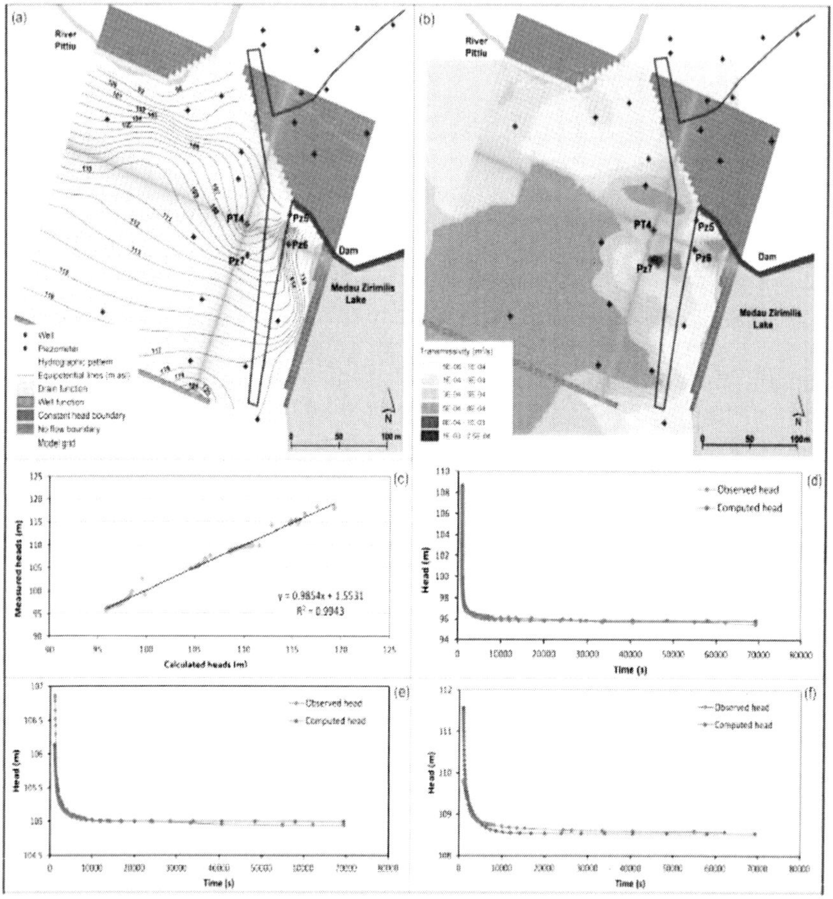

Figure 7. Results of the transient-state calibrated model showing (a) the potentiometric contour map; (b) the transmissivity distribution; (c) the scatter plot of the measured and calculated heads; (d) the comparison of the observed and calculated heads for the PT4 tested well; (e) for the Pz5 piezometer and (f) for the Pz7 piezometer.

Table 5. Groundwater budget resulting from the numerical model compared with the estimated terms from the conceptual model.

Model	Inflow ($\times 10^{-3}$ m^3/s)			Outflow ($\times 10^{-3}$ m^3/s)			% Error
	Lateral Boundaries	Lake	Storage	Well	River	Lateral Boundaries	(in-out)
Numerical Model	6.84	3.10	0.08	2.50	5.69	1.81	0.2
Conceptual model	4–8	4–5	-	2.50	na	3	-

Notes: na: not available.

Composite sensitivities associated with pilot points were calculated for the transient model with reference to the hydraulic conductivity, i.e., the parameter that mainly affects the results. Highest values of composite sensitivity are associated to the pilot points corresponding with the pumping well PT4 and those falling in its surroundings (Figure 8), where the highest number of observations is located. Therefore, we can assume a good representativeness of the calibrated hydraulic conductivity in this zone, where the calibrated parameters give ranges of transmissivity from 5.0 × 10−6 to 2.5 × 10−3 m2/s.

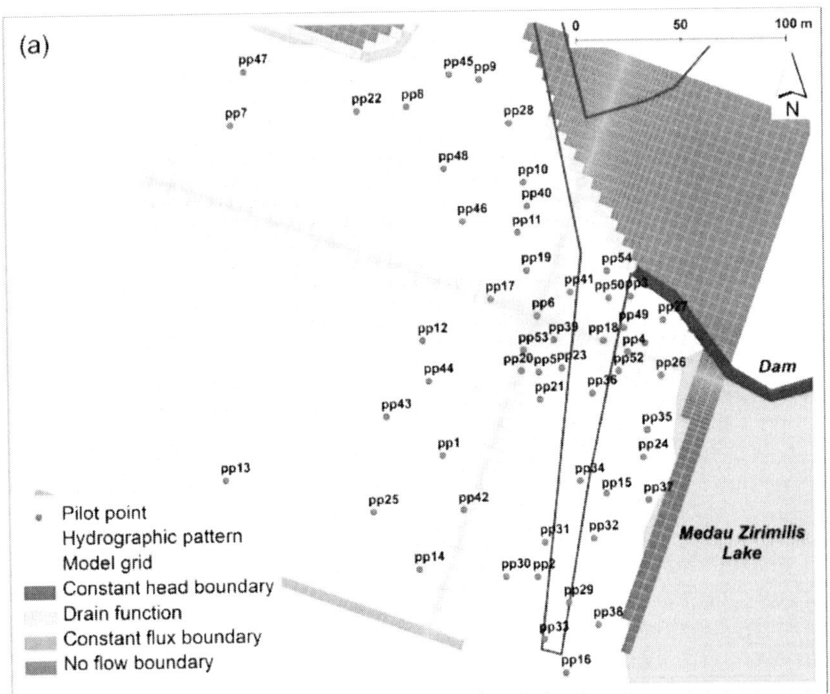

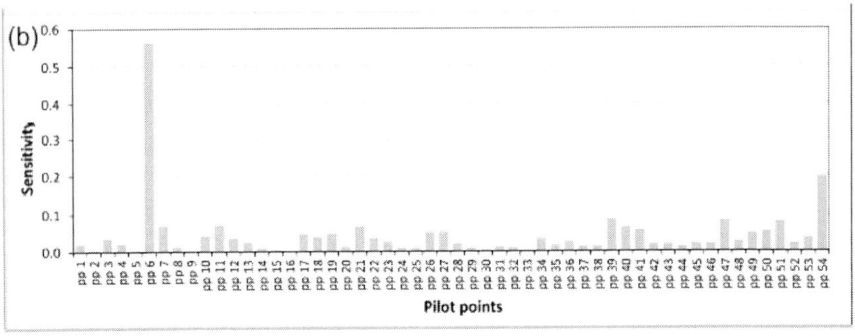

Figure 8. (a) Distribution of the pilot points in the model domain and (b) values of composite sensitivities for each point.

5. DISCUSSION

Investigations concerning fractured metamorphic rocks in a representative area of Sardinia can be used to hydrogeologically characterize these low-permeability rocks and to examine which approach can be used to represent the fractured aquifer. By interfacing the results of surface fracture surveys, injection tests and pumping tests with those of a flow numerical model, we inspected if the porous equivalent approach could (or could not) provide a good approximation of a fractured aquifer.

The surface fracture survey provides a model of rock mass that is characterized by a well-developed discontinuity network (three discontinuity sets), a high frequency of discontinuities (spacing of less than 0.5 m) and comparable values of the average apparent aperture of the different sets (approximately 0.1 mm). Therefore the results of the surface fracture survey implied the possibility of considering the fractured medium as an equivalent porous medium at the outcrop scale, because the discontinuity network gives rise to a reduced representative elementary volume (less than 1 m3) for the rock mass.

The discontinuity network produces a slight anisotropy in the medium, with the major principal component of the hydraulic conductivity directed vertically downward, as expected given the predominance of vertical discontinuities over horizontal ones in the studied area. However, the modules of the hydraulic conductivity determined for the rock mass should be considered in relative terms and not in absolute terms because the apparent aperture does not correspond to the hydraulic aperture. A greater reliability can be assigned to the orientation of the three principal components, independent from the aperture.

The injection tests clearly showed the extreme heterogeneity of the rock mass, as highlighted by the wide range of values determined for the hydraulic conductivity (10–8–10–6 m/s). This can be interpreted by considering the variability of the rock mass fracturing in relation to the proximity to the faulted zone in the valley of the dam site. The highest values of hydraulic conductivity were found for the inclined borehole, which was close to the fault zone and, thus, intercepted a great number of vertical fractures. With the exception of the inclined borehole close to fault zone, the hydraulic conductivity was observed to diminish below a depth of 25 m.

The hydraulic conductivity derived from the pumping tests covered a lower range of values (10–6–10–5 m/s) than did those calculated using the injection tests. No particular trend was evident from the comparison of the values of hydraulic conductivity obtained for piezometers located in different directions from the pumped well. The variation in the parameter is therefore attributable to the heterogeneity of the aquifer rather than to its horizontal anisotropy. Moreover, if we consider the simultaneous response to pumping and the similar shape of the drawdown curves for the different piezometers when the pumping was performed from the same well, the porous medium approximation appears appropriate.

The comparison between the hydraulic conductivity determined by the injection tests and the pumping tests revealed values that were substantially lower in the first case (Figure 9). This can be attributed to the limited volume of rock mass affected by the injection tests compared to that affected by the pumping tests. In both cases, the hydraulic heterogeneity of the rock mass was evident. The injection tests and the pumping tests also converged and indicated higher hydraulic conductivities near the fault that runs in the valley of the dam site (Figure 3 and Table 4). Instead, the limit for the determination of the aperture by the surface survey did not permit a direct comparison of the hydraulic conductivity values determined using this method. However, if the average hydraulic conductivity (i.e., the geometric mean) calculated from the pumping tests and the average spacing of the three sets determined from the surface survey are introduced in Equation (1), the average aperture value would be 0.16 mm, which is comparable with those measured in outcrop. However, the range of hydraulic conductivity resulting from the surface survey (10–6–10–5 m/s) is consistent with that evaluated by pumping tests (Figure 9).

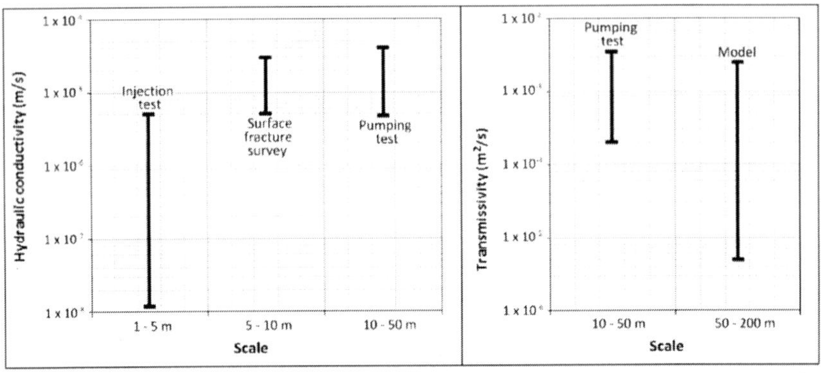

Figure 9. Hydraulic parameters and scale of measurement of the different methods used.

This scale effect of the parameter is expected in heterogeneous media [44,45,46,47] because the volume investigated by the injection tests was significantly smaller than that of the pumping tests. In the graph of Figure 9 one can see that the range of hydraulic conductivity values obtained from the surface fracture survey is consistent with the increase of hydraulic conductivity with the increase of the measuring scale. This implies that the results of surface fracture survey can be considered in the characterization of hard rocks when the discontinuity apertures are measured on the unweathered portions of discontinuity, at least to identify the order of magnitude of the parameter.

The numerical model considered one representative pumping test and provided additional information on the hydrogeological characterization of the hard rocks.

The results of the field investigations together with the model output would encourage the equivalent porous approach, at least in geological situations similar to that here considered. The estimate of transmissivity resulting from the model highlighted a range of values higher than those resulting from the pumping tests. However, the transmissivity values determined in the pumping tests (from 10−4 to 10−3 m2/s) fell in the range of the calibrated values of the model (from 10−6 to 10−3 m2/s), with a common upper limit of the parameter (Figure 9). This still seems to be consistent with what is reported in the literature, i.e., that the increase in average permeability apparently does not continue from the pumping test to a regional scale [44,45]. On the other hand, four orders of magnitude of transmissivity obtained from the model compared with two orders of magnitude from the pumping tests, highlight a high hydraulic heterogeneity of the aquifer, this time in agreement with the results of the injection tests.

6. CONCLUSIONS

The multi-scale characterization of the metamorphic rocks of the area studied herein demonstrated that the continuum medium approach can be used when a dense network of discontinuities exists (representative elementary volume of few cubic meters), as in the case of the fractured aquifer examined. What strongly affects this approach is the hydraulic heterogeneity of the medium. The multi-scale approach is thus recommended for investigating the hydraulic heterogeneity, and it can help to manage uncertainty by placing greater trust in the obtained results. In particular, it is evident that the surface fracture survey enhances the hydraulic characterization of fractured rocks, when the fracturing affects the layers closest to the surface, as the general case of the hard-rock aquifers.

Regarding the transmissivity of the examined hard-rock aquifer, it falls in the classes III and IV of the classification proposed by Krasny [48], corresponding to rocks from low to intermediate transmissivity magnitude. This aquifer, commonly believed to be not productive, may be of interest for the local water supply, especially if one considers the large extent of these rocks and the scarcity of water resources in Sardinia. The same concept could be applied in other regions of the world where these aquifers outcrop over very large areas. In this regard, our findings confirm that the most promising areas for well location correspond to valleys close to faulted zones where transmissivity increases; in these specific zones a relative higher yield of wells is expected, especially in case of boreholes inclined according to an angle which can be determined after the fracture characterization.

ACKNOWLEDGMENTS

The authors wish to thank the two anonymous reviewers for their constructive comments and suggestions. Vincenzo Piscopo want to remember recently deceased Pietro Celico to have introduced him with passion and dedication to hydrogeology.

AUTHOR CONTRIBUTIONS

All of the authors contributed extensively to the work. Vincenzo Piscopo led the research. Antonella Baiocchi, Francesca Lotti and Simone Piacentini performed the data analysis and numerical modeling. Walter Dragoni contributed to the manuscript review. All authors contributed to the manuscript preparation, discussing and sharing the conclusions.

REFERENCES

1. Banks, D.; Banks, S. Hydrogeology of Hard Rocks. In Proceedings of the Memoires of the 24th Congress, International Association of Hydrogeologists, Oslo, Norway, 28th June–2nd July 1993; volume 28, pp. 1–684.
2. Stober, I.; Bucher, K. Hydrogeology of Crystalline Rocks; Kluwer Academic Publishers: Dordrecht, The Netherlands, 2003.
3. Taylor, R.; Howard, K. A tectono-geomorphic model of the hydrogeology of deeply weathered crystalline rock: Evidence from Uganda. Hydrogeol. J. 2000, 8, 279–294.
4. Cho, M.; Ha, K.-M.; Choi, Y.-S.; Kee, W.-S.; Lachassagne, P.; Wyns, R. Relationship between the Permeability of Hard-Rock Aquifers and Their Weathered Cover Based on Geological and Hydrogeological Observation in South Korea. In Proceedings of the IAH Conference on "Groundwater in fractured rocks", Prague, Czech, 15–19 September 2003.
5. Dewandel, B.; Lachassagne, P.; Wyns, R.; Maréchal, J.C.; Krishnamurthy, N.S. A generalized 3-D geological and hydrogeological conceptual model of granite aquifers controller by single or multiphase weathering. J. Hydrol. 2006, 320, 260–284.
6. Krasny, J.; Sharp, J.M. Groundwater in Fractured Rocks; Taylor & Francis: London, UK, 2007. [Google Scholar]
7. Foster, S. Hard-rock aquifers in tropical regions: Using science to inform development and management policy. Hydrogeol. J. 2012, 20, 659–672.
8. Sharp, J.M. Fractured Rock Hydrogeology; Taylor & Francis: London, UK, 2014. [Google Scholar]

9. Uhl, V.W.; Sharma, G.K. Results of pumping tests in crystalline-rock aquifers. Ground Water 1978, 16, 192–203.

10. Howard, K.W.K.; Hughes, M.; Charlesworth, D.L.; Ngobi, G. Hydrogeologic evaluation of fracture permeability in crystalline basement aquifers of Uganda. Hydrogeol. J. 1992, 1, 55–65.

11. Chilton, P.J.; Foster, S.S.D. Hydrogeological characteristics and water-supply potential of basement aquifers in tropical Africa. Hydrogeol. J. 1995, 3, 3–49.

12. Maréchal, J.C.; Dewandel, B.; Subrahmanyam, K. Use of hydraulic tests at different scales to characterize fracture network properties in the weathered-fractured layer of a hard rock aquifer. Water Resour. Res. 2004, 40.

13. Banks, D.; Gundersen, P.; Gustafson, G.; Makela, J.; Morland, G. Regional similarities in the distributions of well yield from crystalline rocks in Fennoscandia. Norges Geol. Unders. Bull. 2010, 450, 33–47. [Google Scholar]

14. Lee, C.H.; Chang, J.L.; Deng, B.W. A continuum approach for estimating permeability in naturally fractured rocks. Eng. Geol. 1995, 39, 71–85.

15. Scanlon, B.R.; Mace, R.E.; Barrett, M.E.; Smith, B. Can we simulate regional groundwater flow in a karst system using equivalent porous media models? Case study, Barton Springs Edwards aquifer, USA. J. Hydrol. 2003, 276, 135–158.

16. Lemieux, J-M.; Therrien, R.; Kirkwood, D. Small scale study of groundwater flow in a fractured carbonate-rock aquifer at the St-Eustache quarry, Québec, Canada. Hydrogeol. J. 2006, 14, 603–612.

17. Long, J.C.S.; Remer, J.S.; Wilson, C.R.; Witherspoon, P.A. Porous media equivalents for networks of discontinuous fractures. Water Resour. Res. 1982, 18, 645–658.

18. Hsieh, P.A.; Neuman, S.P.; Stiles, G.K.; Simpson, E.S. Field determination of the three-dimensional hydraulic conductivity tensor of anisotropic media. 2 Methodology and application to fractured rocks. Water Resour. Res. 1985, 21, 1667–1676.

19. Neuman, S.P. Stochastic Continuum Representation of Fractured Rock Permeability as an Alternative to the Rev and Fracture Network Concepts. In Proceedings of the 28th US Symposium on Rock Mechanics, Tucson, AZ, USA, 29 June–1 July 1987; pp. 533–561.

20. Bradbury, K.R.; Muldoon, M.A.; Zaporec, A.; Levy, J. Delineation of Wellhead Protection Areas in Fractured Rocks; US EPA Technical Guidance Document, EPA 570/9-91-009; EPA: Washington, DC, USA, 1991. [Google Scholar]

21. Nuclear Regulatory Commission (NRC). Rock Fractures and Fluid Flow: Contemporary Understanding and Applications; National Academy Press: Washington, DC, USA, 1996.

22. Neuman, S.P. Trends, prospects and challenge in quantifying flow and transport through fractured aquifer. Hydrogeol. J. 2005, 13, 124–147.

23. Baiocchi, A.; Dragoni, W.; Lotti, F.; Piscopo, V. Sustainable yield of fractured rock aquifers: The case of crystalline rocks of Serre Massif (Calabria, Southern Italy). In Fractured Rock Hydrogeology; Sharp, J.M., Ed.; Taylor & Francis Group: London, UK, 2014; pp. 79–97.

24. Dragoni, W. Some considerations on climatic changes, water resources and water needs in the Italian region south of the 43° N. In Water, Environment and Society in Times of Climatic Change; Issar, A., Brown, N., Eds.; Kluwer Academic Publishers: Dordrecht, The Netherlands, 1998; pp. 241–271.

25. Intergovernmental Panel on Climate Change (IPCC). Summary for Policymakers. In Climate Change 2013: The Physical Science Basis; Cambridge University Press: Cambridge, UK, 2013.

26. Carmignani, L.; Cocozza, T.; Gandin, A.; Pertusati, P.C. Lineamenti della geologia dell'Iglesiente-Sulcis [Outlines of the geology of the Iglesiente-Sulcis Region]. In Guida alla Geologia del Paleozoico Sardo; Società Geologica Italiana, Guide Geologiche Regionali: Roma, Italy, 1982; pp. 65–77.

27. Istituto Superiore per la Protezione e la Ricerca Ambientale (ISPRA). Notes to the Geological Map of Italy 1:50,000 "Sheet 566–Assemini"; Servizio Geologico d'Italia: Roma, Italy, 2009. (In Italy)

28. Barca, S.; Cocozza, T.; del Rio, M.; Pittau Demelia, P. Discovery of lower Ordovician Acritarchs in the "Postgotlandiano" sequence of southwestern Sardinia (Italy): Age and tectonic implications. Boll. Soc. Geol. Ital. 1981, 100, 377–392.

29. Carmignani, L.; Conti, P.; Barca, S.; Cerbai, N.; Eltrudis, A.; Funedda, A.; Oggiano, G.; Patta, E.D. Notes to the Geological Map of Italy

1:50,000 "Sheet 549–Muravera"; Servizio Geologico d'Italia: Roma, Italy, 2001. (In Italy)

30. Celico, P.; Piscopo, V.; Berretta, G. Influence of a reservoir on groundwater flow in metamorphic rocks. Geol. Appl. Idrogeol. 1993, 28, 253–261. (In Italy).

31. Kruseman, G.P.; de Ridder, N.A. Analysis and Evaluation of Pumping Test Data; ILRI Publication 47; International Institute for Land Reclamation and Improvement: Wageningen, The Netherlands, 1994. [Google Scholar]

32. Allmendinger, R.W.; Cardozo, N.C.; Fisher, D. Structural Geology Algorithms: Vectors & Tensors; Cambridge University Press: Cambridge, UK, 2013.

33. Cardozo, N.; Allmendinger, R.W. Spherical projections with OSX Stereonet. Comput. Geosci. 2013, 51, 193–205.

34. Snow, D.T. Anisotropic permeability of fractured media. Water Resour. Res. 1969, 5, 1273–1289.

35. Oda, M. Permeability tensor for discontinuous rock masses. Geotechnique 1985, 35, 483–495.

36. Moye, D.G. Diamond drilling for foundation exploration. Civil Eng. Trans. Inst. Eng. 1967, 9, 95–100.

37. Theis, C.V. The relation between the lowering of the piezometric surface and the rate and duration of discharge of a well using groundwater storage. Am. Geophys. Union Trans. 1935, 2, 519–524.

38. Hantush, S.M.; Jacob, C.E. Non-steady radial flow in an infinite leaky aquifer. Am. Geophys. Union Trans. 1955, 36, 95–100.

39. Warren, J.E.; Root, P.J. The behavior of naturally fractured reservoirs. One Petro 1963, 3, 245–255.

40. Regione Sardegna (2010) Annali Idrologici. Available online: www.regione.sardegna.it (accessed on 25 October 2014).

41. Doherty, J. PEST, Model-Independent Parameter Estimation–User Manual, 5th ed.; Watermark Numerical Computing: Brisbane, Australia, 2010.

42. De Marsily, G.; Lavedan, C.; Boucher, M.; Fasaniso, G. Interpretation of interference tests in a well field using geostatistical techniques to fit the permeability distribution in a reservoir model. In Geostatistics for Natural Resources Characterization; NATO

Advanced Study Institute: Dordrecht, The Netherlands, 1984; pp. 831–849.

43. Doherty, J. Groundwater model calibration using pilot-points and regularization. Ground Water 2003, 41, 170–177.

44. Clauser, C. Permeability of crystalline rocks. EOS 1992, 73, 233–240.

45. Schulze-Makuch, D.; Carlson, D.A.; Cherkauer, D.S.; Malik, P. Scale dependency of hydraulic conductivity in heterogeneous media. Ground Water 1999, 37, 904–919.

46. Renard, P.; de Marsily, G. Calculating equivalent permeability: A review. Adv. Water Res. 1997, 20, 253–278.

47. Neuman, S.P.; di Federico, V. Multifaceted nature of hydrogeologic scaling and its interpretation. Rev. Geophys. 2003, 41.

48. Krasny, J. Classification of transmissivity magnitude and variation. Ground Water 1993, 31, 230–236.

7

CHAPTER

CALIBRATION OF A NEW DEVICE TO MEASURE WATER CONTENT OF ROCKS

Maria Clementina Caputo[1] and Rita Masciale[1]

[1] Water Research Institute (IRSA), National Research Council (CNR), Bari, Italy

1. INTRODUCTION

The vadose zone, which extends from the soil–atmospheric interface to the capillary fringe of the water table, is a fundamental part of the hydrologic cycle. It controls how the precipitation splits into infiltration, surface runoff, evapotranspiration, groundwater recharge; it regulates the storage, transfer, filtering, adsorption, retard and attenuation of solutes and contaminants before these reach the groundwater.

Until the last decades, the vadose zone was considered as a black box that merely connect surface water and groundwater. As consequence an incomplete understanding of the complex dynamics of the vadose zone existed.

Nowadays, instead, the monitoring of the hydrological processes that occur through the vadose zone, are receiving increased attention mainly with refer to contaminant transport processes. Fluxes of water and solutes strongly depend on water content thus its monitoring and estimation becomes an important issue.

A wide range of methods, sensors and technologies are available for the measure of soil water content, mainly used in management of precision farming.

The traditional standard method for direct measurement of soil water content is the gravimetric method [1]. This method implies the sampling of the soil to carry out laboratory measurements, modifying the natural condition. As gravimetric method is destructive, labour intensive, not timely and costly, many alternative non-destructive methods for measuring water content have been developed. However, none of all methods measure the water content directly but only some properties, named "surrogate measure", that changes as the soil water content changes. By measuring the value of the surrogate parameter, it is possible to estimated the value of water content, using a calibration curve that represents the relationship between the surrogate measure and the soil water content values [2].

Neutron Thermalization method [3, 4], Capacitance method [5-8], Time Domain Reflectometry (TDR) method [9-13], Frequency Domain Reflectometry (FDR) method [14, 15], Impedance method [16, 17], Electrical Restistance method [18, 19] and Tensiometer method [20-22] are among the well-known technologies utilized to develop different kind of devices and probes for measuring moisture content and soil water potential, respectively.

More details on these technologies and devices, advantages and limitations of each one, are given in several references [23, 24, 2].

Recently, remote sensing technologies, like passive or active radiometry, have been applied in order to provide wide-area indications of surface soil water content [25, 26]. However, the great influence of vegetation and surface structure on the quality of received signal, restricts the sampling depth to the uppermost part (2–5 cm) of soil [27], strongly limiting the usefulness of this technology in hydrological application.

The use of all these methods and devices in the soil is customary by now but when the vadose zone consists of rocks the monitoring of water content become more difficult for several aspects. The main difficulty regards the installation of the probes that are often very delicate and cannot be hammered or screwed into the rock. Furthermore, after their insertion, a good contact between the rock and the sensor must be ensured in order to minimize the gap effect which causes significant errors in the investigated properties, especially for dielectric ones [28, 29].

At present, very few applications of these techniques to rocks, by creating new devices or by adapting the existing ones, are documented in literature.

The first measurements in sandstone and tuff blocks by means of penetration type probes using TDR are reported in [30]. Topp's equation [9] was used to convert the dielectric constant (K) values into water content (θ) highlighting an overestimation of water content. In fact, especially for rocks with low effective porosity, an individual calibration was needed.

In [31] TDR was applied in granodiorite both in the laboratory and in the field. Employing surface probes, it was showed that K and θ were almost linearly related, revealing that TDR was capable to measure the volumetric water content changes.

Using brass rods, individual K - θ calibration functions were developed in [32] for nine different rock types with porosity values ranging from 1 to 54%. However, the experimentally determined K - θ relationships showed an abrupt jump near fully saturation of samples, suggesting that TDR would be accurate for applications in rocks only when the gap effect is negligible.

By applying TDR in seven different types of rocks [33], the authors demonstrated that conductive silicone fillings, carefully applied, can successfully eliminate gap effect in the case of penetration probes. Moreover, they compared the K-θ relationships obtained with surface probes against that obtained using penetration probes. The results showed systematic differences between two types of probes, explained on the basis of their geometry.

In this chapter, we intended primarily to give further contribute to the knowledge on the applicability on rocks of methods developed for measuring water content in the soils by focusing on the Electrical Impedance Spectrometry (EIS) method. In particular, this work presents the results obtained to calibrate a new device, called Z-meter 2, based on EIS method. Samples of calcarenite have been used for the calibration procedure carried out in laboratory under controlled condition.

The electrical impedance in complex form and its changes over time have been measured with the aim to:

a. verify the suitability of the device for water content estimation in rocks;

b. characterize the effects on the measured values of the electric frequency applied and of the electrical conductivity (EC) of the solution used to saturate samples;

c. determine specific calibration curves for the investigated lithotype.

2. MATERIALS AND METHODS

2.1. Electrical Impedance Spectrometry (EIS) method

The basic principle of the EIS method [34] is the measurement of the electrical impedance $(\hat{Z})$ which provides a measure of medium's opposition during the passage of a alternating current (AC). The impedance is described not only by the amplitudes of the voltage and the current, but also by their relative phase-displaced.

The impedance of a circuit element can be defined as the frequency domain ratio of the voltage phasor $\hat{U}$ to thecurrent phasor $\hat{I}$ across the element:

$$\hat{Z} = \frac{\hat{U}}{\hat{I}} \quad (1)$$

Impedance is represented as a complex quantity. In the polar form, it is

$$\hat{Z} = |Z| e^{j\theta}$$

where $|Z|$ is the magnitude, the argument θ is the phase difference between voltage and current and j is the imaginary unit.

In cartesian form, it is

$$\hat{Z} = R + jX \quad (3)$$

where R is the resistance, that forms the real part of impedance, independent from the frequency, and X is the reactance, that forms the imaginary part of impedance, dependent from the frequency. The values of impedance are expressed, as those of resistance and reactance, in ohms [Ω].

The impedance depends on the characteristics of AC electric circuit which can be made by combination of resistor, inductor and capacitor, either in series or parallel mode.

For a real resistor-capacitor (RC) circuit and resistor-inductor (RL) circuit, the impedance can be expressed by the following relations

$$\hat{Z}_L = R + j \cdot X_L = R_L + j\omega L \quad (4)$$

and

$$\hat{Z}_C = R - j \cdot X_C = R_C - \frac{j}{\omega C} \quad (5)$$

where

R = resistance

$X_L = 2 \cdot \pi \cdot f \cdot L =$ = inductive reactance

$X_C = -\frac{1}{2 \cdot \pi \cdot f \cdot C} =$ = capacitive reactanc

f = frequency (Hz)

$\omega = 2\pi f =$ = angular frequency

C = capacitance or electrical capacity (F)

L = inductance (H)

The porous medium (e.g. soil and rock) can be described by the impedance of an equivalent electric circuit. It always has the character of a resistor, so that the real component of the total impedance $\hat{Z}$ of the circuit is always present.

In a dry porous medium the resistance is so high that it can be considered as a non-conductor (dielectric). Therefore it is possible to assume that dry porous medium is polarized by the electric field and thus it is charged as a capacitor characterized by its capacitance.

Using this assumption, the capacitance of medium is substantially higher than the inductance so it can be neglected and the dominant component of imaginary part of the impedance $\hat{Z}$ is the capacitive reactance (X_C).

As for mentioned above, the porous medium can be described as an equivalent RC circuit and its impedance, expressed by the equation (5), describes primarily electromagnetic properties of medium.

The solid part (grains) is formed by insulating materials, characterized by their dielectric constants, and hence it represents the imaginary part (X_C) of the measured impedance. If the material is in non-zero moisture condition, it can be considered as a conductor because water always contains a certain quantity of mineral salts (conductive material). In this case the resistance is lower and the degree of saturation of the material strongly influences the real part of the measured impedance.

2.2. Z-meter 2 device

The Z-meter 2 is a new device developed, constructed and realized within the European International Project E!3838 of EUREKA Program [35]. It's based on EIS method and it is able to measure the electrical impedance in complex form. Z-meter 2 was specifically designed to carry out long-term field experiments.

According to the equivalent diagram of the circuit in Figure 1, the resulting measured value of impedance is represented by a parallel combination of resistor and capacitor.

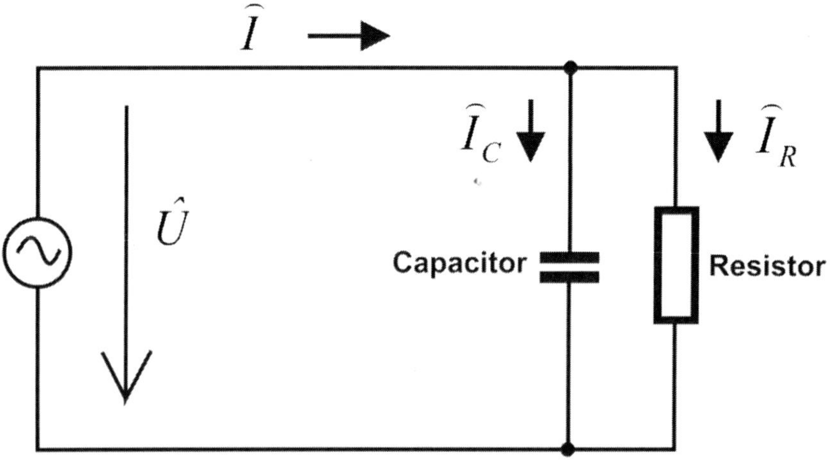

Figure 1. Equivalent diagram of measured impedance.

The electric current is divided into two orthogonal components, real $\hat{I}_R$ flowing through the real component of the equivalent resistor, and imaginary $\hat{I}_C$ flowing through the equivalent capacitor. In this way the Z-meter 2 allows measurement of both parts of the impedance: resistance and reactance. The basic parameters of the device are summarized in Table 1.

Many different kinds of probes can be used with Z-meter 2 for measuring the impedance of soils, either in the field or in laboratory: rod probes, modular probes with variable length constituted by stainless electrical conductive parts alternated with non conductive parts (polyamide), aluminium strip probes with a self-adhesive layer on one side that ensures perfect contact with the medium.

The kind of probe needs to be chosen in relation to the monitored medium, the environmental condition and the aim of the measurements. In this sperimentation, stainless steel road probes, called picket type probe, were used. These probes were chosen because they can be easily installed in the rock without be damaged.

Table 1. Basic parameters of a Z-meter 2 device.

PARAMETERS	Z-METER 2
Range of impedances	10 Ω – 1M Ω
Frequency range	100 Hz – 20 KHz
Accuracy of modulus of impedance	+/- 2% of the range
Accuracy of phase	+/- 2°
Level of measuring voltage	500mV – 5V with an interval of 500 mV
Communication with a PC bus	USB (COM port)
Integrated electric switcher	128 measuring points, each with 4 electrodes

2.3. Rock type and samples preparation

The rock used for the calibration tests was a marine sedimentary rock of Plio-Pleistocene age, known as "Calcarenite di Gravina" Formation. It is a porous rock with granular skeleton consisting of irregularly rounded lithoclasts of limestones and bioclasts, fragments of marine organisms, embedded in a fine, granular, carbonatic matrix, cemented to some degree.

The matrix and the sparry cement, made up predominantly of mineral calcite, are entirely subordinate to the granular skeleton. The calcarenite's composition mostly consists of calcium carbonate (>95%) with an insoluble residue characterized mainly by clay minerals. Previous laboratory tests assessed that porosity ranges between 40% and 47% and the hydraulic conductivity (K) is about $6 \div 7 \times 10^{-5}$ m/s for the coarse calcarenite and 3×10^{-5} m/s for the medium calcarenite [36-38]. The calcarenites are diffused in Apulia Region and frequently constitute part of the vadose zone or form significant porous aquifers in the region, playing an important role in drinking supply. Particularly, the samples of calcarenite used for the tests came from an active quarry, located in Canosa di Puglia, a city in the South of Italy (Figure 2).

Three blocks of calcarenite, hereafter called E1, E2 and E3, cut from one large block of rock, were used in order to verify the repeatability of measurements. The rock samples appeared relatively homogeneous at the sample scale. The samples were oven dried in order to obtain the dry weight. The dimensions of samples and their main physical properties are summarized in Table 2.

The lateral surface of the samples was sealed using a bi-component epoxy resin in order to ensure a one-dimensional water flow during the evaporation process.

The picket type probes, 10 mm diameter, 120 mm long with a conical bit 15 mm long, were used as electrodes in the tests (Figure 3). Two guide holes, 9 cm apart, with approximately 10 mm diameter, were drilled into rock samples for the installation of the probes. These ones were then slowly inserted into the sample with twisting motion and, alternatively, with the help of hammer, leaving 2 cm protrusion from the top edge of sample. In this way a good contact between the rock and the sensor was ensured.

Table 2. Physical characteristics of the calcarenite samples tested.

	Sample E1	Sample E2	Sample E3
Dimensions (cm)	25x16x15	25x15x15	24.9x15x14.7
Bulk density (g/cm³)	1,525	1,515	1,500
Particle density (g/cm³)	2,71	2,71	2,71
Porosity (%)	43,73	44,11	44,66

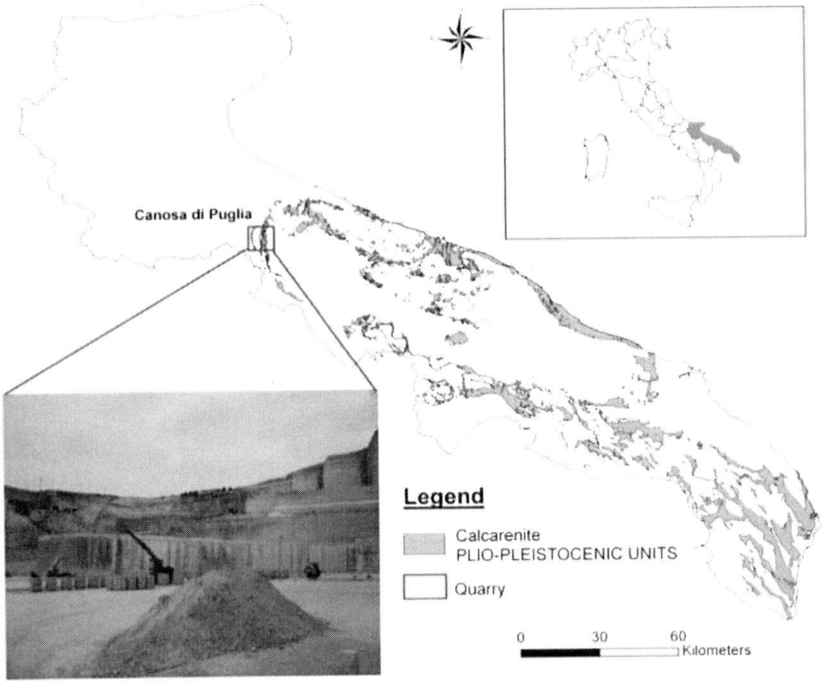

Figure 2. Regional distribution of calcarenites and location of quarry from which the rock samples have been collected.

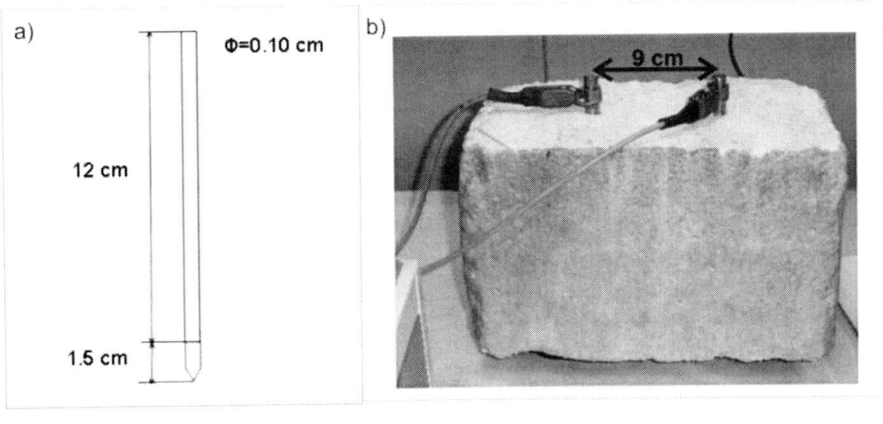

Figure 3. a) Scheme of picket type probe; b) probes position.

After the probes installation, the sample was placed in a water-filled vacuum bell jar, under confining pressure of 1 bar for at least 24 hours, in order to achieve almost full saturation. The saturation process took place from bottom to up by avoiding both air entrapment and uneven distribution of water. Three different solutions were used to saturate the samples in order to establish the effect of electrical conductivity (EC) on the output of the device: deionized water, 0,01 M and 0,03 M KCl solutions. The EC value as well as the KCl concentrations and the density of each water solution are reported in the Table 3.

Table 3. Physical properties of water solutions used to saturate the samples.

KCl solution molarities	EC values (mS/cm)	Salt concentration g/L	Water solution density g/cm³(at 25°C)
deionized water	$3{,}4 \cdot 10^{-3}$	0	0.9971
0.01	1,75	0.745	0.9976
0.03	4,70	2,235	0.9988

2.4. Measuring procedures

The entire cycle of measurements, comprising the measurements for all chosen frequencies, was repeated over time: 2-3 times per day in the first 15 days, 1 time in the following days, according to the expected rate of the evaporation process.

Contemporarily the weight loss of the rock sample, due to evaporation, was measured by electronic balance. The whole experimental set up is shown in Figure 4.

The tests have been performed in a thermostatic room at temperature of 25 ±1°C provided with a ventilation system. This is because temperature variations cause errors in all indirect moisture measurement methods due to the dependence of the dielectric properties from the temperature [39,40]. The tests were stopped when the sample's weight reached approximately it's dry weight.

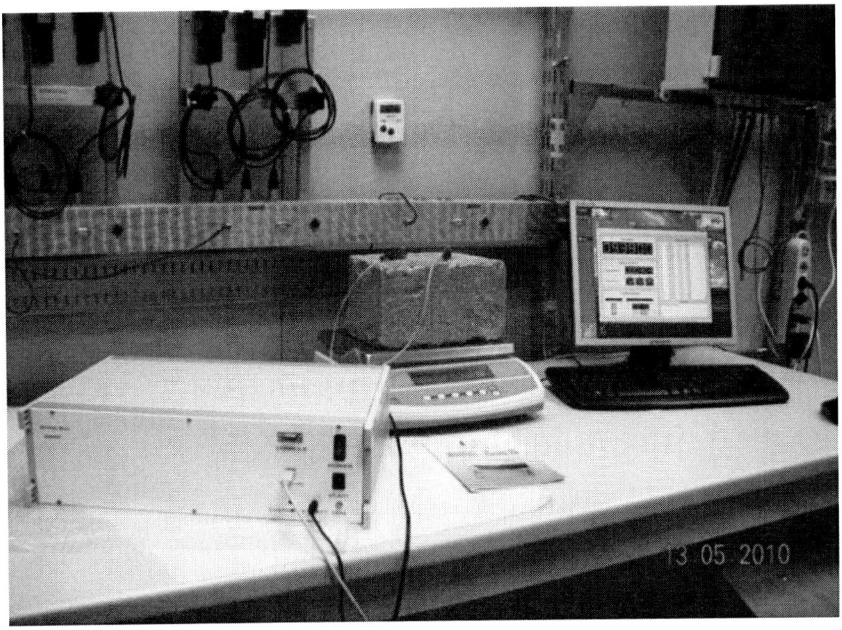

Figure 4. Experimental set up.

3. RESULTS AND DISCUSSION

3.1. Volumetric Water Content (VWC)

The first stage was to obtain the VWC of samples starting from the weight recorded at different times.

The rock VWC was determined, from the gravimetric one, in a standard way by the following equation (6) [41]:

$$\theta_v = \theta_g \left(\frac{\rho_b}{\rho_w} \right) \tag{6}$$

where

θ_v (cm^3 cm^{-3}) = VWC volumetric water content,

θ_g (g g^{-1}) = gravimetric water content,

ρ_b (g cm^{-3}) = rock bulk density,

ρ_w (g cm^{-3}) = water solution density.

The dynamics of the weight loss of the sample during the test is shown in Figure 5, in which the data for sample E2, saturated with 0.01 M KCl solution, are reported as an example.

The weight of the sample decreases over time in different way so that the evaporation from a soil or rock may be modelled as a two-stages process. In the first stage, corresponding to high water contents, the evaporation is governed by environmental conditions; in the second, when the water content decreases beyond a critical value, the evaporation is controlled by the ability of the medium to transmit water or, in other words, by its hydraulic properties [42, 43].

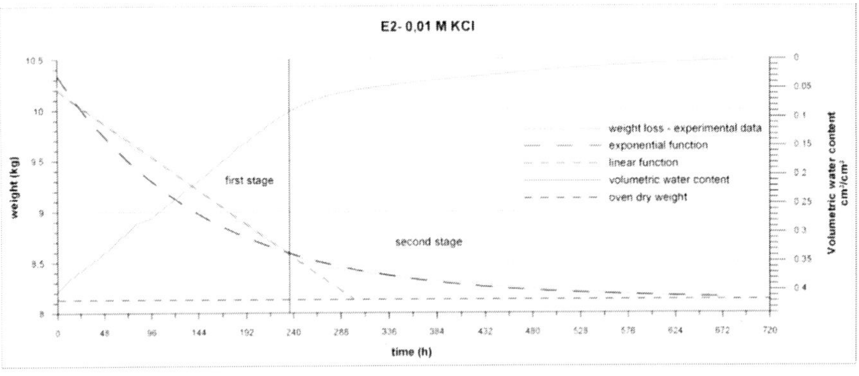

Figure 5. Changes in the weight and volumetric water content vs. time.

This different behavior is reflected in the different equations describing the process: the first part of data, corresponding to high water contents, is described by a linear relation while the second is well described using the following exponential function:

$$y = a + b\exp\left(-\frac{x}{c}\right) \tag{7}$$

where y (kg) is sample weight over time, a (kg) is its dry weight, b (kg) is the weight of water required to saturate the sample, x is the time (h) and c (h) is the only fitting parameter. The transition from the first to the second part occurs at a water content of about 6÷9%.

3.2. Resistance

The use of Z-meter 2 allows to investigate both the real (resistance) and the imaginary component (capacitive reactance) of impedance, being the two components recorded separately. In this study only the resistance was investigated because of its direct correlation with the water content.

The effective independence of the resistance (hereafter called Rx) from the frequency applied was checked and verified. Figure 6 shows the Rx output probe for all the tested measurement frequencies relative to sample E2 saturated with 0.01 M KCl solution.

The Rx values, measured for the same VWC, are very similar to each other for all tested frequencies, except that for VWC close to dry condition (less than 0.02). These means that, for VWC > 0.02, the device has a good accuracy and precision and that changing the frequencies has no effect on the Rx values that are therefore related with the VWC only. This results permits to use the complete dataset, consisting of medium Rx values recorded for each frequencies and the same VWC, in order to obtain the best fit curve, for each solution. For the reason mentioned above, all Rx data corresponding to VWC less than 0.02 are not included in the dataset.

The complete dataset of Rx values, recorded for the sample E2, are shown in Figure 7. Unlike what happens for the frequency, Rx values are strongly affected by changing the molarity, and hence EC value, of the solution used for the sample saturation. Moreover, as expected in a drying experiment, the VWC decreases over the course of the experiment and the corresponding Rx measurements increase significantly.

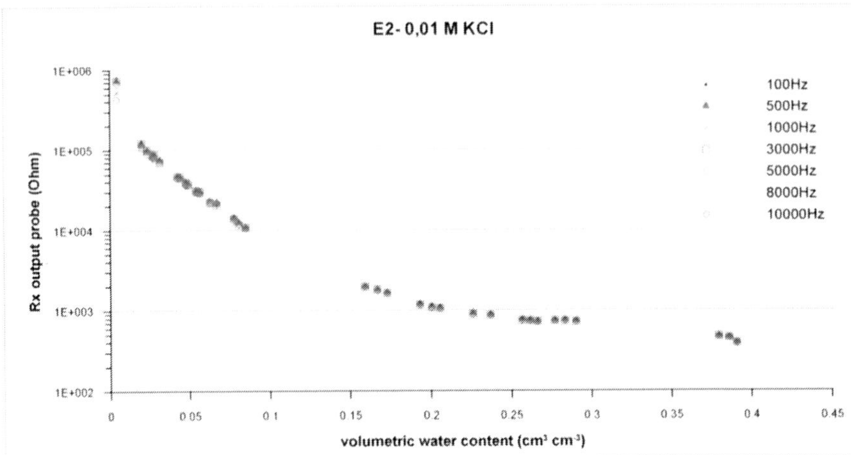

Figure 6. Rx values vs volumetric water content at all tested frequencies for sample E2.

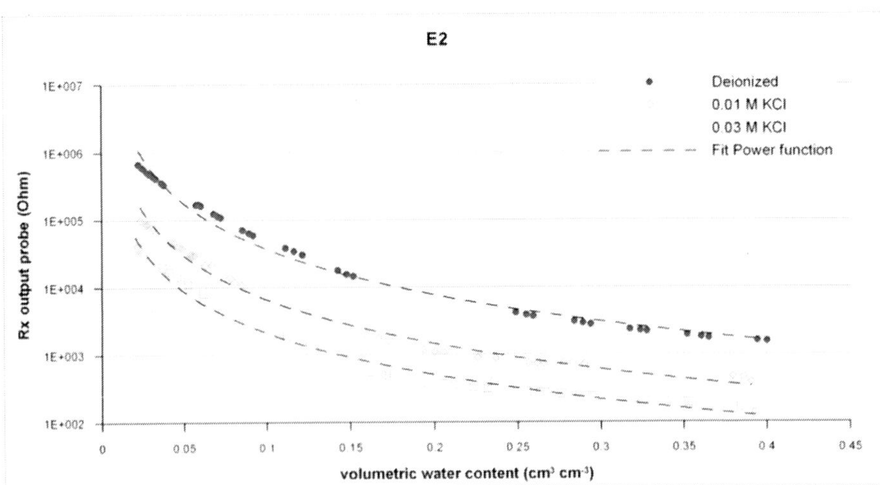

Figure 7. Rx complete dataset for the three solutions used to saturate sample E2.

The best interpolation function of dataset is represented by a power function of the type:

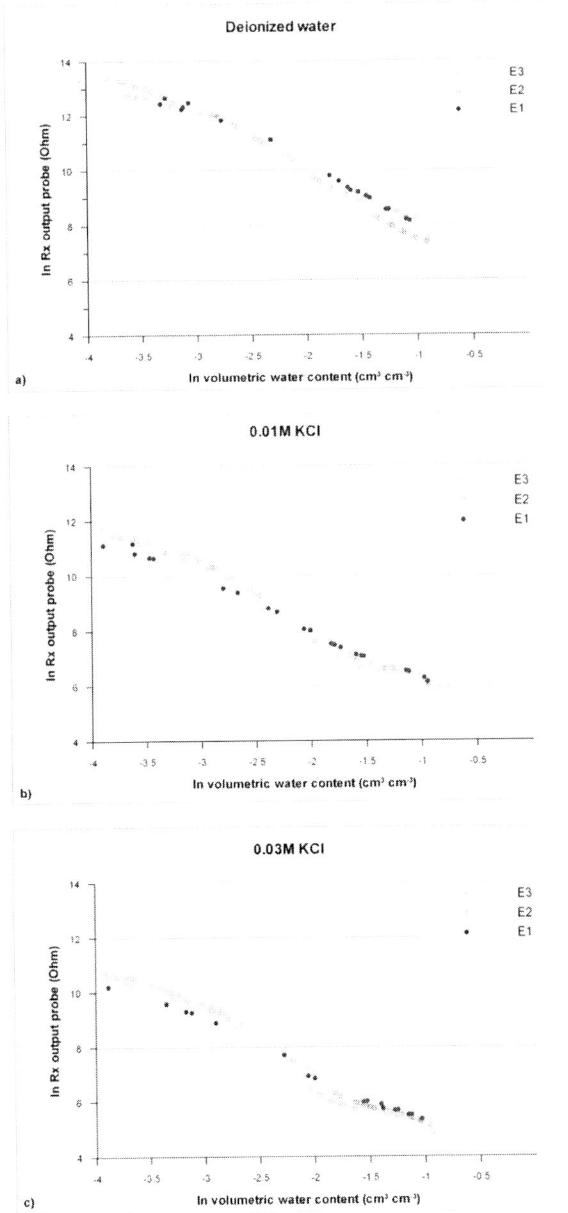

Figure 8. Three replicates (E1, E2, E3) for each solution (a, b, c).

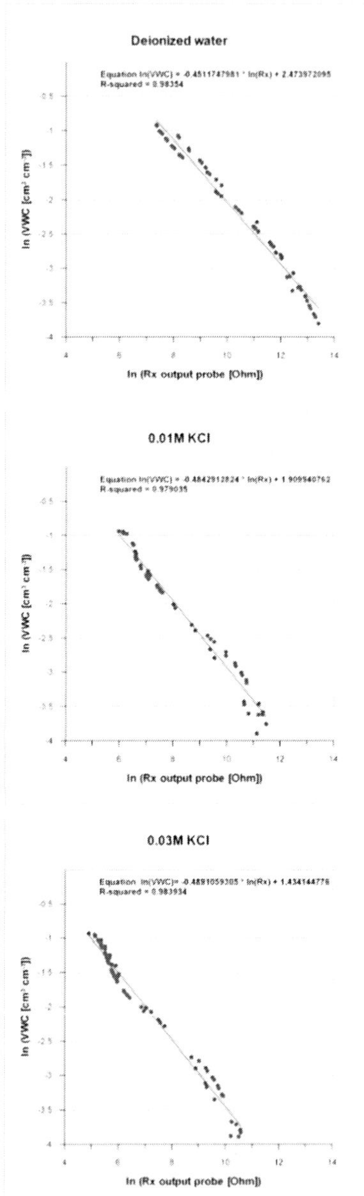

Figure 9. Calibration curves for each solution used to saturate samples.

$$y = a \cdot x^{b} \tag{8}$$

This model can be linearized by taking logarithms:

$$\ln(y) = \ln(a) + b \cdot \ln(x) \tag{9}$$

The linearized experimental data, obtained for each solution used to saturate samples E1, E2 and E3, are shown in Figure 8.

Small differences are found in the three replicates obtained for the same solution that can be explained with the natural variability among different samples of the same rock type (see Table 2). The two replicates E1 and E2 have been used, as a unique dataset, to obtain the calibration function that describes the VWC-Rx relationships, for each solution, while the third replicate (E3) has been used for the validation test. In Figure 9, for each solution, the logarithms of calculated VWC are plotted against the logarithms of measured Rx coming from replicates E1 and E2. The regression curves are also reported in the graphs. The changes in Rx due to the changes in VWC are significant and the relationships show a negative slope. The information provided from the linear regression, as indicated by the value of R-squared coefficient, suggests good degree of correlation of the datasets to these functions.

The calibration function is $VWC = a \cdot Rx^{b}$, by converting the logarithmic function into power function. The values of parameters a and b, for each solution, are reported in Table 4. In order to check the performance of the calibration function obtained by regression of experimental data, a validation test was performed utilizing the data of the sample E3. The actual values of VWC, measured during the test E3, against the predicted values, calculated using the calibration function, are plotted in the graph of Figure 10. The 1:1 line represents the perfect correspondence between the actual and the predicted value. The points below this line indicate an underestimation of the model while the points above indicate an overestimation. Two absolute error lines of ±0.03 and ±0.06 respectively, are also plotted in the graph. Only the 10% of the data fall outside prescribed limits while the 23% of data fall between the ±0.03 and the ±0.06 error lines. For the remaining 66% of the data, the model provides an estimate of the VWC with an error less than ±0.03. The correlation between the actual and predicted values is much higher for low values of VWC (less than 0.07) while tends to decrease for greater values of VWC.

Table 4. Parameters of the calibration curve for each solution used to saturate samples.

KCl solution molarities	a	b
deionized water	2.4739	-0.4512
0.01	1.9099	-0.4843
0.03	1.4341	-0.4891

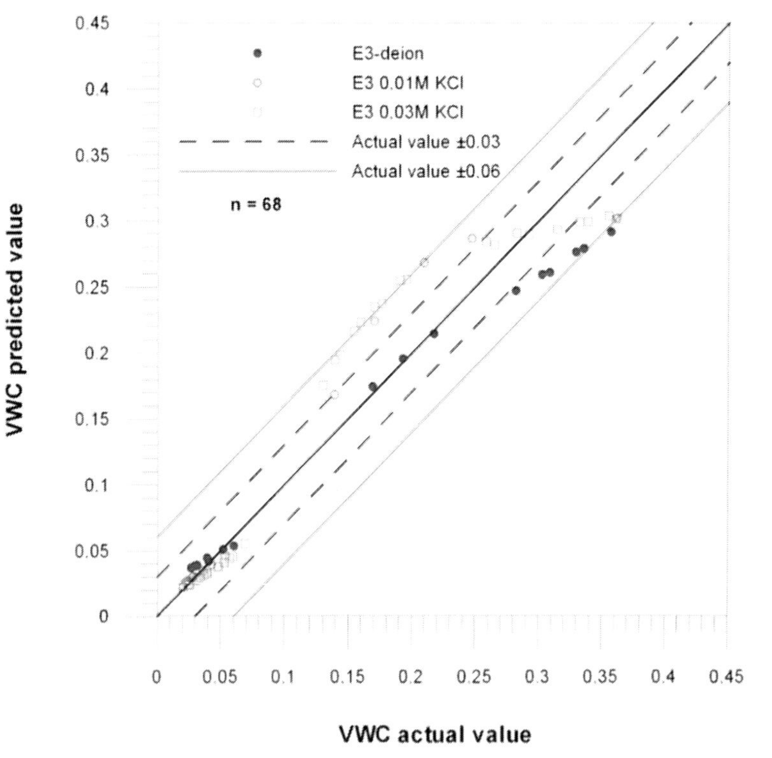

Figure 10. Validation test.

4. CONCLUSIONS

In this chapter has been described the experimental procedure used to obtain the calibration curves for a new device Z-meter 2, to measure water content in porous rock. Experimental tests have been carried out in laboratory, using samples of sedimentary rock, known as calcarenite, collected in a quarry located in Canosa, a city in South of Italy.

The device, based on the EIS method, gives as output both the resistance and the reactance values separately; this study considered the resistance because of its high correlation with the water content.

The tests were aimed at understanding the effects of the applied frequency and EC of KCl solution used to saturate the samples, on the device outputs and, consequently, on the calibration functions. For this reason a wide range of frequencies, from 100Hz to 10000Hz, was investigated and three KCl solutions with different EC values were used to saturate the rock samples.

The probes used were short stainless steel road probes, called picket type probes, chosen because they can be easily installed in the rock without be damaged. By analyzing the different data sets, none abrupt change in resistance values was observed during the evaporation process. This means that the probes work well in the calcarenite and that are able to record measurements of the impedance without any gap effects. The evaporation process monitored during the test showed how the weight of the sample decreases over time with different rates leading to model the process as a two-stages process. The first, corresponding to high water contents, was described by a linear relation while the second by an exponential function. The independence of the resistance from the frequency applied was observed. The resistance values obtained for the same water content at all tested frequencies were almost coincident, except close to dry condition (less than 2%). This means that, for water content higher than 2%, the device has a good accuracy and precision and that changing the frequencies has no effect on the resistence values.

Unlike the frequency, the resistance values were strongly affected by the molarity of the solution used for the sample saturation. For this reason the calibration curves have been obtained considering the data sets recorded for each solutions, separately. The calibration functions, that describe the relationships between the water content and resistance values, are of power type with a good degree of correlation as shown by value of R-squared of about 0.98.

5. ACKNOWLEDGEMENTS

We would like to thank the Project E!3838 "Research, design and realization of the automatized system of the observation of soil moisture changes" and the Project E!4981 "Computerised Measuring System for Analysis of Chosen Characteristics and Processes in Porous Environment by EIS Method" of the Europe International Program EUREKA. A particularly special thanks to Ing. Jana Pařílková professor at Brno University of Technology, Faculty of Civil Engineering and co-applicant of the EUREKA program in the Czech Republic, in order to have permitted free use of the Z-meter 2 device for the laboratory tests. Thanks to Francesco De Benedictis also for his technical support during the experimentations in the laboratory.

REFERENCES

1. Gardner, W.H. Water Content. In: Klute A. (ed.) Methods of Soil Analysis, Part I, Physical and Mineralogical Methods. Madison, Wisconsin, USA: American Society of Agronomy and Soil Science Society of America, Inc; 1986. p493-544
2. Evett, S.R., Heng, L.K., Moutonnet, P., Nguyen, M.L., editors. Field Estimation of Soil Water Content: A Practical Guide to Methods, Instrumentation, and Sensor Technology. IAEA-TCS-30. Vienna, Austria: International Atomic Energy Agency; 2008.http://www-pub.iaea.org/MTCD/Publications/PDF/TCS_30_web.pdf
3. Greacen, E.L., editor. Soil Water Assessment by the Neutron Method, Melbourne, Australia: CSIRO; 1981.
4. Evett, S.R. Measuring soil water by neutron thermalization. In: Stewart B.A. and Howell T. A. (eds.). Encyclopedia of Water Science. New York: Marcel Dekker; 2003. p889–893.
5. Dean, T.J., Bell, J.P., Baty, A.J.B.,. Soil moisture measurement by a capacitance technique, part 1. Sensor design and performance. J. Hydrology 1987; 93, 67–68.
6. Robinson, M. and T.J. Dean. Measurement of near surface soil water content using a capacitance probe. Hydrol. Proc. 1993;7, 77–86.
7. Baumhardt, R.L., Lascano R.J. and Evett S.R. Soil Material, Temperature, and Electrical Conductivity Effects on Calibration of

Multisensor Capacitance Probes. Soil Science Society of American Journal 2000; 64, 1940-1946.

8. Kizito F., Campbell C.G., Cobos D.R., Teare B.L., Carter B. and Hopmans J. W. Frequency, Electrical Conductivity and Temperature Analysis of a Low-Cost Capacitance Soil Moisture Sensor. Journal of Hydrology 2008; 352, 367-378.

9. Topp G.C., Davis J.L. and Annan A.P. Electromagnetic determination of soil water content: Measurement in coaxial transmission lines. Water Resour. Res. 1980; 16, 574–582.

10. Topp G.C. and Davis J.L. Measurement of soil water content using time-domain reflectometry in field evaluation.Soil Science Society of America 1985; 49(1) 19–24

11. Dalton F.N. and van Genuchten M.Th. The time-domain reflectometry method for measuring soil water content and salinity. Geoderma 1986; 38, 237-250.

12. Jones S.B. and Or D. Frequency Domain Analysis for Extending Time Domain Reflectometry Water Content Measurements in Highly Saline Soils. Soil Science Society of American Journal 2004; 68, 1568–1577.

13. Robinson D.A., Jones S.B., Wraith J.A., Or D. and Firedmena S.P. A Review of Advances in Dielectric and Electrical Conductivity Measurements in Soils Using Time Domain Reflectometry. Vadose Zone Journal 2003; 2, 444-475.

14. Veldkamp E. and O'Brien J.J. Calibration of a frequency domain reflectometry sensor for humid tropical soils of volcanic origin, Soil Sci. Soc. Am. J. 2000; 64, 1549–1553.

15. Bilskie J. Using dielectric properties to measure soil water content. Sensors Magazine 1997; 14, pp.26–32.

16. Gaskin G.J. and Miller J.D. Measurement of Soil Water Content Using a Simplified Impedance Measuring Technique. J. Agric. Eng. Res. 1996;63, 153 - 160.

17. Seyfried M.S. and Murdock M. D. Measurement of Soil Water Content with a 50-MHz Soil Dielectric Sensor. Soil Science Society of American Journal 2004; 68, 394-403.

18. McCann I.R., Kincaid D.C. and Wang D. Operational characteristics of the watermark model 200 soil water potential sensor for

irrigation management. Applied Engineering in Agriculture 1992;8 (5) 603–609.

19. Shock C.C. Soil water potential measurement by granular matrix sensors. In Stewart B.A. and Howell T. A. (eds.). Encyclopedia of Water Science. New York, NY: Marcel Dekker; 2003. p899–903

20. Richards L.A. The usefulness of capillary potential to soil moisture and plant investigators. J. Agric. Res. (Cambridge) 1928;37,719–742.

21. Mullins C.E., Mandiringana O.T., Nisbet T.R. and Aitken M.N.,. The design limitations and use of a portable tensiometer. Journal of Soil Science 1986; 37,691–700

22. Cassel D.K. and Klute A., 1986. Water potential: Tensiometry. In: Klute A. (ed.) Methods of Soil Analysis, Part I, Physical and Mineralogical Methods. Madison, Wisconsin, USA: American Society of Agronomy and Soil Science Society of America, Inc; 1986. p563–596.

23. Dane, J.H., and G.C. Topp editors. Methods of Soil Analysis. Part 4-Physical Methods. Madison, Wisconsin, USA: American Society of Agronomy and Soil Science Society of America, Inc; 2002.

24. Evett, S.R. Soil water and monitoring technology. In: Lascano R.J. and Sojka R.E. (eds.) Irrigation of Agricultural Crops. Madison, Wisconsin, USA: American Society of Agronomy and Soil Science Society of America, Inc; 2007. p25–84.

25. Wigneron, J.-P., Calvet, J.-C., Pellarin, T., Van de Griend, A.A., Berger, M., Ferrazzoli, P. Retrieving near-surface soil moisture from microwave radiometric observations: current status and future plans. Rem. Sens. Environ. 2003;854, 489–506.

26. Löw, A., Ludwig, R., Mauser, W.,. Derivation of surface soil moisture from ENVISAT ASAR wide swath and image mode data in agricultural areas. IEEE Trans.Geosci.Remote Sens. 2006; 44, 889–899

27. Walker, J.P., Houser, P.R., Willgoose, G.R. Active microwave remote sensing for soil moisture measurement: a field evaluation using ERS-2. Hydrol. Proc. 2004;1811, 1975–1997

28. Annan A. P. Time domain reflectometry - Air-gap problem for parallel wire transmission lines. In: Rep. Activ., Part B, Pap. 77–1B. Ottawa, Ont., Canada: Geol. Surv. of Canada; 1977. p. 59–62.

29. Knight, J.H., Ferré P.A,. Rudolph D.L., and Kachanoski R.G. A numerical analysis of the effects of coatings and gaps upon relative dielectric permittivity measurement with time domain reflectometry, Water Resour. Res. 1997; 33,1455–1460.

30. Hokett, S. L., Chapman J. B. and Russell C. E. Potential use of time domain reflectometry for measuring water content in rock, J. Hydrol. 1992;138, 89–96.

31. [31]Schneebeli M., Fluhler H., Gimmi T., Wydler H., Laser H.P. and Bear T. Measurements of water potential and water content in unsatured crystalline rock. Water Resour. Res. 1995;31,1837-1843.

32. Sakaki T., Sugihara K., Adachi T., Nishida K. and Lin W. R. Application of time domain reflectometry to determination of volumetric water content in rock, Water Resour. Res. 1998;34, 2623–2631.

33. Sakaki T. and Rajaram H. Performance of different types of time domain reflectometry probes for water content measurement in partially saturated rocks. Water Resour. Res. 2006; 42, W07404, doi 10.1029/2005WR004643.

34. Barsoukov E. and Macdonald R. J., Editors. Impedance Spectroscropy Theory, Experiment, and Applications. Hoboken, New Jersey, USA: John Wiley and Sons, Inc.; 2005.

35. Parilkova J., Veselý J., PavlíkJ., Stoklásek R. Monitoring of the soil status using electrical impedance spectrometry method developed in Project E!3838 of the Europe International Program EUREKA. Proceedings of the XIX IMEKO World Congress Fundamental and Applied Metrology, 6–11 September 2009, Lisbon, Portugal. http://www.imeko2009.it.pt/Papers/FP_443.pdf

36. Iannone, A. and Pieri, P. Considerazioni critiche sui "Tufi Calcarei" delle Murge – Nuovi dati litostratigrafici e paleoambientali, Geografia Fisica e Dinamica Quaternaria 1979; 21,33–58.

37. Andriani G.F. and Walsh N. Fabric, porosity and permeability of calcarenites from Apulia (SE Italy) used as building and ornamental stones. Bull. Eng. Geol. Env. 2003; 62, 77-84;

38. Andriani G.F. and Walsh N. - Physical properties and textural parameters of calcarenitic rocks: qualitative and quantitative evaluation. Bull. Eng. Geol. Env. 2002; 67, 5-15.

39. Ghretli M.,. Khalid K, Grozescu I. V., Sahri H. and Abbas Z. Dual frequency microstrip antenna sensor for water content

measurements independent of temperature variation. Measurement Science and Technology 2007;18, 1054–1060.

40. Evett, S.R., Tolk J.A., and Howell T.A. Soil profile water content determination: Sensor accuracy, axial response, calibration, temperature dependence and precision. Vadose Zone J. 2006; 5, 894–907.

41. Black C.A., editor. Methods of Soil Analysis, Part I, Physical and Mineralogical Methods. Madison, Wisconsin, USA: American Society of Agronomy and Soil Science Society of America, Inc; 1965.

42. Brutsaert W., editor. Evaporation into the Atmosphere: Theory, History and Applications, Hingham, Massachusetts: Kluwer Academic; 1982.

43. Jury W., Gardner W. R. and Gardner W. H., editors. Soil Physics, 5th ed. New York: John Wiley and Sons, Inc.; 1991.

8

CHAPTER

HYDRAULIC CONDUCTIVITY ESTIMATION TEST IMPACT ON LONG-TERM ACCEPTANCE RATE AND SOIL ABSORPTION SYSTEM DESIGN

Jakub Nieć * and Marcin Spychała

Department of Hydraulic and Sanitary Engineering, Poznan University of Life Sciences, Piątkowska St. 94A, Poznań 60-649, Poland

ABSTRACT

The aim of this paper was to verify the common methods of hydraulic conductivity estimation for soil assessment in respect to wastewater disposal. The studies were conducted on three types of sandy soils. Hydraulic conductivity was determined using a scale effect-free laboratory method, empirical equations and compared with measurements estimated from a laboratory infiltration column with identified head loss. Based on the hydraulic conductivity values, the long-term acceptance rates (LTAR) [1] were calculated. The differences in LTAR values were about one order of magnitude smaller than differences in hydraulic coefficient. The study showed a good convergention of the results obtained from the constant head method (CHM) by solving the Glover Equation for medium and coarse sands. In low permeability soil (fine sand), the best result was obtained using CHM-a with a capillary rise consideration (a is a factor included in the flow in the unsaturated and saturated zones calculated from a capillary rise). From a practical point of view the relatively small value of LTAR underestimation (20%-for constant head method) is responsible for the extended surface area of the system and provides a security margin (the avoidance of clogging risk). The use of the falling head method, based on the Van Hoorn equation, can be said to be highly overestimated. For medium and coarse sandy soils the underestimation of LTAR calculated and based on CHM test determination is 14%–18%. The total cost of soil absorption system (SAS) designed-based on CHM in comparison to that designed-based on real hydraulic conductivity value in Poland is only about 7%–9% higher.

Keywords: hydraulic conductivity; long term acceptance rate; LTAR; percolation test; soil absorption system; SAS; soil permeability

1. INTRODUCTION

Hydraulic conductivity is the main parameters used for estimation water and wastewater applicability in saturated soils.

Based on the determination of hydraulic conductivity the long-term acceptance rate can be calculated. The term "long term acceptance rate" (LTAR) is the daily volume of wastewater that can be applied over an indefinite period of time to a unit of soil surface area. LTAR values were determined using the equation [1]:

$$LTAR = 5K_s - \frac{1.2}{\log_{10} K_s} \tag{1}$$

where,

LATR–long term acceptance rate, gal·ft−2·d−1, the utilization range between 0.32 and 0.80 gal·ft−2·d−1 (1.3 and 3.2 cm·d−1);

Ks–hydraulic conductivity, ft·min−1.

Several studies can be found in the literature and related papers were published in accordance with LTAR [2,3,4].

There are many different methods to estimate hydraulic conductivity of soils under various conditions, which may have impact on hydraulic conductivity. For example, some methods are used to estimate hydraulic conductivity to obtain contamination migration velocity and others, to estimate hydraulic conductivity of soil for sub-surface disposal of waste-water. Some authors [5] have utilized the hydraulic conductivity decrease as a factor of biofilm growth (clogging factor).

Methods of estimation of hydraulic conductivity can be mainly divided into field and laboratory. The former are conducted in natural conditions where factors, such as temperature, pressure and the level of the water table, may affect the result. An additional advantage of field methods in comparison to laboratory ones is the opportunity to take into account the local conditions in a given research context. The most popular method among field methods is a percolation test which can be conducted in situ, where a given subsurface disposal system will be completed [6]. During the recent past, the percolation tests were modified and now can be divided into tests conducted in steady-state (constant head-CHM) and unsteady-state (falling head-FHM) conditions.

Laboratory methods can be divided into direct and indirect ones. Direct methods use soil samples with an unbroken or broken structure taken from where a subsurface drainage is planned. Hydraulic conductivity is often estimated using a filtration column where measurements are conducted in steady or unsteady conditions. This research is simple and inexpensive but the problem is how to obtain a representative size of soil sample. Indirect methods are based on information about the physical characteristics of investigated soils, e.g., bulk density, effective porosity, grain size distribution, and solid matrix properties, such as grain shape, pore shape, specific surface, tortuosity and organic matter content, to estimate hydraulic conductivity [7,8,9].

One of the first researchers who described the relationship between porosity and hydraulic conductivity was Slichter [10]. Aronovici [11] showed a correlation between the content of silts and clay in a cohesive soil and its impact on the soil hydraulic conductivity estimated in laboratory conditions. Oosterbaan and Nijland [12] reported that Smedema and Rycroft in 1983 presented the hydraulic conductivity in a range of soils as a function of grain size distribution. Additionally, the authors pointed out that in the case of soils with the same texture but different structure, the results of hydraulic conductivity can be differentiated a great deal more. De Ridder and Wit [13] described a method to estimate hydraulic conductivity using the ratio of irregular particles to regular ones contained in a sample measuring 1 cm in diameter. In the authors' opinion, the range of method validity is limited only to homogeneous and isotropic soils, which are rather rare in nature.

Knowledge about grain size distribution can be used to calculate hydraulic conductivity using one of the well-known empirical formulae [14,15].

The empirical equations can be divided into two groups depending on the input data as follows:

utilizing information about grain size distribution only, e.g., USBR (The United States Bureau of Reclamation) or Seelheim formula, demanding more information about the soil porosity or specific surface area, etc. as per the modified version of Hazen, the Krüger or Kozeny—Carman formula.

During the last few years the methods used to estimate hydraulic conductivity based on drainage porosity [16] have gained more popularity. In this paper the methods based on drainage porosity have not been verified by reason of difficulties in obtaining drainage porosity values.

There is no simple relation between wastewater infiltration rate and Ks decrease due to the complex nature of the clogging process and numerous related conditions and terms affecting such a process. There are many experimental research results showing that the infiltration rates through the biomat or sub-biomat soil layer in permeable, sandy soils can be between 1.4 and 7.5 cm/d [17,18,19]. The hydraulic gradient in fine sand below the biomat (clogging layer) was measured and its value came approximately to 7. For the minimum value of LTAR: 1.3 cm/d the corresponding K value is 8.64 cm/d and calculated (taking into consideration hydraulic gradient equal 7) permeability is about 61 cm/d. This value is almost 50 times higher than the LTAR value (1.3 cm/d). This shows the relation between LTAR value and soil permeability potentially clogged by wastewater. Such a small LTAR value compared to partly clogged soil permeability means that using the recommended daily dosing rate (LTAR) the possibility of full clogging is very low. In this paper, the methods used to estimate hydraulic conductivity for the assessment of soil in respect to wastewater subsurface disposal (soil absorption system) are described in brief and subsequently compared.

The main aim of this paper is the verification of commonly used methods to estimate hydraulic conductivity. The results of hydraulic conductivity measured, using a scale effect-free laboratory method were compared with the estimation from the infiltration column with a constant head of water. In addition, hydraulic conductivity was obtained using an empirical equation. The reason for choosing IC data as reference Ks values was the need to take into account the given water flow according to Darcy's law. This was made possible because during research head loss on sand filter depth was measured. Taking into consideration that the result of hydraulic gradient was close to unity, the use of Darcy law was seen to be appropriate.

2. EXPERIMENTAL PROCEDURES

2.1. Characteristic of Investigated Soils

Measurements were conducted on soils, which potentially can be used as a soil absorption system with different physical parameters (Table 1).

Table 1. Comparison of physical properties of investigated soils.

Soil type	Soil texture				Soil grain diameter		Bulk density	Porosity
	Percentage							
	Gravel	Sand	Silt	Clay	d_{10}	d_{20}	P	n
	%				mm		$g \cdot cm^{-3}$	%
fine sand	0.1	99	0.9	-	0.08	0.14	1.6	39
medium sand	0.5	98	1.5	-	0.14	0.20	1.8	32
coarse sand	15	85	-	-	0.27	0.30	1.85	30

The investigated soils were taken from pits without initial processing, thus the research was conducted on materials with almost natural properties. Before the soil samples were taken, the porosity and humidity were measured. The soil samples were placed in the model with density and moisture predetermined. The samples were disturbed (broken) in structure, then, the research set-up was filled with the layers of sand with compaction to achieve natural soils density identified before samples taken. During filling the porosity and moisture were verified.

2.2. Laboratory Measurements in Infiltration Column

The investigations were conducted in an infiltration column of a diameter of 15 cm (Figure 1).

Before measurements the soil samples were saturated. The saturation was conducted by an upwards flow of water (in the opposite direction) to push all the air trapped in the soil. The water used for research purposes was de-aired. During the measurements with constant head pressure the water level was maintained using a Mariotte bottle with outflow above the soil sample. In the case of measurements conducted with a falling water level (falling head pressure) the inflow to the column was stopped and the corresponding change of water level in time was recorded. Calculation of hydraulic conductivity was according to the well-known equation:

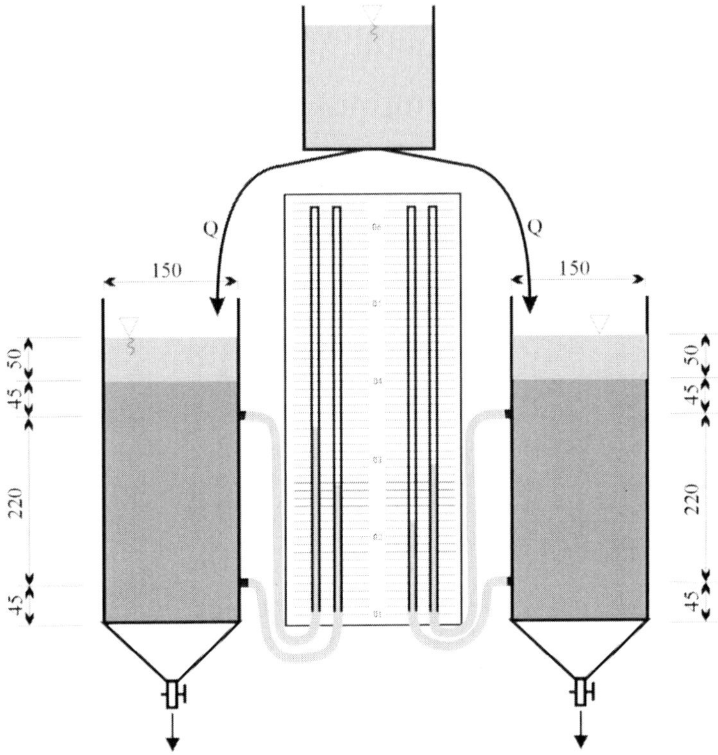

Figure 1. Scheme of infiltration columns (IC) for samples with disturbed structure.

$$K_s = \frac{V}{I} = \frac{Q}{A} \cdot \frac{L}{\Delta h_l} \tag{2}$$

where,

K_s–hydraulic conductivity, m·s−1;

A–cross-section area of infiltration column, m2;

Δh_l–differenced in piezometer heights (head loss), m;

L–length of the sample, m;

V–seepage velocity, m·s−1;

Q–discharge of flow, m3·s−1;

I–hydraulic gradient;

Hydraulic conductivity was calculated using measurements made in the infiltration column. Additionally, relevant values of hydraulic conductivity were calculated, using well-known formulae based on physical parameters of the investigated soils. The first was the Hazen Formula in the form presented by Lange [20] where value of hydraulic conductivity also depends on porosity:

$$K_s = [400 + 40 \cdot (n - 26)] \cdot d_{10}^2 \tag{3}$$

where,

K_s–hydraulic conductivity, m·d−1;

d_{10}–effective grain size, mm;

n–porosity, %.

The sand mine specification indicated d10 as 0.1 mm, however, the d10 estimated using sieving analysis was 0.08 mm on average. Taking into consideration this divergence (related probably to measurement precision) the d10 value was approximated to 0.1 mm.

Hydraulic conductivity was also calculated using the USBR Formula as follows:

$$K_s = 0.36 \cdot d_{20}^{2.3} \tag{4}$$

where,

K_s–hydraulic conductivity, cm·s−1;

d_{20}–effective grain size, mm.

2.3. Scale Effect-Free Laboratory Method Conducted in Controlled Conditions

2.3.1. Research Set-Up Description

Measurements were conducted on a research set-up_presented in Figure 2. The front sidewall of the tank was built out of a transparent plexiglass. The tank bottom with a centrally designed outflow was covered by a 0.1 m layer of gravel. The tank was filled with the investigated soil with a density corresponding to natural occurrence. After filling, at the soil surface a hole had been prepared according to the Amoozegar [21] instruction. Only one half of the hole of diameter of 0.15 m and depth 0.30 m was created at the transparent wall to observe water level and streamlines. The wall of the hole was protected against collapse by a brass mesh screened by a fabric filter [22]. Water used for the experiments was deaerated, as a direct application of tap water in this type of research may give the wrong results [23].

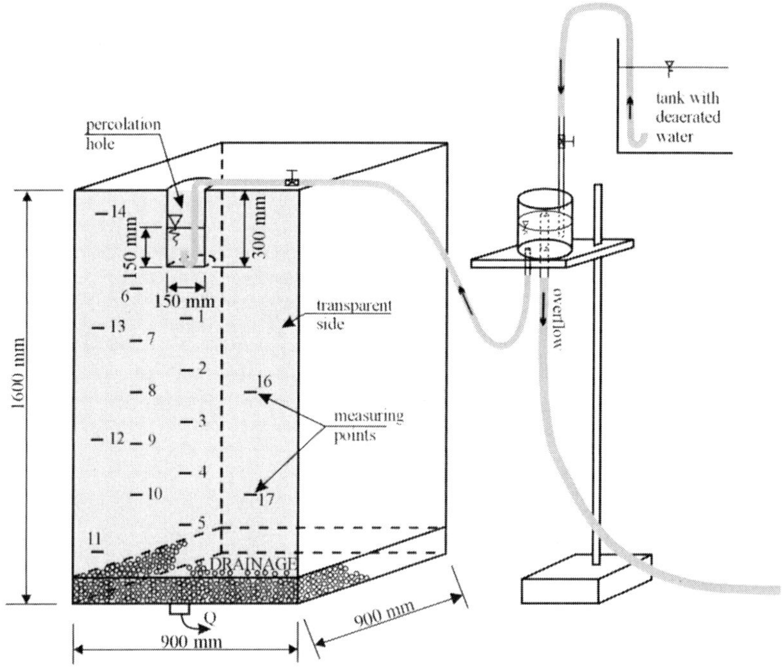

Figure 2. Scheme of the research set-up of the scale effect-free laboratory method.

2.3.2. Calculations of Hydraulic Conductivity

Soil saturation was measured using the TDR method (Time Domain Reflectometry). During and after full saturation, percolation tests were conducted according to schedule. Two kinds of tests were used. The first consisted of maintaining a constant level of water equal to 0.15 m in the hole (steady-state condition). The second was a falling head test, where the time taken to the drop of water level from 0.15 m to zero (bottom) was recorded in each test. Calculations of hydraulic conductivity in falling head tests were conducted using the Van Hoorn equation [1]:

$$K_s = 1.15r \frac{\log\left(H_0 + \frac{r}{2}\right) - \log\left(H_1 + \frac{r}{2}\right)}{t} \tag{5}$$

where,

H_0–initially water level in the hole, m;

H_1–final water level in the hole, m;

r–radius of the hole, m;

t–water falling time, s.

The above-presented equation is used to calculate the value of hydraulic conductivity in American tests known as the inversed auger hole method. In the French literature, this test is known as the Porchet method [12].

In steady-state tests the values of hydraulic conductivity were obtained by the Glover equation [24]. In a steady-state condition for a hole of cylindrical shape it holds:

$$K_s = Q \cdot \frac{\sinh^{-1}\left(\frac{H}{r}\right) - \left[\left(\frac{r}{H}\right)^2 + 1\right]^{\frac{1}{2}} + \left(\frac{r}{H}\right)}{2 \cdot \pi \cdot H^2} \tag{6}$$

where,

K_s–hydraulic conductivity, m·s−1;

Q–quantity water needed to hold a constant water level at H, m3·s−1;

H–constant water level in the hole, m.

In this paper additionally, the Philip equation was used to estimate hydraulic conductivity [24]:

$$K_s = \frac{Q}{C_p \cdot r^2} \tag{7}$$

$$C_p = \left(\frac{H^2}{r^2}-1\right)^{\frac{1}{2}} \cdot \left\{ \frac{4.117 \cdot \left(\frac{H}{r}\right) \cdot \left(1-\frac{r^2}{H^2}\right)}{\ln\left[\frac{H}{r}+\left(\frac{H^2}{r^2}-1\right)^{\frac{1}{2}}\right] - \left(1-\frac{r^2}{H^2}\right)^{\frac{1}{2}}} + \frac{\left(8.056+5.034 \cdot \frac{r}{H}\right)}{a \cdot r \cdot \ln\left[\frac{H}{r}+\left(\frac{H^2}{r^2}-1\right)^{\frac{1}{2}}\right]} \right\} \quad (8)$$

where,

Cp-factor, which depends on the shape of the hole, and takes into consideration both saturated and unsaturated components of water flow around the auger hole.

This equation allows to calculate the value of hydraulic conductivity for both saturated and unsaturated flow of water out of the hole. Coefficient a must be determined independently or to be estimated for a given type of soil from values reported by Erlic but presented by Amoozegar [21]. The a depends on the type of soil, as follows:

a = 1–compacted clays;

a = 4–unstructured fine-textured soils;

a = 12–most structured soils from clays to loam and unstructured medium and fine sand and sandy loam;

a = 36–coarse and gravelly sands.

3. RESULTS AND DISCUSSION

Laboratory Measurements

Hydraulic conductivity was calculated using measurements made in the infiltration column. The results for a water temperature of 10 °C are presented in Table 2. Additionally, relevant values of hydraulic conductivity were calculated using the Hazen and the USBR formulae.

Table 2. Results of hydraulic conductivity obtained by different methods.

Soil type	Method	Mean value of K_{10}	N number of measurements	LTAR *	Percentage variation comparing to IC
		(m·d^{-1})	-	(cm·d^{-1})	%
fine sand	IC	0.80 ± 0.01	85	1.79	
	FHM	11.66 ± 0.11	45	3.60	101%
	CHM	2.03 ± 0.03	47	2.15	20%
	CHM a = 4	0.83 ± 0.01	47	1.80	1%
	Hazen	5.9	-	2.85	59%
	USBR	3.4	-	2.44	36%
medium sand	IC	1.93 ± 0.12	25	2.13	
	FHM	7.65 ± 0.57	104	3.09	45%
	CHM	0.90 ± 0.03	31	1.83	−14%
	CHM a = 36	0.90 ± 0.03	31	1.83	−14%
	Hazen	12.5	-	3.70	73%
	USBR	7.7	-	3.10	45%
coarse sand	IC	5.27 ± 0.1	102	2.75	
	FHM	6.02 ± 0.39	147	2.86	4%
	CHM	2.46 ± 0.08	295	2.25	−18%
	CHM a = 36	2.55 ± 0.09	295	2.13	−23%
	Hazen	40.8	-	6.57	139%
	USBR	19.5	-	4.47	62%

Notes: Not measured; ** LTAR is calculated using the formula presented by Laak-scale effect-free laboratory tests conducted in controlled conditions.

During research using the FHM in gravel and medium sand higher variations of the results were noted. Infiltration rates were found to be gradually decreasing during percolation tests.

The results of hydraulic conductivity obtained using the CHM showed a relatively low variation.

In the Philips equations, the parameter a was calculated using capillary rise measurements (hk = 0.48 m) for fine sand. Assuming that the value of capillary rise is equal to the matrix potential the latter was calculated as a reciprocal of a, as follows [25]:

$$a = 2\left|\psi\right|^{-1} = 2h_k^{-1} \tag{9}$$

where,

h_k–capillary rise in m H2O;

ψ–matrix potential m H2O.

The value of a calculated using the Equation (9) a = 4.17 was close to the assumed value of the parameter (a = 4.0). In the calculation of a, the value was estimated and based on literature data. In the case where capillary rise is less then about 6 cm the value of a being equal to 36 has been found to be correct. Calculations of hydraulic conductivity in fine sand using Glover equations, comparing to those which neglect capillary rise, gave overestimated results.

The results of hydraulic conductivity obtained by empirical formulae based only on grain size distribution should be treated carefully, because the formulae are oversimplified. The results can be only used as a preliminary assessment of hydraulic conductivity.

Analysis of variance was completed with a significance level set on $\alpha = 0.05$ confirmed significant differences in the compared tests. Therefore, it was important to conduct post hoc tests that can explain which values were responsible for rejecting the null hypothesis. The Tukey method was chosen to find which means are significantly different from one another. This test compare every group mean with every other group mean and typically incorporate some method of controlling for Type I errors.

Although Pedescol [26] did not explain the obtained result, the author reported that hydraulic conductivity of gravel soil obtained using the falling head method was about 38% higher than that obtained using a constant head method. However, the results of hydraulic conductivity obtained by both methods in a coarse sand were not significant ($p < 0.05$) despite the fact that the mean value estimated by the falling head method was several times higher than that obtained using the constant head method.

Comparison of hydraulic conductivities obtained by the falling head method and constant head method confirmed significant differences in the parameter ($p < 0.05$) between investigated soils obtained earlier by the author [27]. Comparable results were obtained by Spychała and Nieć [28], values obtained using the FHM test were higher than using CHM in the same conditions (geo-textile supplied with wastewater) but the difference was not as high (0.057 m/d and 0.047 m/d, respectively).

For fine and medium sand (most common soil types in practice) the relatively small values of the variation coeffcient (20% for fine sand and −14% for medium sand) for the 47 and 31 measurements respectively were made.

For the same types of soil the impact of differences between column measurements and constant water head test (CHM) measurement mean values on longterm acceptance rate (LTAR) [1] were relatively small. The hydraulic coefficient measurement points were located on the plotted LTAR function line (Figure 3). The significant differences in results for the same type of ground but using different methods (CHM, CHM-a and FHM) can be observed. These differences in hydraulic coefficient values implicates much smaller differences in LTAR values.

The best agreements were given by the method of constant head measurement for the calculation of hydraulic conductivity using the Glover equation, which gave similar results obtained, using the Philip equation for a parameter (which was equal to 36) as presented in Table 2. The differences between both variants of this test (without and with regard to capillary rise), for all types of sand, and infiltration column mean values were lower than 25%.

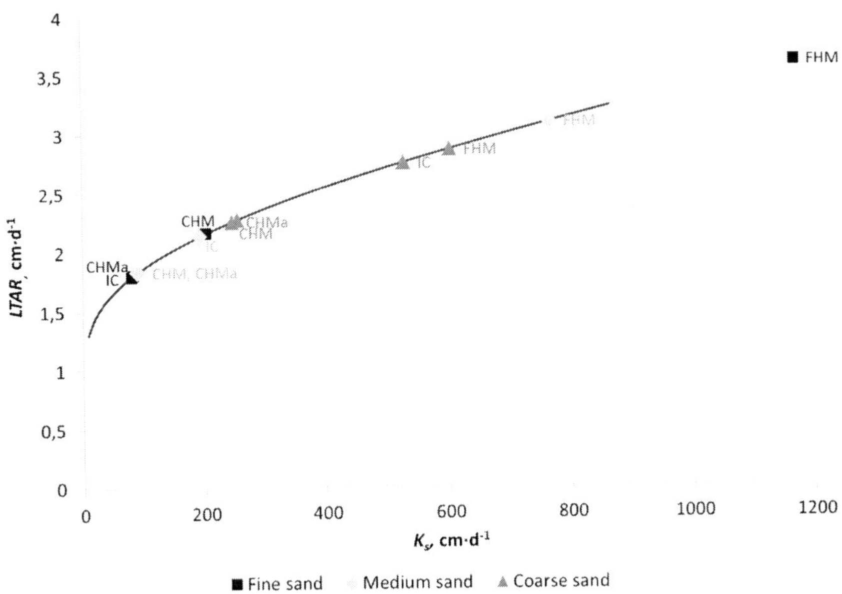

Figure 3. The hydraulic coefficient measurement points plotted on the LTAR function line.

Taking into consideration that even filter column tests values were relatively highly differentiated (about 16%–31%) it seems that the differences between CHM test values and infiltration column mean values of about 20%–25% could be accepted. Values obtained from the CHM test were acceptable for medium and coarse sand (−14% and −18%, respectively) but less acceptable for fine sand (20%). The values with a minus sign mean that assessed hydraulic loading is lower than probably more certain values obtained from an infiltration column test. This is the reason that a potentially designed system will have a surface area of about 20% larger than optimal. From a practical point of view the 20% is a relatively small value and designing a larger surface area system gives some security margin, allowing for a decrease in clogging risk. In the case of fine sand the CHM-a test is preferred by authors for applications in practice than the CHM test, as it gives results in the long-term acceptance rate (calculated and based on the hydraulic coefficient) overestimated in comparison to the infiltration column. The value is not high but the sign is adverse due to the resulting smaller infiltration surface area than optimal (calculated based on infiltration column measurement). The difference between CHM-a test mean values and infiltering column test measurement values has the same sign but is very small—1% only, so it can be assumed that a measurement using this test is almost optimal. Taking this into consideration the authors of this paper suggest using CHM tests for medium and coarse sand and CHM-a for fine sand, or the CHM-a test for all types of soil.

For medium and coarse sandy soils (medium are the most common as a system location) the underestimation of calculated LTAR based on CHM test determination gives 14%–18% value of system infiltration surface area. In Poland the full cost of the most common type of on-site wastewater system (septic tank with soil absorption system in sandy soil for one four-person family) is about 2,000 EUR and soil absorption systems cost is about a half of the total cost (1,000 EUR). The cost of SAS designed systems, based on CHM in comparison to those designs based on real hydraulic conductivity value, is up to 7%–9% of total higher than on-site wastewater plant cost. This relatively low cost (70–90 EUR) gives some security value of infiltration surface area (14%–18%), preventing the system from clogging. For large surface area systems other methods or apparatuses should be recommended and used to determine K in the field—e.g., infiltrometers.

The FHM and well-known equations are not recommended by the authors for soil permeability and infiltration systems assessment due to the high difference compared to infiltration columns measurement results. The above can be used, however, for investigations of particular factors and their impact on soil permeability, indicating, e.g., wastewater solids [29].

4. CONCLUSIONS

Based on this study the following conclusions can be drawn:

the differences in long-term acceptance rate values were about one magnitude smaller than differences in the hydraulic coeffcient (Figure 3), which is related to plotted function line slope; the research on medium and coarse sand showed that a similar result for hydraulic conductivity to that obtained from infiltration columns, which can be obtained using constant head tests; for the same types of soil the impact of difference between column measurements and constant water head test (CHM) measurement mean values on long-term acceptance rate value were relatively small–up to 14%–18%; studies conducted in fine sand confirmed a good convergence of hydraulic conductivity, using the infiltration column with constant head tests when the value of hydraulic conductivity is calculated with regard to capillary rise-the best agreements were given by the method of constant head measurement for the calculation of hydraulic conductivity, using the Philip equation for a parameter of 4; In the authors' opinion the CHM tests measurements are acceptable for small systems (e.g., one family household). This test gives some overestimation, however it can be treated as a security factor preventing clogging risk. For calculating LTAR for low permeable soils (fine sand) the authors of this paper suggest using the CHM-a test where the unsaturated flow of water (common in small grain and pore diameter soils) can be determined thanks to a capillary rise determined by the test. For large surface area systems other, more precise methods or tests should be used to determine K in the field, e.g., infiltrometers; the cost of SAS designed based on CHM in comparison to designed based on real hydraulic conductivity value is higher up to 7%–9% of total on-site wastewater plant cost only; the result of hydraulic conductivity obtained in the falling head test can be overestimated due to the higher value of hydraulic gradient than unity assumed in development of the calculation formulae; the authors of this paper suggest using CHM-a especially for low permeable soils (e.g., fine sand) where the unsaturated flow of water occurs with regard to capillary rise so as to calculate hydraulic load as a LTAR.

ACKNOWLEDGMENTS

The authors gratefully acknowledge the funding of this investigation by the National Science Centre (grants: No. N305316740).

AUTHOR CONTRIBUTIONS

Jakub Nieć has formulated the problem and accomplished laboratory tests. Marcin Spychała has contribiuted to discussions and writing of the manuscript.

REFERENCES

1. Laak, R. Subsurface Soil System. In Wastewater Engineering Design for Unsewered Areas; Technomic Publishing Company: Lancaster, PA, USA, 1986.
2. Amoozegar, A. Report No.316, Impact of Wastewater Quality on the Long-Term Acceptance Rate of Soils for On-Site Wastewater Disposal Systems; WRRI Project No. 70140; Water Resources Research Institute of the University of North Karolina, North Carolina State University: Raleigh, NC, USA; July; 1998.
3. Beal, C.D.; Gardner, E.A.; Kirchhof, G.; Menzies, N.W. Long-term flow rates and biomat zone hydrology in soil columns receiving septic tank effluent. Water Res. 2006, 40, 2327–2338.
4. Radcliffe, D.E.; West, L.T. Spreadsheet for converting saturated hydraulic conductivity to long-term acceptance rate for on-ste wastewater systems. Soil Surv. Horiz. 2009, 50, 20–24.
5. Pedescoll, A.; Uggetti, E.; Llorens, E.; Granés, F.; García, D.; García, J. Practical method based on saturated hydraulic conductivity used to assess clogging in subsurface flow constructed wetlands. Ecol. Eng. 2009, 35, 1216–1224.
6. Scott Purdy; Vector Engineering Inc.; Joel Peters; Vector Chile Ltda. Comparison of Hydraulic Conductivity Test Methods for Landfill Clay Liners. In Proceedings of the Beacon Conference "Sanitary Landfills for Latin America", Buenos Aires, Argentina, 8–10 March 2004.
7. Sobieraj, J.A.; Elsendbeer, H.; Vartessy, R.A. Pedotransfer functions for estimating saturated hydraulic conductivity: Implications for modeling storm flow generation. J. Hydrol. 2001, 251, 202–220.
8. Islam, N.; Wallander, W.W.; Mitchell, J.P.; Wicks, S.; Howitt, R.E. Performance evaluation of methods to the estimation of soil hydraulic parameters and their suitability in a hydrologic model. Geoderma 2006, 134, 135–151.
9. Ghanbarian-Alavijeh, B.; Liaghat, A.M.; Sohrabi, S. Estimating saturated hydraulic conductivity from soil physical properties using neural networks model. World Acad. Sci. Eng. Technol. 2010, 4, 108–113. [Google Scholar]

10. Slichter, C.S. Theoretical Investigations of the Motion of Ground Waters; 19th Annual Report, Part II; U.S. Geological Survey: Lei Sidui, VA, USA, 1899.

11. Aronovici, V.S. The mechanical analysis as an index of subsoil permeability. Soil Sci. Soc. Am. J. 1947, 11, 137–141.

12. Oosterbaan, R.J.; Nijland, H.J. Determining the saturated hydraulic conductivity. In Drainage Principles and ApplicationsILRI Publication 16, 2nd ed.; Ritzema, H.P., Ed.; ILRI Publication: Wageningen, the Netherlands, 1994.

13. De Ridder, N.A.; Wit, K.E. A comparative study on the hydraulic conductivity of unconsolidated sediments. J. Hydrol. 1965, 3, 180–206.

14. Odong, J. Evaluation of empirical formulae for determination of hydraulic conductivity based on grain-size analysis. Available online: http://www.engineerspress.com/pdf/IJAE/2013-01/a1%20_IJAE-133101_.pdf (accessed on 18 September 2014).

15. Vuković, M.; Soro, A. Determination of Hydraulic Conductivity of Porous Media from Grain-size Composition; Water Resources Publications: Littleton, CO, USA, 1992; p. 83.

16. Spychalski, M.; Kaźmierowski, C.; Kaczmarek, Z. The possibilities of indirect filtration coefficient assessment in soil. Polski Towarzystwo Geologiczne Oddział w Poznaniu 2004, Tom XIII, 48–56. (In Polish).

17. Bouma, J. Unsaturated flow during soil treatment of septic tank effluent. J. Environ. Eng. 1975, 101, 967–983.

18. Kropf, F.W.; Laak, R.; Healy, K.A. Equilibrium operation of subsurface absorption systems. J. Water Pollut. 1977, 49, 2007–2016.

19. Beach, D.; McCray, J.; Lowe, K.; Siegrist, R. Temporal changes in hydraulic conductivity of sand porous media biofilters during wastewater infiltration due to biomat formation. J. Hydrol. 2005, 311, 230–243. [Google Scholar]

20. Lange, O.K. Basic Hydrogeology (Osnovi Gidrogeologii); Moskovskii Gosudarstvenii Univerzitet: Moscow, Russia, 1958. (In Russian)

21. Amoozegar, A. Comparison of saturated hydraulic conductivity and percolation rate: Implication for designing septic system. In proceedings of Symposium on the Site Characterization and Design

of On-Site Septic Systems, New Orleans, LA, USA, 16–17 January 1997.

22. Bagarello, V.; Giordano, G. Comparation of procedures to estimate steady flow rate in field measurement of saturated hydraulic conductivity by the Guelph Permeameter method. J. Agric. Eng. Res. 1999, 74, 63–71. [Google Scholar]

23. Hayashi, M.; Quinton, W.L. A Constant-head well permeameter method for measuring field-saturated hydraulic conductivity above impermeable layer. Can. J. Soil Sci. 2004, 84, 255–264.

24. Bronders, J. Field and Laboratory Measurements to Determine Hydraulic Conductivity; Faculty of Applied Science Laboratory of Hydrology, Free University Brussels: Brussels, Belgium, 1991. [Google Scholar]

25. Elrick, D.E.; Reynolds, W.D.; Tan, K.A. Hydraulic conductivity measurements in the unsaturated zone using improved well analyses. Ground Water Monit. Remediat. 1989, 9, 184–193.

26. Pedescoll, A.; Samo, R.; Romero, E.; Puigagut, J.; Garcia, J. Reliability, repeatability and accuracy of the falling head method for hydraulic conductivity measurements under laboratory conditions. Ecol. Eng. 2011, 37, 754–757.

27. Nieć, J. Comparison of hydraulic conductivity determination methods for wastewater drainage. Seria Melior. Inż. Środ. 2005, 26, 293–304. (In Polish).

28. Spychała, M.; Nieć, J. Impact of septic tank sludge on filter permeability. Environ. Prot. Eng. 2013, 39, 77–89.

29. Spychała, M.; Nieć, J.; Pawlak, M. Preliminary study on filamentous particle distribution in septic tank effluent and their impact on filter cake development. Environ. Technol. 2013, 34, 2829–2837.

9 CHAPTER

DEVELOPMENT OF PEDOTRANSFER FUNCTIONS FOR SATURATED HYDRAULIC CONDUCTIVITY

John P. O. Obiero[1], Lawrence O. Gumbe[1], Christian T. Omuto[1], Mohammed A. Hassan[2], Januarius O. Agullo[1]

[1]Department of Environmental and Biosystems Engineering, University of Nairobi, Nairobi, Kenya; [2]Department of Earth and Environmental Science and Technology, Technical University of Kenya, Nairobi, Kenya.

ABSTRACT

The purpose of the study was to develop pedotransfer functions for determining saturated hydraulic conductivity (K_s). Pedotransfer functions (PTFs) for predicting soil physical properties used in determining saturated hydraulic conductivity, based on moisture retention characteristics, were developed. The van Genuchten moisture retention equation was fitted to measured moisture retention properties obtained from International Soil Reference and Information Centre (ISRIC) soils data base in order to determine parameters in the equation i.e. saturated soil moisture content (θ_s), residual soil moisture (θ_r), air entry parameter (α) and the pore size distribution parameter (n). 457 samples drawn from the data base were used to be the maximum possible sample size that contained the measured soils characteristics data required. Using statistical regression, mathematical relationships were developed between moisture retention parameters (response variables) and appropriately selected transformed basic soil properties (predictor variables). The developed PTFs were evaluated for accuracy and reliability. It was found that pedotransfer functions developed for θ_s produced the best performance in reliability compared to the remaining parameters yielding a correlation coefficient value of coefficient of determination (R^2 = 0.76), RMSE = 2.09, NSE = 0.75 and RSR = 0.5 indicating good performance. Relatively poorest performance was obtained from the pedotransfer function developed for α which yielded a correlation coefficient, R^2= 0.06, RMSE = 0.85 and a NSE of 0.02 reflecting the best possible equation derived for the parameter for use in predicting hydraulic conductivity. Out of the pedotransfer functions developed for each of the moisture retention parameters, the best performing PTF was identified for each parameter. The accuracy of the pedotransfer functions assessed based on R^2 were for θ_s (R^2 = 0.80), θ_r (R^2 = 0.42), α (R^2 = 0.04) and for n (R^2 = 0.30), when the variables were expressed directly in terms of the selected transformations of the basic soil properties.

Keywords: Pedotransfer Functions; Development; Evaluation

1. INTRODUCTION

Hydraulic conductivity is an important parameter that influences hydrological processes which affect flow in rivers. For instance, ground water flow is determined by saturated hydraulic conductivity (K_s). A number of formulations have been developed over the years to predict soil hydraulic conductivity from readily measurable soil properties. The need for mathematical modeling of K_s arises from the fact that insitu or laboratory measurements of hydraulic conductivity are time consuming, labour intensive and expensive as noted by [1], making it practically unlikely in reality to collect permeability data. Besides, direct measurements are also unreliable for site specific applications. The use of pedotransfer functions is an alternative to determining hydraulic conductivity and involves relationships that enables K_s to be predicted using measurable soil physical and hydraulic properties. K_s serves as an input parameter in many hydrological models. There is no specific method considered most accurate in determining K_s as the method used depends on soil and environmental conditions. Furthermore, data on K_s is not readily available in most local soil survey data bases and even where they are available, the reliability is not guaranteed. There is uncertainty in the prediction of K_s using existing pedotransfer functions. Methods to develop PTFs are also unknown and not well understood especially in Africa. Well known moisture characteristic functions are based on parameters that can be determined using moisture retention characteristics. Such data is not readily available in most soil data bases. It is essential to estimate such parameters using basic and easily available soil properties like bulk density, texture etc. Attempts to develop such equations is rare.

[2] notes that modern simulation and analysis of hydrologic processes relies heavily on the accurate and reliable description of soil water holding and transmission characteristics. Predicting K_s from basic soil properties using developed pedotransfer functions is based on analysis of the existing data sets, an approach applied in agricultural hydrology and watershed management. This research demonstrates how pedotransfer functions can be developed from a data base, with limited soils data, in predicting van Genuchten moisture retention parameters, a step in predicting K_s from readily available data on basic soil properties available from soil survey data. K_s can be indirectly predicted from soil properties using moisture retention equations like van Genuchten, Brooks Corey etc. where data on moisture retention curves is available. This technique was used in the study. Readily available data on measured K_s is rare and characterized with uncertainty on their accuracy as different methods in different environments yield varied results as observed by [3] therefore necessitating use of the indirect methods.

Three types of pedotransfer functions are recognized [4]. One category predicts soil moisture retention from basic soil properties. The other is based on point prediction of water retention characteristic as used by [5] to predict points along the moisture retention curve on Iranian soil. Another category involves prediction of parameters in models describing the θ-h-k relationship. This latter approach used in the study involved pedotransfer functions that relate the simple and easy to measure soil properties to van Genutchen moisture retention parameters, a notable gap in the science of development of pedotransfer functions. Most established pedotransfer functions for predicting soil hydraulic characteristics from continuous soil properties are based on statistical regression [6] in which the response variable, y, predicted from a number of n predictor variables, x_i yields the statistical multiple linear regression tool expressed as

$$y = a + \sum_{i=1}^{n} b_i x_i + \varepsilon, \quad i = 1, \cdots, n \tag{1}$$

where "a" is the intercept, b_i is the regression coefficient and ε is the error. The procedure to develop the pedotransfer functions involves steps which include:

The first step of statistical analysis is intended to test if each response variable distribution may be considered as normal distribution [7]. The normality of a distribution is tested by certain statistics like the Shapiro-Wilk (Wvalue), Skewness coefficient and Kurtosis. The shape of the distribution is determined by a measure of its Skewness which is given by the equation,

$$S = \left[\frac{n}{(n-1)(n-2)} \right] \sum_{i=1}^{n} \frac{(x_i - \bar{x})^3}{s^3} \tag{2}$$

where S is Skewness coefficient, s the standard deviation, x_i the observation value, $\bar{x}$ the mean value and n the number of observations. For a normally distributed data, the skewness = 0, hence the closer the skewness coefficient is to zero (0), the better the normality of the distribution. The level of Kurtosis also provides information about the extent of the normality of the distribution and is given by,

$$K = \left[\frac{n(n+1)}{(n-1)(n-2)(n-3)} \right] \times \sum_{i=1}^{n} \left[\frac{(x_i - \bar{x})^4}{\left[s^4 - (3n-1)^2 \right] / (n-2)(n-3)} \right] \tag{3}$$

The measure of Kurtosis is zero (0) for a normally distributed population. Graphical aids used to check normality includes the normal probability plot.

Mathematical relationships between potential predictor variables and the response variables are checked by detecting exponential, logarithmic, or square root tendencies. Important to eliminate is the problem of redundant information in the predictor set brought about by linear dependence between predictor variables [6]. This problem of multicollinearity can be examined by means of correlation matrix. Common statistics used in evaluating pedotransfer functions include [4]:

Multiple determination coefficient given by

$$R^2 = 1 - \frac{\sum_1^N (y_i - \hat{y}_i)^2}{\sum_1^N (y_i - \bar{y})^2} \tag{4}$$

Root mean square error given by

$$RMSE = \sqrt{\frac{\sum_1^N (y_i - \hat{y})}{N}} \tag{5}$$

Mean Error given by

$$ME = \sum_1^N \frac{(y_i - \hat{y})}{N} \tag{6}$$

Mean Absolute Error given by

$$ME = \sum_1^N \frac{|y_i - \hat{y}_i|}{N} \tag{7}$$

where y_i denotes the actual value, $\hat{y}_i$ the predicted value, and $\bar{y}$ the average of the actual value and N is the total number of observations. [8] evaluated performance of different pedotransfer functions for estimating certain soil hydraulic properties that include Available Water Capacity (AWC) using R^2, RMSE and ME. [9] used only RMSE in assessing pedotransfer functions developed for predicting hydraulic conductivity and water retention.

The van Genuchten water retention equation which relates moisture content (θ) to the soil matric suction (h) and is expressed as follows;

$$h(\theta) = \frac{1}{\alpha}\left[S_e^{n-1} - 1\right]^{\frac{1}{n}} \tag{8}$$

where

$$S_e = \left[\frac{\theta - \theta_r}{\theta_s - \theta_r}\right] \tag{9}$$

To determine K_s, the hydraulic conductivity function used is expressed as

$$K(h)=K_s\left[\frac{\left(1-(\alpha h)^{n-1}\left(1+(\alpha h)^n\right)^{-m}\right)^2}{\left(1+(\alpha h)^n\right)^{m/2}}\right] \tag{10}$$

To determine K_S using the above equation, a relationship between hydraulic conductivity and matric suction, h is used and can be obtained through equations that relate hydraulic conductivity to permeability [10].

2. METHODOLOGY

2.1. Data Collection and Analysis

[4] noted that data from existing international soil data bases having measured soil data can be analyzed to enable prediction of hydraulic characteristics from measured soil data. Availability of measured soil hydraulic characteristics for a wide range of soils and from a large and reliable international data bases are considered prerequisite for development of pedotransfer functions. The ISRIC soils database was used to obtain measured data on soil hydraulic characteristics and basic soil properties. It was possible to obtain all the required measured data on 457 soil samples from various parts of the world. The data contained measured moisture retention curves as well as data on texture (Sand%, Clay% and Silt%). The ISRIC dataset [11] was used for this test. This dataset contains Water Retention Curve (WRC) for both topsoil (between 0 and 20 cm depth) and subsoil (between 20 and 100 cm) from many countries in the world. It has 8 levels of measured suction potential (at 0, 0.1, 0.312, 1.0, 1.99, 5.01, 2.5, and 15.8 m) (Figure 1). The lowest suction potential was taken as 0.01 to avoid errors with logarithmic transformations. The observed moisture retention curve obtained from measured data on moisture retention characteristics was fitted to the van Genuchten moisture retention Equation (11)

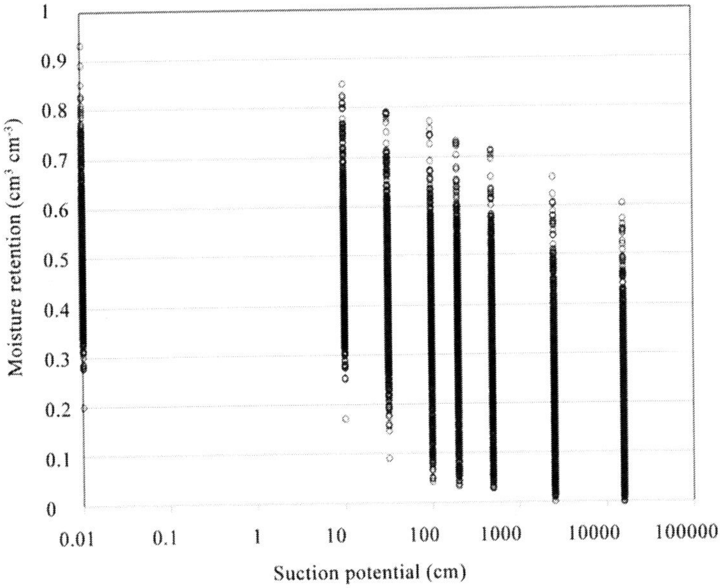

Figure 1. Measured water retention characteristics from ISRIC database.

$$\theta(h) = \theta_r + (\theta_s - \theta_r)\left[1 + (\alpha h)^n\right]^{-m} \quad (11)$$

where q(h) is the soil moisture content at h suction potential, q_r is the residual moisture content, q_s is the saturated moisture content, a is the inverse of air-entry potential, and n and m are the shape parameters for the water retention curve representing pore-size distribution index [12]. HydroMe computer programme was used to determine parameters of the van Genuchten equation (http://cran.r-project.org/ web/packages/HydroMe/index.html) and is executable in R programme The pedotransfer functions were then developed to estimate these parameters from readily available measured basic soil properties. Relating the moisture retention parameters to the basic soil properties then makes it possible to predict the moisture retention characteristics from the basic properties and hence enabling prediction of the saturated hydraulic conductivity. It would therefore not be necessary to have measured data on moisture retention characteristics to determine K_s but instead, the basic soil properties would be used in determining the equations for moisture retention which can then be used to estimate K_s. Measurement of moisture retention characteristics is quite elaborate, tedious, time consuming and also expensive especially if a large number of sites is involved. From the sample data set of 457 samples, a sub dataset consisting of 342 (75%) samples was randomly selected for use in calibration of the pedotransfer functions to be developed while the remainder portion of 115 samples (25%) of the data was used in the validation process. In development of the pedotransfer functions, the following procedures were undertaken.

2.1.1. Evaluation of the Distribution of the Moisture Retention Parameters

A requirement in developing pedotransfer functions is that the response (dependent) variables should be normally distributed. As a result, the moisture retention parameter of the van Genuchten equation and their possible transformations were evaluated to establish which would best approximate normal distribution. To check the extent to which the response variables would be normally distributed, statistical measures of Shapiro Wilk (W-value), Skewness Coefficient, and measures of Kurtosis were used in the assessment.

2.1.2. Statistical Regression Analysis

Statistical regression was performed between each transformed response variable and each of the basic properties and their transformations so as to establish the goodness of fit in each case based on the measure of the coefficient of determination (R^2). The purpose of this was to determine, for each response variable, the predictor variable or its transformation that gives the best quality of fit between the two when a regression is performed. This would give an indication of which predictor variable or its transformation correlates best with the response variable or it transformation.

2.1.3. Analysis of Cross Correlations

To establish the level of dependence between the selected predictor variables for each response variable best correlated to them, a cross correlation was performed between the independent variables by determining the correlation coefficient (r) between each pair of the vari ables. This kind of analysis is performed to determine if there is any correlation between two predictor variables among the set chosen for determining the multiple regression equations. It is preferable that the independent variables in a multiple linear regression equation be independent among themselves.

2.1.4. Multiple Linear Regression and Validation of Developed Equations

This was carried out between each selected transformed response variable and selected combination of predictor variables/transformation consisting of the independent variables for which the response variable best correlated as determined by the measure of coefficients of determination. The multiple coefficient of determination was determined for each equation developed to assess the accuracy of the equation. The multiple regression was carried out between response variables and the predictor variables or transformations consisting of independent variables that are not themselves significantly correlated ($r < 0.5$). After development of the pedotransfer functions using the given data set, their reliability is then tested by comparing the values of predicted parameters using the developed equations with the observed data using an independent data set. An independent data set consisting of 115 samples was used in the validation process. The evaluation statistic used included the correlation coefficient among others.

3. RESULTS AND DISCUSSION

3.1. Evaluation of the Distribution of the Moisture Retention Parameters

Table 1 shows the values of statistical parameters used for evaluating normality of distribution for the response variables being the van Genuchten moisture retention parameters and their proposed transformations for the data set of 457 samples. The transformations of van Genuchten parameters (response variables) that produced the best approximation to normal distribution are θ_s^{-1}, e^{θ_r}, $\sqrt{\alpha}$ and $\sqrt{n}$. This was based on measures of the ShapiroWilk (W-value), Skewness coefficient and Kurtosis. For illustration, Figure 2 shows the histogram and normality plot in the normal distribution check for the saturated soil moisture content transformation (θ_s).The transformations of van Genutchen parameters (response variables) that produce the best approximation to normal distribution are θ_s^{-1}, e^{θ_r}, $\sqrt{\alpha}$ and $\sqrt{n}$. This was based on measures of the Shapiro-Wilk (W-value), Skewness coefficient and Kurtosis.

3.2. Statistical Linear Regression Analysis

The quality of fit for the linear regression between each

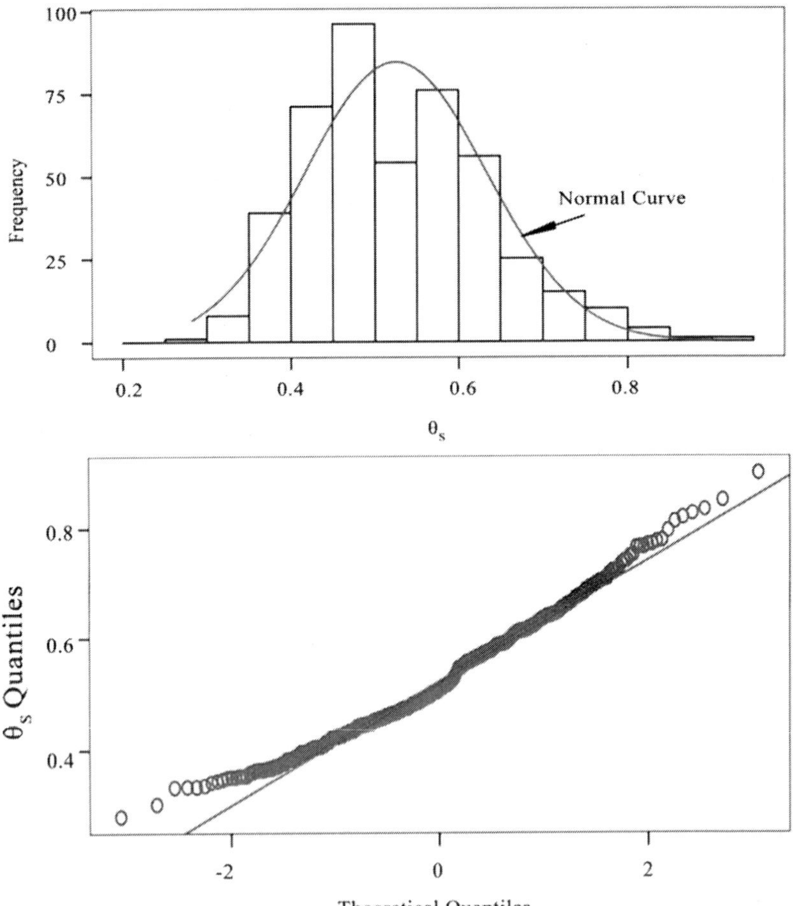

Figure 2. Check for normality of distribution of the saturated soil moisture content (θ_s) showing the histogram and normality plot.

	Kurtosis	W-value	Skewness
θ_r	0.99	0.09	−0.21
α	0.67	4.08	29.47
n	0.78	2.65	11.13
$\ln(\theta_s)$	0.99	0.04	−0.39
$\sqrt{n}$	0.99	0.28	−0.29
θ_s^{-1}	0.98	0.06	0.06
e^{θ_s}	0.96	0.82	0.66
θ_s^2	0.93	1.05	1.25
$\ln(\theta_r)$	0.75	−4.16	38.34
$\sqrt{\theta_r}$	0.93	−1.09	1.77
$1/\theta_r$	0.03	21.09	444.94
θ_r^2	0.89	1.34	2.29
e^{θ_r}	0.99	0.39	0.12
$\theta_r^{1/E}$	0.83	−2.02	5.78
$\ln(\alpha)$	0.72	−2.94	11.8
$1/\alpha$	0.03	21.17	450.93
e^{α}	0.62	5.01	44.15
α^2	0.20	13.89	238.13
$\sqrt{\alpha}$	0.95	0.91	2.27
$\mathrm{Log}(n)$	0.98	0.49	0.94
$\sqrt{n}$	0.91	1.45	3.67
n^2	0.46	6.23	56.04
$1/n$	0.93	1.32	3.93
e^{n}	0.24	12.69	201.48

Table 1. Normality of distribution check for parameters in the van Genuchten water retention characteristic and their transformations.

selected transformed dependent variables (moisture rention parameters) and the transformed independent variables (basic soil properties) was notably different for each predictor variable transformation. For instance, the linear regression performed between θ_s and sand yielded a quality of fit, measured by the value of R^2, to be 0.10 while that between θ_s and e^{sand} yielded $R^2 = 0.01$ (Table 2). The same observation was made in the case of regreson performed with silt, clay, bulk density and organic carbon. Also each selected transformed variable (i.e. $\ln(\theta_s)$, e^{θ_r}, $\sqrt{\alpha}$, etc.) produced different values of R^2 when a linear regression was performed between each of the response variables and a transformed predictor varile for each of the variables sand, silt, clay bulk density and organic carbon. For example, a regression performed between the independent variable $sand^2$ and θ_s^{-1} yielded $R^2 = 0.14$ while that between $sand^2$ and e^{θ_r} produced $R^2 = 0.20$, hence no particular transformed variable could be said to generally give the best quality of fit with all the dependent variables whether it is sand, silt, clay, bulk density or organic carbon. Also, there is no particular response variable that could be said to yield the best quality of fit compared to others for all the independent variables or transformations to which regression was performed. The purpose of regression analysis was to identify the pair of transformed dependent and independent variables that yield the best quality of fit between them as measured by coefficient of determination (R^2) so that for each dependent variable, one is able to identify which transformations of each of the dependent variables would be used in developing the pedotransfer functions during multiple regression equation to relate each response variable and selected predictor variables and appropriate transformations. This would help establish the best possible mathematical relationship between the moisture retention parameters (response variables) and selected basic soil properties (independent variables) in the pedotransfer function. A number of observations on the regression relationships can be made for each response variable and how it relates with each of the predictor variables and their transformations. The transformed response variable θ_s^{-1} yielded the best quality of fit ($R^2 = 0.14$) with the independent variables sand or $sand^2$ compared to the other transformations when a regression analysis was performed. In the case of regression with silt, the best possible quality of fit was obtained with the transformation $silt^{-1}$ ($R^2 = 0.04$) and so was with bulk density, BD^3 ($R^2 = 0.60$), $\sqrt{clay}$ or $clay^2$ ($R^2 = 0.12$), organic carbon, $\ln(orgC)$ ($R^2 = 0.27$). In general, the linear regression

Table 2. Measurement of the quality of fit based on coefficient of determination (R^2) between the moisture retention parameters (response variable) and sand with its transformation (predictor variables) in the linear regression.

	Sand	$Sand^2$	$Sand^3$	$Sand^5$	$Sand^{10}$	$Sand^{15}$	$\sqrt{Sand}$	ln(sand)	$sand^{-1}$	e^{Sand}
θ_r	$R^2 = 0.10$	$R^2 = 0.10$	$R^2 = 0.08$	$R^2 = 0.06$	$R^2 = 0.03$	$R^2 = 0.03$	$R^2 = 0.09$	$R^2 = 0.06$	$R^2 = 0.01$	$R^2 = 0.01$
θ_r^{-1}	$R^2 = 0.14$	$R^2 = 0.14$	$R^2 = 0.12$	$R^2 = 0.08$	$R^2 = 0.05$	$R^2 = 0.04$	$R^2 = 0.12$	$R^2 = 0.08$	$R^2 = 0.01$	$R^2 = 0.02$
$\ln(\theta_r)$	$R^2 = 0.12$	$R^2 = 0.12$	$R^2 = 0.10$	$R^2 = 0.10$	$R^2 = 0.07$	$R^2 = 0.04$	$R^2 = 0.12$	$R^2 = 0.08$	$R^2 = 0.01$	$R^2 = 0.02$
θ_s	$R^2 = 0.21$	$R^2 = 0.21$	$R^2 = 0.19$	$R^2 = 0.15$	$R^2 = 0.04$	$R^2 = 0.03$	$R^2 = 0.11$	$R^2 = 0.07$	$R^2 = 0.01$	$R^2 = 0.02$
e^{θ_s}	$R^2 = 0.20$	$R^2 = 0.20$	$R^2 = 0.18$	$R^2 = 0.13$	$R^2 = 0.07$	$R^2 = 0.05$	$R^2 = 0.18$	$R^2 = 0.13$	$R^2 = 0.02$	$R^2 = 0.03$
$\sqrt{\alpha}$	$R^2 = 0.01$	$R^2 = 0.01$	$R^2 = 0.02$	$R^2 = 0.02$	$R^2 = 0.01$	$R^2 = 0.01$	$R^2 = 0.00$	$R^2 = 0.00$	$R^2 = 0.00$	$R^2 = 0.02$
$\ln(\alpha)$	$R^2 = 0.01$	$R^2 = 0.03$	$R^2 = 0.05$	$R^2 = 0.07$	$R^2 = 0.09$	$R^2 = 0.09$	$R^2 = 0.01$	$R^2 = 0.00$	$R^2 = 0.00$	$R^2 = 0.08$
$Log(n)$	$R^2 = 0.01$	$R^2 = 0.13$	$R^2 = 0.15$	$R^2 = 0.15$	$R^2 = 0.16$	$R^2 = 0.15$	$R^2 = 0.15$	$R^2 = 0.08$	$R^2 = 0.02$	$R^2 = 0.08$
$\sqrt{n}$	$R^2 = 0.11$	$R^2 = 0.15$	$R^2 = 0.18$	$R^2 = 0.21$	$R^2 = 0.20$	$R^2 = 0.12$	$R^2 = 0.08$	$R^2 = 0.05$	$R^2 = 0.01$	$R^2 = 0.10$
n^{-1}	$R^2 = 0.07$	$R^2 = 0.08$	$R^2 = 0.09$	$R^2 = 0.08$	$R^2 = 0.09$	$R^2 = 0.06$	$R^2 = 0.06$	$R^2 = 0.04$	$R^2 = 0.01$	$R^2 = 0.03$

performed between θ_s^{-1} and the bulk density transformations produced best quality of fit as compared to those of silt, sand, organic carbon and clay, showing that the best relationship was obtained with bulk density followed by organic carbon, sand, clay and silt in that order. In the case of residual moisture content, θ_r, the selected transformation variable e^{θ_r} related best with clay with a quality of fit $R^2 = 0.32$ obtained when regression was performed between e^{θ_r} and $\sqrt{\text{clay}}$. The poorest fit was obtained when the regression was performed with transformations of organic carbon e.g. ln(orgC) with $R^2 = 0.01$. α yielded the best possible quality of fit when a regression was performed with $\sqrt{\text{clay}}$ or 1/clay. $\sqrt{\alpha}$ generally showed poor quality of fit with most of the independent variables or transformations. The transformation n^{-1} also showed poor quality fit with most of the selected response variables when regression analysis was performed. The best quality fit was obtained with the transformation of sand when the linear regression was performed between n^{-1} and $sand^3$ ($R^2 = 0.09$). Figures 3(a)-(d) illustrates the regression between selected transformations of response variables and the predictor variables to which they best fit.

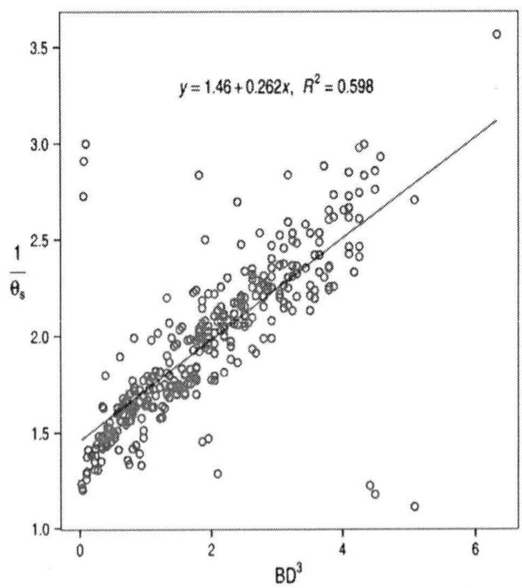

(a)

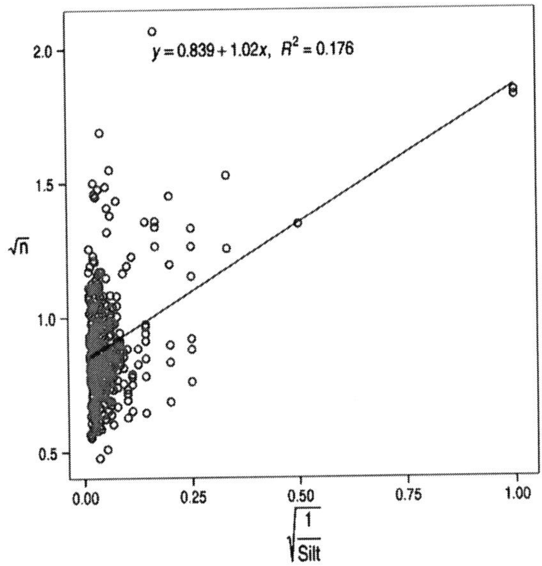
y = 0.839 + 1.02x, R² = 0.176
2.0
1.5
1.0
0.5
√n
0.00
0.25
0.50
0.75
1.00
√(1/Silt)

(b)

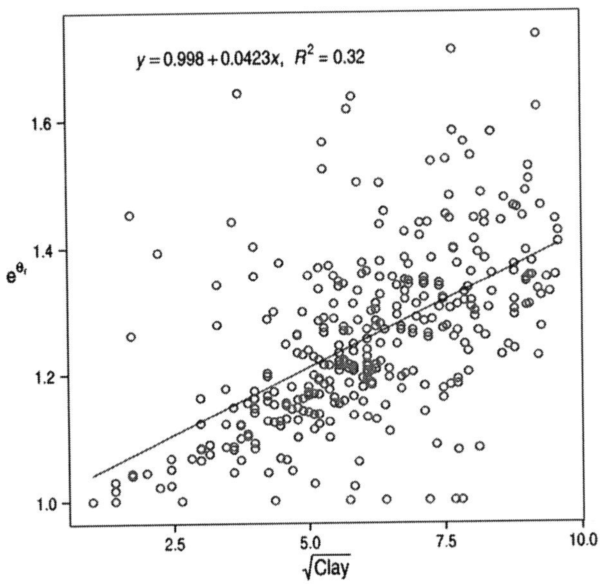
y = 0.998 + 0.0423x, R² = 0.32
1.6
1.4
1.2
1.0
e^θr
2.5
5.0
7.5
10.0
√Clay

(c)

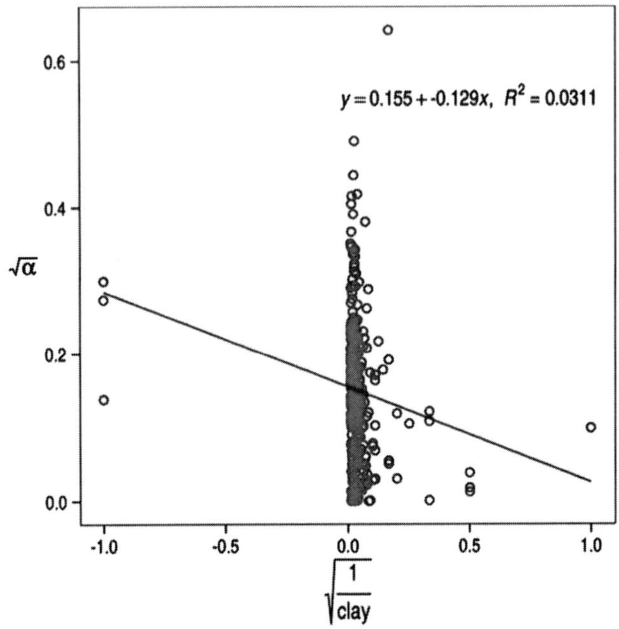

(d)

Figure 3. Response versus predictor variables correlations with corresponding best possible quality fit.

After performing linear regression between two variables involving the selected transformed response variables, associated with the moisture retention parameters, and the predictor variables associated with the basic soil properties, the following conclusions were arrived at. For the transformed response variable θ_s^{-1} the predictor variables that yielded the best possible quality of fit with it when linear regression was performed were sand, silt^{-1}, BD^3, clay^2, and ln(orgC). These variables would then be used in developing the multiple linear regression between the said response variable and the predictor variables. In the case of e^{θ_r}, the best possible fit (measured by the value of R^2) was obtained from linear regression with sand, silt^{10}, BD^2, $\sqrt{clay}$, and ln(orgC), while in the case of $\sqrt{\alpha}$ the best possible fits were obtained with e^{sand}, Silt^{-1}, BD, Clay^{-1}, and ln(orgC). In the case of $\sqrt{n}$, the best possible quality of fit was obtained with the predictor variables Sand^5, silt^{-1}, ln(clay), (orgC^2)$^{-1}$.

3.3. Multiple Regression Analysis and Accuracy of the Pedotransfer Functions Developed

Table 3 shows the quality of fit obtained when multiple linear regression was performed between each selected response variable and the transformed predictor variables to which it relates best e.g. a multiple linear regression performed between the response variable θ_s^{-1} and sand, silt^{-1}, BD^3, clay^2, ln(orgC), yielded the quality of fit meas ured by $R^2 = 0.63$ which is considered acceptable and therefore accurate. The resulting equation is illustrated below Equation (12). The predictor variables indicated above that related best in linear regression with the selected response variables were the ones considered in developing the pedotransfer functions using multiple linear regression.

$$\theta_s^{-1} = 1.447 + 0.003045\text{sand} - 0.1505\text{silt}^{-1}$$
$$+0.2126BD^3 - 006\text{clay}^2 - 0.04542\ln(\text{orgC}) \quad (12)$$

where θ_s is the saturated soil moisture content (cm^3/cm^3), silt the % silt, sand the % sand, BD the bulk density (g/cm^3), clay the % clay, and orgC the Organic carbon content (g/kg). The qualities of fit for the remaining transformed response variables are illustrated in Table 3 for each of e^{θ_r}, $\sqrt{\alpha}$ and $\sqrt{n}$ when the simple linear regression was done with the selected predictor variable transformations.

3.4. Analysis of Cross Correlation and Correlation Matrix

Cross correlation was done for the independent variables (transformations) used in developing multiple linear regression for each of the response variables θ_s^{-1}, e^{θ_r}, $\sqrt{\alpha}$ and $\sqrt{n}$. Tables A1-A4 (Appendix) shows the relationships, based on correlation coefficient, between the various predictor variables used for each of the response variable transformations.

In the case of θ_s^{-1} the predictor variables sand and $\sqrt{\text{clay}}$ appear to be highly negatively correlated with r = −0.71 while BD^3 and ln(orgC) are also

Table 3. Measurement of Quality of fit based on coefficient of determination (R^2) between the moisture retention parameters (response variables) and selected predictor variable transformations in the multiple linear regression.

	sand, $silt^{-1}$, BD^3, $clay^2$, ln(orgC)	sand, e^{BD}, sand	$silt^{-1}$, BD^3	$\sqrt{clay}$, BD^2, $silt^{-1}$	BD^3, ln(orgC)	BD^3, $\sqrt{clay}$	sand, BD^3
θ_r^{-1}	$R^2 = 0.63$	$R^2 = 0.63$	$R^2 = 0.63$	$R^2 = 0.61$	$R^2 = 0.59$	$R^2 = 0.61$	$R^2 = 0.63$
	sand, $silt^{10}$, BD^2, $\sqrt{clay}$, ln(orgC)	sand, $\sqrt{clay}$	$\sqrt{clay}$, $silt^{10}$, BD^2	clay, $silt^{10}$, BD^2	$\sqrt{clay}$, $Silt^{10}$, ln(orgC)	$\sqrt{clay}$, BD^2	$\sqrt{clay}$, ln(orgC)
e^{β}	$R^2 = 0.37$	$R^2 = 0.33$	$R^2 = 0.35$	$R^2 = 0.32$	$R^2 = 0.32$	$R^2 = 0.35$	$R^2 = 0.32$
$\sqrt{\alpha}$	e^{sand}, $silt^{-2}$, BD, $clay^{-1}$, ln(orgC)	e^{sand}, BD, $clay^{-1}$	e^{sand}, $clay^{-1}$, ln(orgC)	$clay^{-1}$, $silt^{-2}$, ln(orgC)	$silt^{-2}$, BD, $clay^{-1}$	BD, $clay^{-1}$	$silt^{-1}$, $clay^{-1}$
	0.05	0.05	0.04	0.04	0.05	0.04	0.04
	$sand^5$, 1/silt, ln(clay), $orgC^{-2}$	$sand^5$, $orgC^{-2}$	$silt^{-1}$, ln(clay)	ln(clay), or GC^{-2}	$orgC^{-2}$, BD^3	ln(clay), BD^3	$sand^5$, BD^3
$\sqrt{n}$	$R^2 = 0.24$	$R^2 = 0.21$	$R^2 = 0.23$	$R^2 = 0.12$	$R^2 = 0.04$	$R^2 = 0.11$	$R^2 = 0.20$

fairly correlated in a negative sense (r = −0.59). For the dependent variable e^{θ_r}, the response variables BD^3 and ln(orgC) are also closely correlated in the negative sense with r = −0.62. Other reasonable correlations include that between e^{sand} and $silt^{-2}$ (r = 0.89), BD and ln(orgC) (r = −0.64) in the case of predictor variables for $\sqrt{\alpha}$. The other significant correlations observed is that between $sand^5$ to $(silt)^{-1}$ (r = 0.74) and $sand^5$ to ln(Clay) (r = −0.69) in the predictor variables for $\sqrt{n}$.

3.5. Multiple Linear Regression Results

For each of the selected response variables, Table 3 shows the various possible combinations of the predictor variables that were used in performing the multiple linear regression and also indicates the qualities of fit obtained based on the coefficient of determination R^2.

The various possible multiple linear regression equations obtained for the various independent variables are underlisted (Equations (13)-(27)). The choice of the independent variables or their transformations was based on combinations that would yield the highest possible quality of fit i.e. the predictor variable combinations chosen include the variable that relate very well with the response variable, hence the equations are the best possible equations that could be obtained using the variables indicated. Equations developed involving θ_s^{-1} are as follows with the measures of accuracy based on coefficient of determination (R^2) indicated:

$$\frac{1}{\theta_s} = 0.813 + 0.003534\text{sand} + 0.3099e^{BD} \quad (R^2 = 0.63) \tag{13}$$

$$\frac{1}{\theta_s} = 1.397 + 0.00317\text{sand} + 0.2451BD^3 \quad (R^2 = 0.63) \tag{14}$$

$$\frac{1}{\theta_s} = 1.397 + 0.003385\text{sand} - 0.1433\frac{1}{\text{silt}} + 0.246BD^3 \quad (R^2 = 0.63) \tag{15}$$

$$\frac{1}{\theta_s} = 1.477 + 0.003045\text{sand} - 0.1505\frac{1}{\text{silt}} + 0.212BD^3 - 9.3857E - 0.006\text{clay}^2 - 0.0452\ln(\text{orgC}) \quad (R^2 = 0.63) \tag{16}$$

The set of possible multiple linear regression equations developed involving e^{θ_r} are listed as follows:

$$e^{\theta_r} = 1.083 + 0.0382\sqrt{\text{clay}} - 5.0290E - 021\text{silt}^{10} - 00392BD^2 \quad (R^2 = 0.35) \tag{17}$$

$$e^{\theta_r} = 1.062 - 0.0007465\text{sand} + 0.03532\sqrt{\text{clay}} \quad (R^2 = 0.33) \tag{18}$$

$$e^{\theta_r} = 1.009 + 0.04099\sqrt{\text{clay}} - 4.9035E - 021\text{silt}^{10} + 0.004877\ln(\text{orgC}) \quad (R^2 = 0.32) \tag{19}$$

$$e^{\theta_r} = 1.007 + 0.03896\sqrt{\text{clay}} - 0.03928BD^2 \quad (R^2 = 0.35) \tag{20}$$

The set of possible multiple linear regression equations developed involving $\sqrt{\alpha}$ are listed as follows:

$$\sqrt{\alpha} = 0.18 - 1.0440E - 043e^{\text{sand}} - \frac{0.071381}{\text{silt}^2} - 0.02068BD - \frac{0.088661}{\text{clay}} + 0.0008447\ln(\text{orgC}) \quad (R^2 = 0.05) \tag{21}$$

$$\sqrt{\alpha} = 0.1847 - 2.0502E - 043\exp(\text{sand}) - 0.02556BD - 0.09316\frac{1}{\text{clay}} \quad (R^2 = 0.05) \tag{22}$$

$$\sqrt{\alpha} = 0.1564 - 2.0817E - 043e^{\text{sand}} - 0.08877\frac{1}{\text{clay}} + 0.004372\ln(\text{orgC}) \quad (R^2 = 0.04) \tag{23}$$

The set of possible multiple linear regression equations developed involving $\sqrt{n}$ are listed as follows:

$$\sqrt{n} = 0.9871 + 4.0051E - 01\,1\text{sand}^5 + 0.623\frac{1}{\text{silt}} - 0.03977\ln(\text{Clay}) - \frac{0.00006391}{\text{orgC}^2} \quad (R^2 = 0.24) \tag{24}$$

$$\sqrt{n} = 1.101 + 0.8702\frac{1}{\text{silt}} - 0.07228\ln(\text{clay}) \quad (R^2 = 0.24) \tag{25}$$

$$\sqrt{n} = 0.8617 + 8.4062E - 01\,1\text{sand}^5 + 2.2259E - 005\left(\frac{1}{\text{orgC}^2}\right) \quad (R^2 = 0.21) \tag{26}$$

$$\sqrt{n} = 1.22 - 0.09247\ln(\text{clay}) + \frac{0.00011681}{\text{orgC}^2} \quad (R^2 = 0.12) \tag{27}$$

Based on the above indicated equations that produced the best quality of fit, equations relating the moisture retention parameters that are not transformed to the trans

formed basic properties were also compared to the measured (fitted) parameters and yielded the values of R^2 indicated (Equations (28) to (31)) alongside the equations.

$$\theta_s = \left(0.8103 + 0.003534\text{sand} + 0.3099\text{e}^{BD}\right)^{-1} \left(R^2 = 0.77\right) \quad (28)$$

3.6. Validation of the Developed Pedotransfer Functions to Evaluate Reliability

Table 4 shows regression equations developed for each of

$$\theta_r = \ln\left(1.083 + 0.0382\sqrt{\text{clay}} - 5.029E - 0.021\text{silt}^{10} - 0.0329BD^2\right) \left(R^2 = 0.43\right) \quad (29)$$

$$n = \left(0.8617 + 8.4062E - 011\text{sand}^5 + 2.2259E - 0.005\frac{1}{\text{orgC}^2}\right)^2 \left(R^2 = 0.28\right) \quad (30)$$

Table 4. Summary of equations for transformations of van Genuchten parameters and their performance in reliability.

Response variable	Equation	R^2	RMSE	RSR	NSE	ME	PBIAS
θ_s^{-1}	$\theta_s^{-1} = 0.8103 + 0.003534\text{sand} + 0.3099e^{BD}$	0.76	2.09	0.50	0.75	0.001	0.07
θ_s^{-1}	$\theta_s^{-1} = 1.397 + 0.003176\text{sand} + 0.2451\text{BD}^3$	0.76	2.12	0.50	0.75	0.004	0.19
θ_s^{-1}	$\theta_s^{-1} = 1.477 + 0.003045\text{SAND} - 0.15051/\text{silt} + 0.2126\text{BD}^3$ $-9.3857E - 006\text{clay}^2 - 0.04542\ln(\text{orgC})$	0.74	2.11	0.51	0.74	0.009	0.44
θ_s^{-1}	$\theta_s^{-1} = 1.397 + 0.003385\text{sand} - 0.1433\text{silt}^{-1} + 0.246\text{BD}^3$	0.76	2.13	0.51	0.74	0.006	0.31
e^{θ_r}	$e^{\theta_r} = 1.083 + 0.0382\sqrt{\text{clay}} - 5.0290E - 021\text{silt}^{10} - 0.0392\text{BD}^2$	0.41	1.10	0.77	0.41	0.011	0.90
e^{θ_r}	$e^{\theta_r} = 1.062 - 0.0007465\text{sand} + 0.03532\sqrt{\text{clay}}$	0.36	1.14	0.80	0.36	0.008	0.63
e^{θ_r}	$e^{\theta_r} = 1.009 + 0.04099\sqrt{\text{clay}} - 4.9035E - 021\text{silt}^{10}$ $+0.004877\ln(\text{orgC})$	0.40	1.11	0.78	0.39	0.007	0.55
e^{θ_r}	$e^{\theta_r} = 1.007 + 0.03896\sqrt{\text{clay}} - 0.03928\text{BD}^2$	0.41	1.10	0.77	0.40	0.011	0.90
e^{θ_r}	$e^{\theta_r} = 1.003 + 0.0417\sqrt{\text{clay}} + 0.005286\ln(\text{orgC})$	0.38	1.12	0.79	0.38	0.007	0.57
$\sqrt{\alpha}$	$\sqrt{\alpha} = 0.18 - 1.0440E - 043\exp(\text{sand}) - 0.07138\text{silt}^{-1}$ $-0.02068\text{BD} - 0.088661/\text{clay} + 0.0008447\ln(\text{orgC})$	0.04	0.85	1.00	0.01	−0.012	−8.82
$\sqrt{\alpha}$	$\sqrt{\alpha} = 0.1847 - 2.0502E - 043e^{\text{sand}} - 0.02556\text{BD} - 0.09316\text{clay}^{-1}$	0.04	0.85	0.99	0.01	−0.011	−8.29
$\sqrt{\alpha}$	$\sqrt{\alpha} = 0.1564 - 2.0817E - 043e^{\text{sand}} - 0.088771/\text{clay}$ $+0.004372\ln(\text{orgC})$	0.06	0.85	0.99	0.02	−0.013	−9.39
$\sqrt{n}$	$\sqrt{n} = 0.9871 + 4.0051E - 011\text{sand}^5 + 0.623\text{silt}^{-1}$ $-0.03977\ln(\text{Clay}) - 0.00006391/\text{orgC}^2$	0.20	1.89	0.90	0.20	−0.006	−0.62
$\sqrt{n}$	$\sqrt{n} = 1.101 + 0.87021/\text{silt} - 0.07228\ln(\text{Clay})$	0.18	1.91	0.91	0.17	−0.002	0.17
$\sqrt{n}$	$\sqrt{n} = 0.8617 + 8.4062E - 011\text{sand}^5 + 2.2259E - 0051/\text{orgC}^2$	0.20	1.88	0.89	0.20	−0.005	−0.59
$\sqrt{n}$	$\sqrt{n} = 1.22 - 0.09247\ln(\text{Clay}) + 0.00011681/\text{orgC}^2$	0.08	2.02	0.96	0.08	−0.002	0.19

$$\alpha = \left(0.1564 - 2.0817E - 043e^{\text{sand}} - 0.08877\frac{1}{\text{clay}} + 0.004372\text{Ln}(\text{orgC})\right)^2 \quad (R^2 = 0.04) \tag{31}$$

the selected transformations of the moisture retention parameters of the van Genutchen moisture retention equation and measures of their performance in validation based on the statistical measures indicated. The transformed model parameter that yielded the best reliability is the saturated soil moisture content θ_s^{-1} (considering the measure of coefficient of determination ($R^2 = 0.76$). This indicates that θ_s is reasonably well predicted by the pedotransfer function developed. Prediction of transformed variable $\sqrt{\alpha}$ yielded the lowest possible value for the selected measure for the quality of fit (r = 0.25) reflecting the lowest level of reliability compared to the other pedotransfer functions developed for moisture retention parameters under consideration.The transformed variable for residual moisture content e^{θ_r} yielded modest performance with the best possible value of $R^2 = 0.41$.

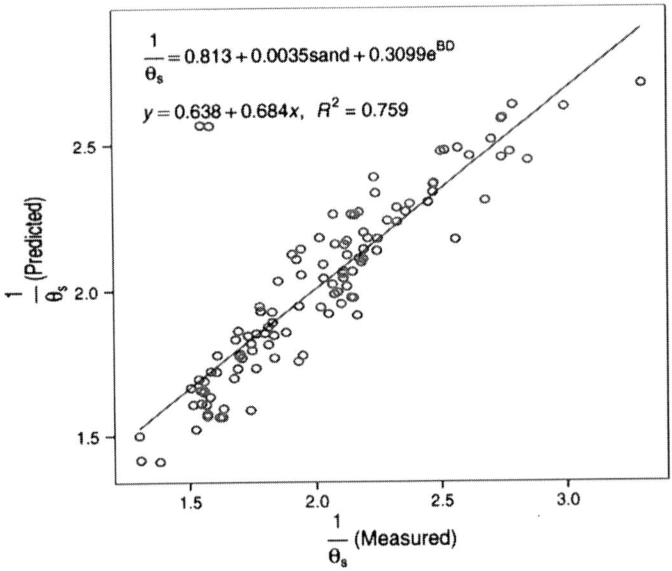

Figure 4. Predicted value of transformed variable θ_s^{-1} versus measured by correlation for a developed equation that yielded the best possible quality of fit.

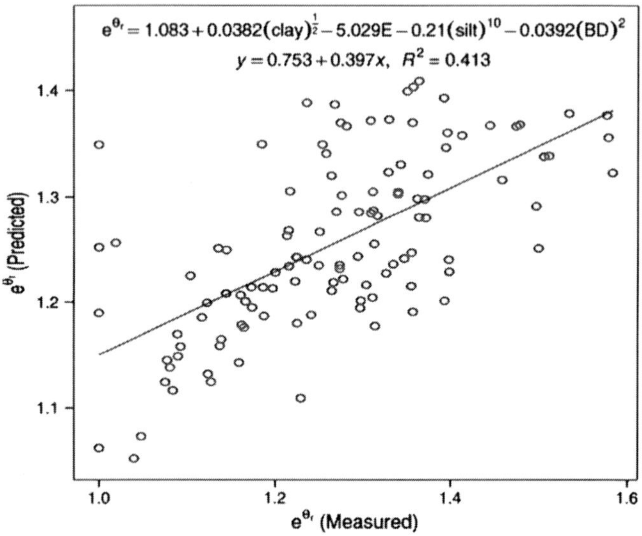

Figure 5. Predicted value of transformed variable e^{θ_r} versus measured by correlation for a developed equation that yielded the best possible quality of fit.

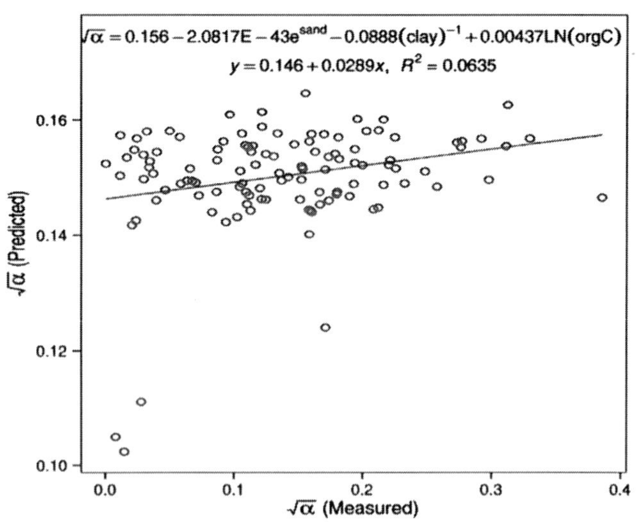

Figure 6. Predicted value of transformed variable $\sqrt{\alpha}$ versus measured by correlation for a developed equation that yielded the best possible quality of fit.

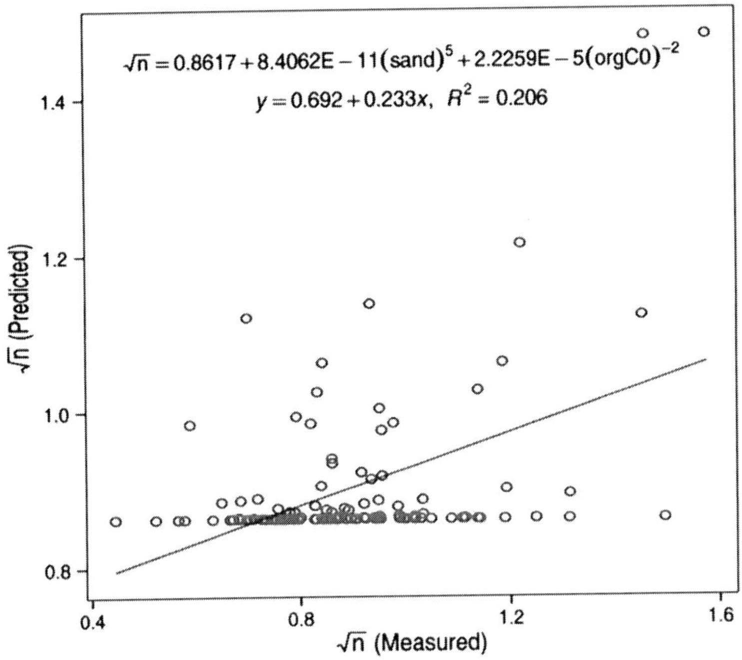

Figure 7. Predicted value of transformed variable $\sqrt{n}$ versus measured by correlation for a developed equation that yielded the best possible quality of fit.

[4] notes that continuous pedotransfer functions can be applied in the case of more site specific applications, where measured data is available, since they do not provide site specific information. The pedotransfer functions predict soil hydraulic characteristics for broadly defined textural classes. Figures 4-7 illustrates the performance of the developed pedotransfer functions for the transformed response variables based on analysis by linear regression.

4. CONCLUSION AND RECOMMENDATION

The performance of developed pedotransfer functions for estimating moisture retention parameters in the van Genuchten moisture retention equation varied for the individual moisture retention parameters. Equations developed for predicting saturated soil moisture content (θ_s) yielded the best performance in model validation with the highest possible value of coefficient of determination $R^2 = 0.76$, RMSE = 2.09 and a NSE of 0.75. The relatively poorest performance was observed with the parameter "α" with the best possible value of R^2 being 0.06 indicating unsatisfactory performance and a RMSE value of 0.85. The remaining parameters (θ_r and "n") registered performance intermediate between the aforementioned two extremes. It is therefore evidently not possible to obtain excellent performance in pedotransfer functions for all the parameters. The equations derived from these parameters would be the best possible obtainable for estimating hydraulic conductivity. Continued effort should be made in developing more pedotransfer functions to provide users with more alternative equations and especially those PTFs that use the very basic soil properties available in many local soils data bases.

5. ACKNOWLEDGEMENTS

I wish to acknowledge the support provided by the Deans Committee Research Grant, University of Nairobi (UoN) who funded the research project, The Principal's office College of Architecture and Engineering (UoN) who funded journal publication fees. I also wish to acknowledge support given by staff of the department of Environmental and Biosystems Engineering (UoN) who contributed useful academic ideas that enriched this document.

REFERENCES

1. H. R. Fooladmad, "Pedotransfer Functions for Point Estimation of Soil Moisture Characteristic Curve in Some Iranian Soils," African Journal of Agricultural Research, Vol. 6, No. 6, 2011, pp. 1586-1591.
2. J. P. O. Obiero, "Evaluation of Infiltration Using Green Ampt Model and Rainfall Runoff Data for Lagan and Sambret Catchments, Kericho, Kenya," M.Sc. Thesis, Department of Agricultural Engineering, University of Nairobi, Nairobi, 1996.
3. A. Rasoulzadeh, "Estimating Hydraulic Conductivity Using Pedotransfer Functions," In: L. Elango, Ed., Hydraulic Conductivity Issues, Determination, and Applications, Intech, 2011, pp. 145-164.

4. K. E. Saxton and W. J. Rawls, "Soil Water Characteristic Estimates by Texture and Organic Matter for Hydrologic Solutions," Soil Science Society of America Journal, Vol. 70, No. 5, 2006, pp. 1569-1578.

5. M. G. Schaap and F. J. Leij, "Using Neural Networks to Predict Soil Water Retention and Soil Hydraulic Conductivity," Soil and Tillage Research, Vol. 47, No. 1-2, 1998, pp. 37-42.

6. J. Tomasella and M. Hodnett, "Pedotransfer Functions in Tropical Soils," In: Y. Pachepsky and W. J. Rawls, Eds., Development of Pedotransfer Functions in Soil Hydrology. Developments in Soil Science, Vol. 30, Elsevier B. V., Amsterdam, 2004, pp. 415-429.

7. H. Vereecken and M. Herbst, "Statistical Regression," In: Y. Pachepsky and W. J. Rawls, Eds., Development of Pedotransfer Functions in Soil Hydrology. Developments in Soil Science, Vol. 30, Elsevier B. V., Amsterdam, 2004, pp. 3-19.

8. R. T. Walczack, F. Moreno, C. Slawinski, E. Fernandez and J. L. Arrue, "Modelling of Soil Water Retention Curve Using Soil Solid Phase Parameters," Journal of Hydrology, Vol. 329, No. 3-4, 2006, pp. 527-533.

9. T. Wells, S. Fityus, D. W. Smith and H. Moe, "The Indirect Estimation of Saturated Hydraulic Conductivity of Soils, Using Measurements of Gas Permeability. I. Laboratory Testing with Dry Granular Soils," Australian Journal of Soil Research, CSIRO Publishing, 2006. http://www.highbeam.com

10. J. H. Wosten, M. Y. A. Pachepsky and W. J. Rawls, "Pedotransfer Functions: Bridging the Gap between Available Basic Soil Data and Missing Soil Hydraulic Characteristics," Journal of Hydrology, Vol. 251, No. 3-4, 2001, pp. 123-150.

11. N. H. Batjes, Ed., "A Homogenized Soil Data File for Global Environmental Research: A Subset of FAO, ISRIC, and NRCS Profiles (Version 1.0)," Working Paper and Preprint 95/10b, International Soil Reference and Information Centre, Wageningen, 1995. http://www.isric.org/NR/exeres/545B0669-6743-402B-B79A-DBF57E9FA67F.htm

12. M. van Genutchen, "A Closed Form Equation for Predicting the Hydraulic Conductivity of Unsaturated Soils," Soil Science Society of America Journal, Vol. 44, No. 5, 1980, pp. 892-898.

10
CHAPTER

A MODIFIED VAN GENUCHTEN-MUALEM MODEL OF HYDRAULIC CONDUCTIVITY IN KOREAN RESIDUAL SOILS

ABSTRACT

According to the Mualem capillary model, hydraulic conductivity (HC) is integrated theoretically from the function related to soil water retention curves (SWRC). On the other hand, based on the smooth type of SWRC, the predicted HC function decreases abruptly near saturation, which often challenges the stability of numerical solutions. To improve the Mualem HC, van Genuchten's function for SWRC was modified within the range of low matric suction. The van Genuchten-Mualem HC was then modified to integrate the proposed SWRC for each interval decomposed by a tangential curve. The analytical solutions of the modified HC were derived to prevent an abrupt decrease near saturation. The SWRC and HC data were acquired from laboratory tests for unsaturated soils sampled from five areas in Korea. The results of the HC tests were compared with the theoretical HC models using both the van Genuchten SWRCs and the modified curves. For fine grained soils, the modified model predicts a saturated HC at very small suctions. Furthermore, the modified model was shown to accurately predict the unsaturated behavior of the HC functions for Korean weathered soils.

Keywords: hydraulic conductivity; soil water retention curve; infiltration; capillary model; unsaturated soils

1. INTRODUCTION

In variably saturated soils, the hydraulic conductivity is a crucial factor for understanding the hydrological behavior due to infiltration or evaporation. Experimental measurements are a very reliable way of estimating the unsaturated conductivity for various saturated conditions but the process is cumbersome. Consequently, for decades, indirect approaches have been developed to predict the hydraulic conductivity functions from the water retention characteristics [1,2,3,4,5]. Among the various soil water retention curves (SWRCs), the van Genuchten's model can describe the normalized hydraulic behavior successfully from two parameters and provide conductivity by a closed-form equation based on Mualem theory [3].

The Mualem conductivity function, however, is extremely sensitive near saturation when smooth retention curves are used [6,7]. Small changes in the shape of the soil water retention curve near saturation can lead to significantly different results regarding the unsaturated hydraulic conductivity (HC) function based on the Mualem model. Furthermore, for fine grained soils, the HC decreases abruptly below a very small matric suction and the HC function becomes extremely steep under conditions close to saturation. From unsaturated flow analysis, the numerical solution can be highly nonlinear because of the steepness of the hydraulic conductivity [8,9]. Very steep functions in the HC can create difficulties with convergence and the solution will diverge and oscillate near saturation [10,11].

To reduce the steepness of the HC, the original fit of the van Genuchten's curve can be modified to a non-smooth curve at a small suction [7,12,13]. Vogel et al. [7] suggested a modified SWRC to van Genuchten's equation, in which a small capillary pressure introduces a break in the SWRC fit. On the other hand, the suggested fit of the SWRC requires additional effort to define the apparent air-entry pressure (e.g., −0.2 kPa according to Vogel et al. [7]) and replace the maximum water content with a fictitious value greater than the saturated one. In this study, to overcome such shortcomings, a modification of the SWRC is proposed near saturation, but the same parameters are used as the original fit of the van Genuchten's model. The Mualem HC is derived as an analytical solution by integrating the proposed SWRC to preserve the stability of the solutions in infiltration analysis.

In the mountainous terrain of South Korea, the rock layer is weathered heavily and the upper layer near the surface is composed of residual soil. Consequently, the hydraulic behavior needs to be assessed frequently in geotechnical planning and construction because of rainfall infiltrating the variably saturated layers in the residual soils [11]. The modified model of the hydraulic conductivities was examined from experimental evidence obtained from residual soils in Korea. Laboratory tests on both the water retention and HC behaviors were carried out for multiple sites. Based on the fit of the water retention curves, the HC was predicted by both the van Genuchten-Mualem model and the modified model. The modified model reduced the slope of HC near saturation and also predicted the actual HC accurately.

2. THE RELATIONSHIP BETWEEN SOIL WATER RETENTION AND HYDRAULIC CONDUCTIVITY

The soil water retention curve defines the relationship between the measures of the pore water volume and suction in the soil matrix. The matric suction p is defined as the difference between the air and water pressures in the pore. The variable for the pore water content is commonly the volumetric water content θ or effective saturation Θ , which is defined as

$$\Theta = \frac{\theta - \theta_r}{\theta_s - \theta_r} \tag{1}$$

where θ_s is the saturated water content and θ_r is the residual water content. According to the van Genuchten (VG) model (1980), the SWRC is defined by the relation of Θ and p as follows:

$$\Theta = \left[\frac{1}{1 + \{\alpha p\}^n} \right]^m \tag{2}$$

$$p = \frac{1}{\alpha} \left[\frac{1}{\Theta^{1/m}} - 1 \right]^{1/n} \tag{3}$$

where α is the inverse of the air-entry pressure, pb, n is the parameter on the pore size distribution and m = 1 – 1/n.

According to Mualem [2], the relative hydraulic conductivity function, K r , can be derived as follows:

$$K_r(\Theta) = \Theta^q \left(\int_0^\Theta \frac{d\Theta}{p} \right)^2 \Big/ \left(\int_0^1 \frac{d\Theta}{p} \right)^2$$

(4)

$$K(\Theta) = K_s K_r(\Theta)$$

(5)

where K is the unsaturated hydraulic conductivity function with respect to Θ , and K s is the saturated hydraulic conductivity. q is an empirical parameter on the pore connectivity, which is commonly assumed to be 0.5 [2].

In the integrand of Equation (4), the inverse of p is a function of Θ , which is obtained from the equation on the SWRC. Substituting Equation (3) into (4), the van Genuchten-Mualem (VGM) model on the hydraulic conductivity can be derived as follows [3]:

$$K_r(\Theta) = \Theta^q\ f(\Theta)^2 / f(1)^2$$

(6)

$$f(\Theta) = \alpha \left\{ 1 - (1 - \Theta^{1/m}) \right\}^m$$

(7)

Figure 1 shows the hydraulic characteristics according to Mualem theory for a van Genuchten's SWRC (pb = 2.43 kPa, n = 1.34, q = 0.5). As shown in Figure 1a, in the domain of a sufficiently small p, the pores are almost saturated and the water content varies insignificantly in both the SWRC data and VG fit. In contrast, the slope of the SWRCs near full saturation has a significant effect on integrating the unsaturated HC in Equation (4). As shown in Figure 1b, when Θ approaches 1, the value of 1/p increases to infinity in the smooth SWRCs, such as the VG curves. According to Equation (4), Kr (Figure 1c) is reduced abruptly in the VGM model and is much less than 1.0 (saturated value) near saturation or at very low suction. In particular, the decrease in Kr is drastic in the VGM model, when n < 1.5 for silty soils [7]. Numerical solutions using such HCs can frequently result in oscillations or divergence according to the steep gradient near saturation. Hence, the VGM model tends to estimate the HCs differently, even with extremely small variations in the shape of the SWRCs near saturation and it does not provide a unique HC for a set of SWRC data [14].

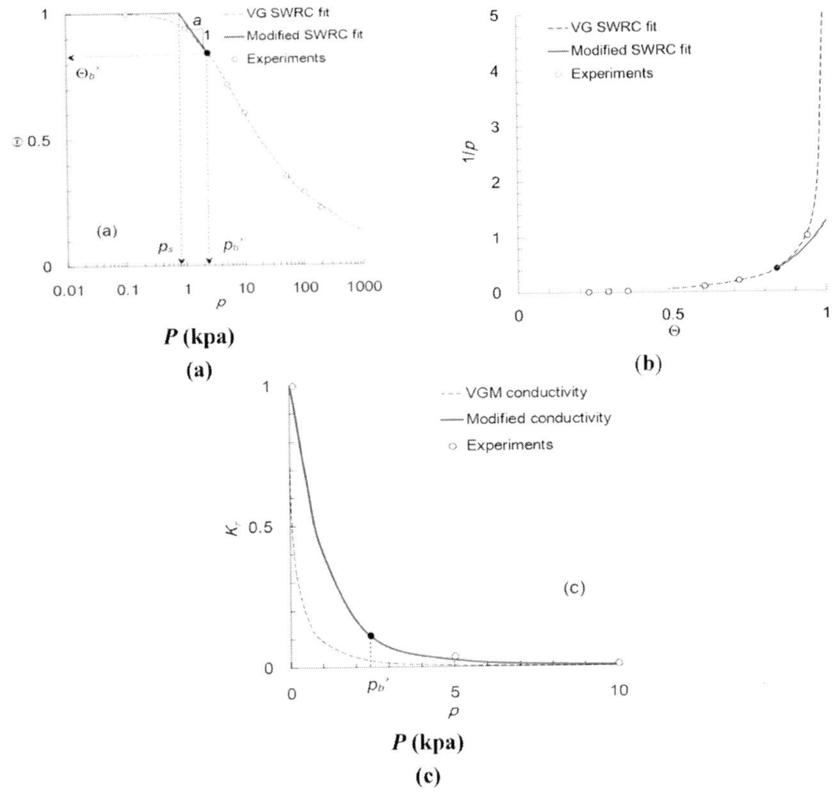

Figure 1. Comparison of the soil water retention curves and hydraulic conductivities in both van Genuchten-Mualem theory and its modification: (a) soil water retention curves; (b) 1/p-Θ relationships; and (c) hydraulic conductivities.

To improve the VGM HC function, Vogel et al. [7] introduced a break near saturation, which serves the same purpose as ps in the non-smooth curve (Figure 1a). In the region where $p > ps$, the SWRC should be fitted again to the experimental results using a fictitious water content greater than the saturated one. Therefore, the parameters α and n of the SWRCs are different from those of the original fit in the VG SWRC. Furthermore, ps is determined to be a small value to retain the nonlinear form of the VG SWRC, but is difficult to assess uniquely.

pb is the inverse of α in the van Genuchten's parametrization, which is evaluated from the curve fit of the SWRC data. In Figure 1a, pb′ is an arbitrary estimate from which the curve is extrapolated tangentially on the log p scale to full saturation, which is less than or equal to pb. Θb′ is the effective saturation corresponding to pb′ in the VG fit of Equation (2). The VG curve is linearized in the logarithmic p axis by the tangential line from pb′ to ps. ps is the value of the matric suction on the tangential line when Θ = 1. The modified VG SWRC has a partially linear shape on the logarithmic scale, where p is between ps and pb′ and follows the VG curve smoothly in Equation (2), where p > pb′. The modified VG model on SWRC in Figure 1a is derived as follows:

$$p = \frac{1}{\alpha}\left[\frac{1}{\Theta^{1/m}} - 1\right]^{1/n},\ \Theta < \Theta_b{}' \qquad (8)$$

$$p = p_b{}'\exp\{a(\Theta - \Theta_b{}')\},\ \Theta_b{}' \le \Theta < 1 \qquad (9)$$

$$0 \le p < p_s,\ \Theta = 1 \qquad (10)$$

where

$$p_s = p_b{}'\exp\{a(1 - \Theta_b{}')\},\ a = \frac{\left\{1 + (\alpha p_b{}')^n\right\}^{m+1}}{-mn(\alpha p_b{}')^n} \qquad (11)$$

Equation (8) is the same as the VG function (2b) when $\Theta<\Theta_b{}'$, and Equation (9) is a linear function in the ln p-Θ axis when $\Theta_b{}' \le \Theta < 1$. At full saturation, Θ = 1 when p ≤ ps. Θb and Θb′ are the effective saturations at pb and pb′, respectively, calculated from Equation (2) of the VG model. In Equations (8–11), the modified SWRC is established by two parameters, α and n, of the VG SWRC, but it also needs to define pb′.

Figure 1b shows the function of 1/p with respect to Θ, in the modified SWRC according to Equations (8)–(11). For $\Theta<\Theta_b{}'$, the function is the same as the VG curve, which leads smoothly to the modified curve in the region, $\Theta_b{}' \le \Theta < 1$, according to Equation (9). When Θ = 1, the function of 1/p is a vertical line according to Equation (10) with a break of 1/ps. In the integration of Equation (4), the function of 1/p is integrated for each interval of Θ. Substituting Equations (8)–(11) into Equation (4), the relative HC is derived as follows:

$$K_r(\Theta) = \Theta^q f(\Theta)^2 / \{f(\Theta_b') + g(1)\}^2,\ \Theta < \Theta_b' \quad (12)$$

$$K_r(\Theta) = \Theta^q \{f(\Theta_b') + g(\Theta)\}^2 / \{f(\Theta_b') + g(1)\}^2,\ \Theta_b' \le \Theta < 1 \quad (13)$$

$$K_r(\Theta) = 1,\ \Theta = 1 \quad (14)$$

where

$$g(\Theta) = \frac{1}{a}(1/p(\Theta) - 1/p_b') \quad (15)$$

Detailed derivation of Equations (8)–(15) is given in the Appendix. Equations (12)–(15) show the analytical solution of HC for the modified VGM model. The function of the modified SWRC is integrated to Equations (12) and (13). For full saturation (Θ = 1), Kr = 1 (Equation (14)) according to the definition of Equation (5). Figure 1c shows that the modified HC varies gradually without any abrupt decreases in the small matric suction near saturation. The difference between the SWRCs in Figure 1a is very small but the value of Kr from the modified SWRC is much greater than that of the VGM model.

According to Equations (12)–(15), the modified Kr is deduced from the parameters, α and n, of the VG SWRC. In addition, pb′ should be determined using a sufficiently small value so that the modified SWRC can fit the actual shape of the SWRC in the experiments equivalently to the original VG model. Figure 2 shows the effect of pb′ on the SWRCs and Kr functions (pb = 1 kPa, n = 1.2, q = 0.5), in which the maximum value of the p axis is equal to pb or 1kPa for comparison in a small p value. In Figure 2a, the modified SWRCs were compared with the VG model for three cases: pb′ = pb (1 kPa), pb/5 (0.2 kPa) and pb/50 (0.02 kPa). Note that ps is 0.29 kPa, 0.081 kPa and 0.0087 kPa, respectively. All modified SWRCs have the same VG curve in the greater matric suction than pb′. As shown in Figure 2b, the Kr for the VGM model was reduced significantly, by 50%, at a matric suction of 0.001 kPa, and the overall values of Kr are less than the modified HCs. In very small matric suction, the modified VGM model prevents abrupt changes in Kr and retains the saturated Kr of 1.0. For the higher pb′, the value of Kr is greater than that for pb′ = 0.02 kPa.

To compare with the Kr function of Vogel et al. [7], ps is estimated by 0.2 kPa, and α and n of the SWRCs are assumed to be the same as the original fit in the van Genuchten's model. The modification by Vogel et al. [7] revealed a non-smooth shape at the break pressure both in the SWRC (Figure 2a) and the HCF (Figure 2b). On the other hand, the proposed Kr functions show a smooth shape, even though the modified SWRCs are not smooth.

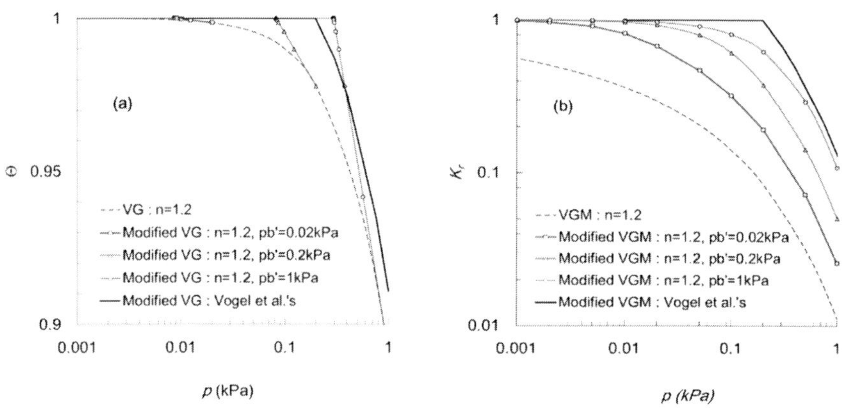

(a) (b)

Figure 2. Effect of linearization by the modified SWRCs on Kr functions (pb = 1 kPa, n = 1.2, q = 0.5): (a) soil water retention curves; (b) hydraulic conductivities.

As pb′ is defined at a tangential point to extrapolate the SWRC, it is far from the physical parameter or variable, and a parametric study is examined to find pb′. To fit the water retention behavior equivalently to the VG model, it is better to establish a smaller pb′. If pb′ is infinitesimal, Kr has the same value as that of the Mualem model based on the VG SWRC, and the numerical stability is no longer preserved. In Figure 2b, Kr at the matric suction of pb or Kr(pb) can be a measure to compare the hydraulic conductivities according to pb′, which decreases from 0.1–0.03 (0.01 for VGM) with decreasing pb′. Figure 3 shows the change in Kr(pb) as a function of pb′ when n = 1.1 and 1.5 (pb = 1 kPa, q = 0.5). When pb′ > pb/50 (or 0.02 kPa), Kr(pb) increases remarkably with respect to the logarithmic pb′. On the other hand, when pb′ < pb/50, Kr(pb) varies steadily and converges slowly to the VGMs for very small pb′(<0.001 pb). Therefore, pb′ can be assumed to be a threshold value of pb/50 to consider both the accurate fit of the experimental SWRC and the numerical stability due to the

Kr function. In the case of n = 1.5, the difference in Kr(pb) between the modified and original VGM models is less than in the case of n = 1.1.

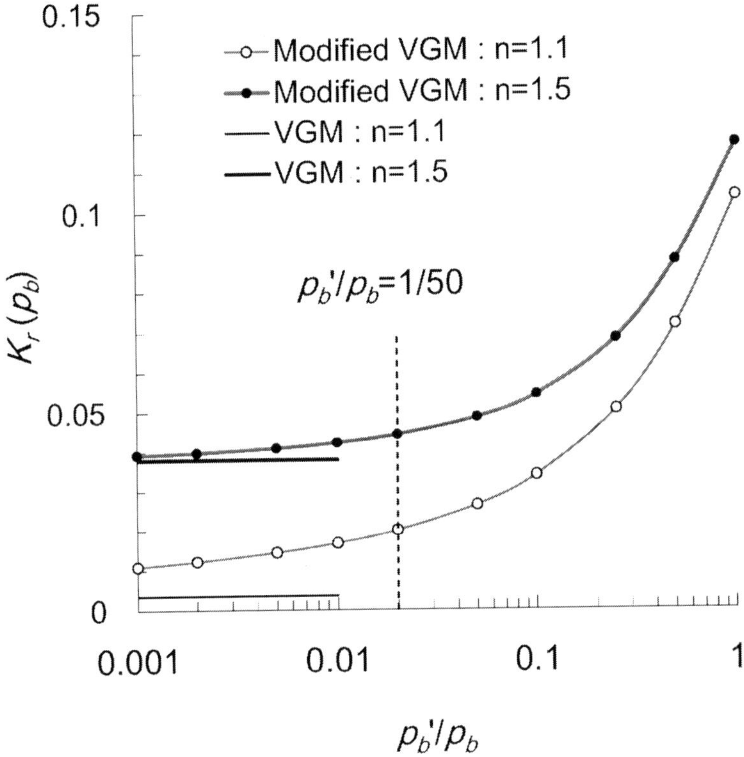

Figure 3. Variation of Kr(pb) with respect to pb′ for two cases of n = 1.1 and 1.5.

3. TEST PROCEDURE AND SAMPLE CONDITIONS

Unsaturated hydraulic conductivity tests were conducted by controlling the air pressure and water pressure independently to apply matric suction into the soil sample using triaxial cell [15,16,17]. Figure 4 shows that the disks with high air entry pressure are located at both the pedestal and cap. The high air entry disk (or ceramic disk) is saturated by submerging and applying pressure in advance. In the compartments of both the cap and pedestal, a half-spiral grooved channel induces a uniform rate of water flow into or out of the soil sample, and also plays a role as a path flushing air bubble. A half-circular groove exists along the perimeter of the cap to distribute the uniform air pressure into the soil sample.

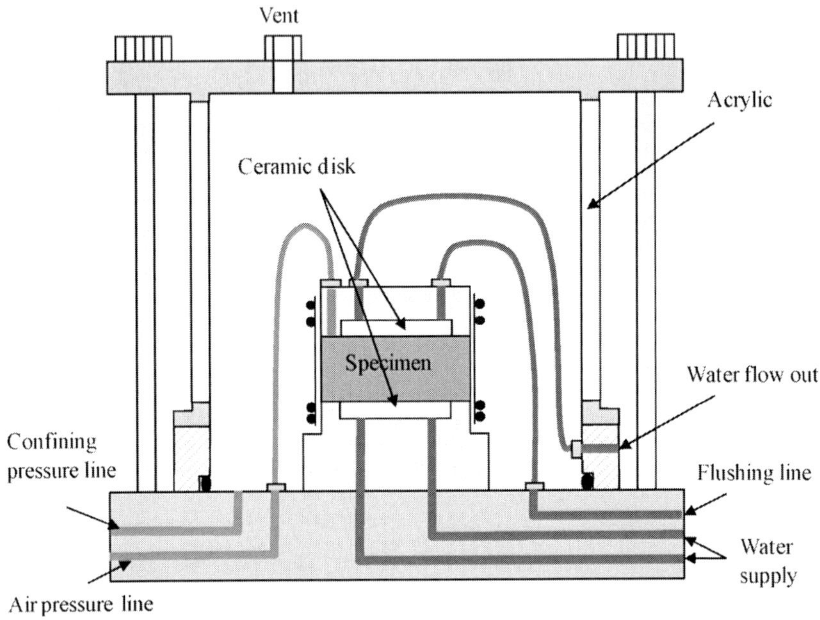

Figure 4. Apparatus to measure the hydraulic conductivities for unsaturated soils.

To saturate the samples initially, the air is removed by suction and the water is supplied to the sample. The test is started from low matric suction values at the wetness near saturation and proceeds through a drying process. The matric suction is set to a specified value by controlling the pore air pressure, whereas the confining pressure is applied using the same air pressure for zero net stress. The difference in water pressure (10 kPa) between the pedestal and cap induces a hydraulic gradient and, hence, upward water flow within the soil sample. The pressure gradient is not so small and the test is limited by the fact that the measured conductivity is unreliable at low suctions. The matric suction in the soil is determined from the air pressure and the mean water pressure. The steady state flow condition is achieved when the outflow rate is equal to the inflow rate, and the volume of water and the period of time flowing across the soil are measured. The hydraulic conductivity is then evaluated for each matric

Table 1. Geotechnical properties.

Region	Void Ratio	Dry Unit Weight γ_d (kN/m^3)	Initial Gravimetric Water Content w (%)	Saturated Hydraulic Conductivity K_s (m/s)	Specific Gravity G_s	Particles Less Than 75μm #200	Unified Soil Classification System USCS **
Asan *	0.45	16.8	–	2.47×10^{-7}	2.58	0	SP
Kimpo *	0.75	14.5	–	2.99×10^{-7}	2.65	19	SM
Okcheon 1	0.57	16.7	19.87	1.29×10^{-8}	2.67	5	SW
Okcheon 2	0.77	14.9	7.93	2.38×10^{-6}			
Seochang	0.61	16.3	20.61	9.13×10^{-7}	2.64	14	SC-SM
Yesan 1	0.58	16	19	1.39×10^{-7}	2.63	2	SW
Yesan 2	0.78	14.2	19	3.70×10^{-7}			
Yesan SP	0.95	13.5	13.84	5.00×10^{-7}	2.68	1	SP

Notes: * Undisturbed samples: ** SP = poorly graded sand. SW = well graded sand. SM = silty sand. SC = clayey sand.

suction according to Darcy's law. The saturated hydraulic conductivity is obtained independently from the falling head tests [18].

In the pressure plate extractor, the soil water retention curves during the drying and wetting procedures were measured to a maximum matric suction of 200 kPa on the pre-saturated high air entry disk. The drying branch of the soil water characteristic curve was first acquired, followed by measurements of the wetting branch. The sample size for the soil-water characteristic curve was 62 mm in diameter and 15 mm in height, and that for the hydraulic conductivity was 72 mm in diameter and 30 mm in height.

The hydraulic behavior of undisturbed and remolded samples of weathered granite was examined. Two undisturbed samples were obtained, as in Asan and Kimpo. The other disturbed samples were obtained in Okcheon, Seochang and Yesan, and compacted statically by a hydraulic jack with a controlled water content and dry density, as shown in Table 1. The Yesan SP sample was obtained from different sites compared to the Yesan 1 and Yesan 2 samples. The soil samples have void ratios varying widely from 0.47 to 0.95 (Table 1). The soils in Seochang and Kimpo contain 14% and 19% fine particles (finer than #200 sieve or 75 μm in diameter), respectively. The other samples have less than 5% fine particles.

4. COMPARISON OF THE HCS OF THE VGM MODEL WITH A MODIFIED MODEL

The parameters for SWRC (θs, θr, pb, n) were evaluated to fit the experimental data by RETC code [19] and an additional parameter q was evaluated to fit the HC data. As listed in Table 2, the hydraulic behaviors of the soils are sorted into three groups. Soils with $n < 1.5$ were classified as group A and the others were classified as group B. Group A was again sorted by two groups of group A-1 and group A-2, in which $pb \geq 5$ kPa. The samples belong to each group due to the presence of some fine-grained soil and residual formation environments. The low values of the tortuosity factor can be related to adsorptive water retention and film conductivity [20,21], which is beyond the scope of this study. For each group, the relative hydraulic conductivity, Kr, for the modified VGM model was first compared with that of the VGM model and subsequently with the measured value.

Table 2. Parameters of the van Genuchten-Mualem model for each SWRC and HC.

Group	Region	θ_s	θ_r	p_b (kPa)	n	q	R^2 for Modified HC Fit	R^2 for VGM HC Fit
A-1 ($n < 1.5$, $p_b \geq 5$ kPa)	Kimpo	0.40	0.00	10.90	1.23	−1.5	0.638	0.702
	Seochang	0.37	0.00	9.48	1.24	−2.5	0.968	0.965
	Yesan 1	0.25	0.00	16.11	1.32	−3.5	0.979	0.972
A-2 ($n < 1.5$, $p_b < 5$ kPa)	Asan	0.28	0.13	2.43	1.34	−3.5	0.990	0.977
	Okcheon 1	0.37	0.06	2.01	1.32	−2.7	0.963	0.951
	Yesan 2	0.25	0.05	1.89	1.33	−4.3	0.990	0.941
B ($n \geq 1.5$)	Okcheon 2	0.33	0.14	7.86	1.50	−3.7	0.916	0.880
	Yesan SP	0.46	0.18	1.14	1.69	−3.6	0.970	0.967

Figure 5 shows the SWRCs and HCs of group A-1 ($n < 1.5$, $pb \geq 5$ kPa): Kimpo, Seochang and Yesan 1 samples. In the VG fits, n = 1.23–1.32 and pb = 9.5–16.1 kPa, while θs = 0.25–0.40 and θr = 0. The three measured SWRCs described by the effective saturation are similar because of the similar n and pb values. In the modified SWRCs, $pb' = pb/50$ or 0.19–0.32 kPa. The modified VG model fitted differently to the VG model just in the region where the matric suction is less than pb′. Figure 5a shows the modified VG fit when $p < pb'$. In the SWRCs of Figure 5a, the VG model fits the measured data quite accurately on three samples of group A-1 and both models have an equivalent ability to fit the experimental SWRCs.

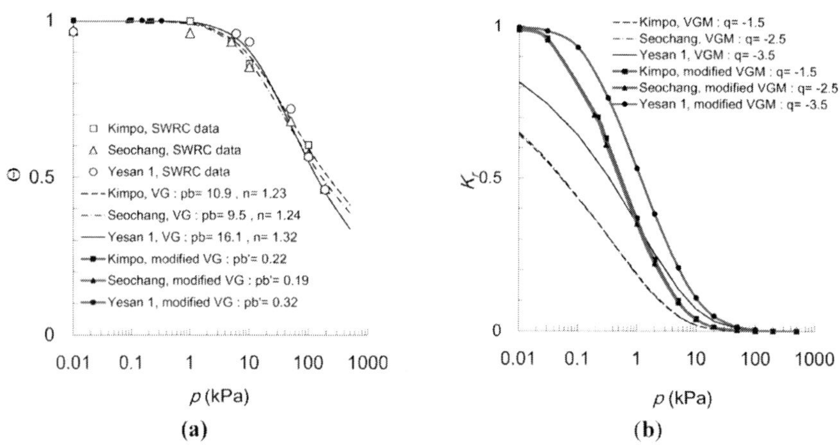

Figure 5. Comparison of the VGM model with the modified model in Group A-1: (a) soil water retention curves (the measured data are shown by points, and the modified VG curve by lines with point when $p < pb'$); (b) hydraulic conductivities.

Kr was calculated using Equations (12)–(15), and the modified models were compared with the VGM HC models on Kimpo, Seochang and Yesan 1 samples (Figure 5b). The Kr function was calculated from the SWRC curve. The parameter q was evaluated to fit HC data as −3.5––1.5. The SWRCs in Figure 5a incorporates the small difference in a suction less than pb′ but Kr for the modified models is clearly greater than that for the VGM models. Near saturation, Kr = 1 in the modified VGM model, whereas Kr is reduced at a suction less than 0.01 kPa in the VGM model. The difference between the two models is 18%–37% at 0.01 kPa, which increases with decreasing n.

As shown in Figure 6, the SWRCs for group A-2 (n < 1.5, pb < 5 kPa) are used to predict the relative HC in Asan, Okcheon 1 and Yesan 2 samples. Both the VG model and the modified VG model can fit the measured SWRCs accurately, in which the parameters of the three samples to fit SWRCs are quite similar; n = 1.32–1.34, pb = 1.9–2.4 kPa and pb′ = 0.04–0.05 kPa. θs, θr and q were evaluated to be 0.25–0.37, 0.05–0.13 and −4.3−−2.7, respectively. In Figure 6b, the VGM models for the three samples also showed an abrupt decrease in Kr, which were less than the modified VGM model by ~25% at suction of 0.01 kPa. Because pb in group A-2 is smaller than that in group A-1, the gradient of the Kr function to the matric suction in Figure 6b is greater than that in Figure 5b.

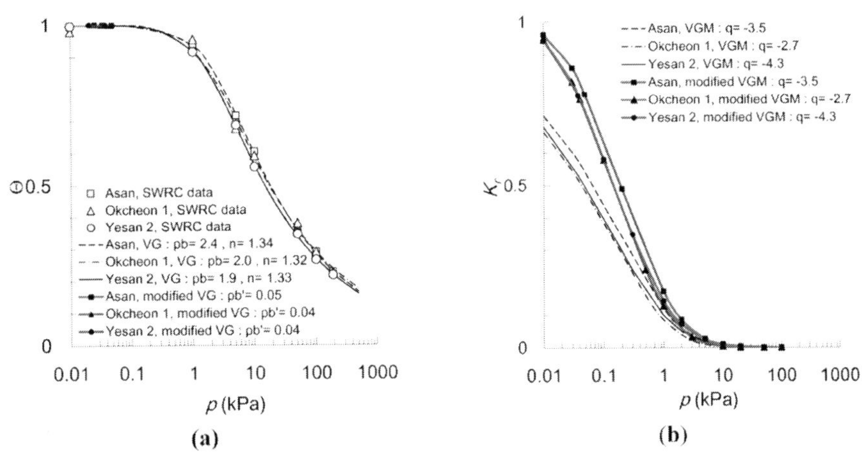

Figure 6. Comparison of the VGM model with the modified model in Group A-2: (a) soil water retention curves (The measured data are shown by points, and the modified VG curve by lines with point when p < pb′); (b) hydraulic conductivities.

As shown in Figure 7, the SWRCs (Figure 7a) for group B (n ≥ 1.5 are shown with the relative HC (Figure 7b) in Okcheon 2 and Yesan SP samples. Both the VG model and the modified VG model can fit the measured SWRCs accurately; n = 1.50 and 1.69, pb = 1.1 and 7.9 kPa, θs = 0.33 and 0.46, θr = 0.14 and 0.18, and q = −3.7−−3.6, respectively. The data on the SWRCs were measured within a matric suction less than 200 kPa, which is far from the actual suction at the residual water content. Therefore, θr (Table 2) increased with increasing gradient of the SWRCs or n to optimize the SWRC fit. Near saturation, Kr ≈ 1 in both the modified VGM model and the VGM model, where Kr is reduced by 10% at most in group B. Therefore, the VGM model estimates Kr similar to the modified VGM model.

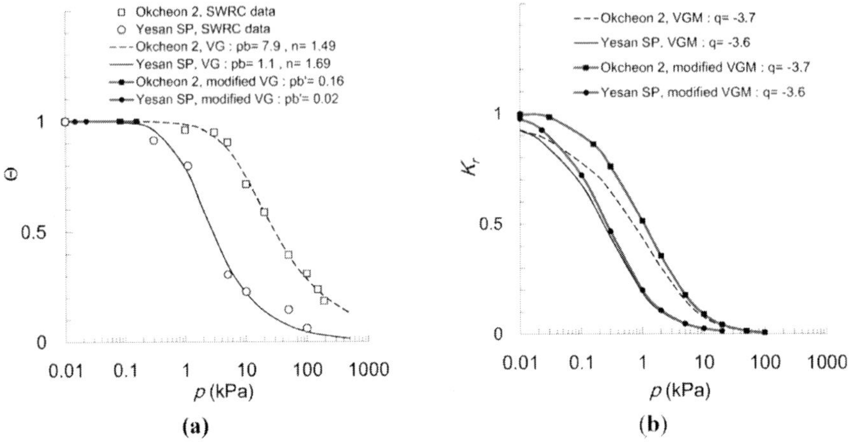

Figure 7. Comparison of the VGM model with the modified model in Group B: (a) soil water retention curves (The measured data are shown by points, and the modified VG curve by lines with point when $p < pb'$); (b) hydraulic conductivities.

In groups A-1 and A-2, the prediction of Kr by the VGM model decreases abruptly near saturation. In such cases, the VGM model induces abnormal oscillation or divergence of the solutions in the numerical analysis near saturation. On the other hand, in group B, i.e., $n > 1.5$, there was a smaller decrease in Kr in the initial part of the VGM model.

5. COMPARISON OF HCS OF THE MODIFIED VGM MODEL WITH THE MEASURED DATA

Mualem's model [2] incorporates an empirical parameter, q, accounting for the tortuosity and connectivity to predict the actual hydraulic conductivity, which is commonly assumed to be 0.5. On the other hand, many studies examined a range of values of the tortuosity parameter in the fit of the actual data [4,22,23,24], which is frequently negative [25]. The parameter, q, was estimated to be −1.5−−4.3 (−3.2 in average) in Table 2, which was obtained directly from the experimental data for Korean residual soils.

Figure 8 shows the Kr predicted by both the original and the modified VGM models for group A-1 (n < 1.5, pb ≥ 5kPa), where q was −2.5 on average. The goodness of fit of both models are compared in terms of the correlation coefficient R2 for log Kr (Table 2). Using the modified model, the measured Kr was predicted quite accurately for Yesan 1 sample (n = 1.32, R2 = 0.98 in Table 2). Because the SWRCs of Kimpo (n = 1.23) and Seochang (n = 1.24) samples are similar, as shown in Figure 5a, the measured data of the unsaturated HC were predicted well as similar curves using the modified VGM model (R2 = 0.64 for Kimpo and R2 = 0.97 for Seochang). The VGM model could predict the measured HC equivalently to the modified model (R2 = 0.70–0.97 for group A-1), but underestimated the nearly saturated Kr (~1.0) and subsequently the unsaturated part of the Kr data compared to the modified model.

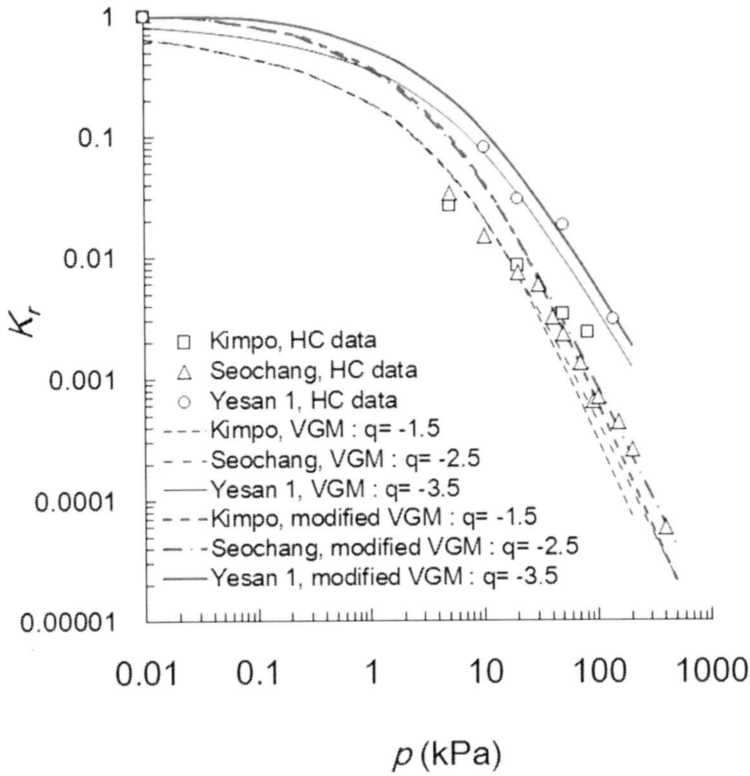

Figure 8. Comparison of the measured hydraulic conductivities with the modified model in Group A-1.

In group A-2 (n < 1.5, pb < 5 kPa) in Figure 9, the measured Kr was also predicted quite well by the modified model for Asan (R2 = 0.99 in Table 2), Okcheon 1 (R2 = 0.96) and Yesan 2 (R2 = 0.99) samples. The VGM model could predict the measured HC fairly (R2 = 0.94–0.98 for group A-2), but underestimated the saturated and the unsaturated parts of the Kr data compared to the modified model. Regarding the effects of q on Kr, Kr increased with decreasing q. Group A-2, which has a smaller pb, q (−3.5 on average), has a much stronger effect on HC than Group A-1. Therefore, q contributes significantly to the prediction of Kr. Yesan 2 sample in Group A-2 (Figure 9) is compacted by the looser conditions and shows a higher Kr for both the measurements and predictions than the Yesan 1 sample in Group A-1 (Figure 8). Okcheon 2 sample is also looser than Okcheon 1 sample in Group A-2 (Figure 9) and belongs to Group B (Figure 10).

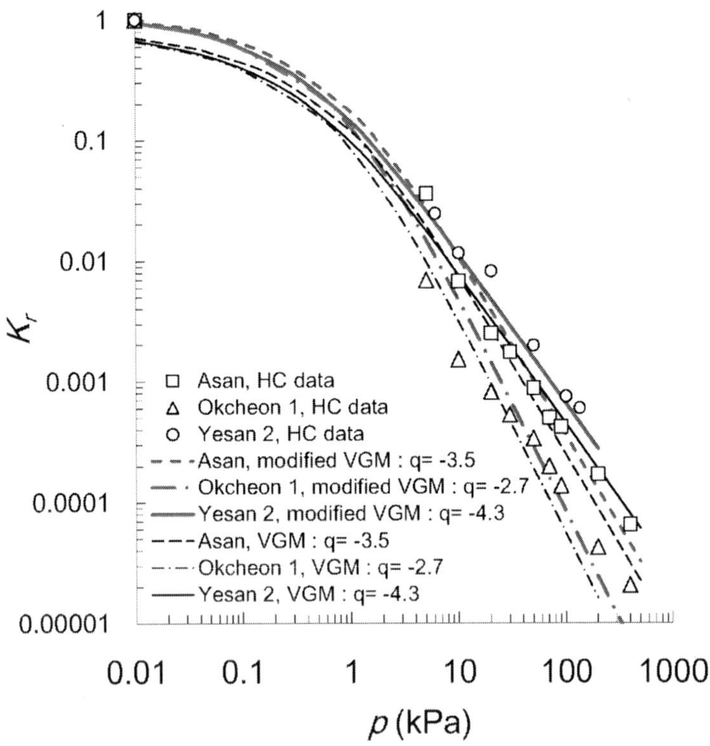

Figure 9. Comparison of the measured hydraulic conductivities with the modified model in Group A-2.

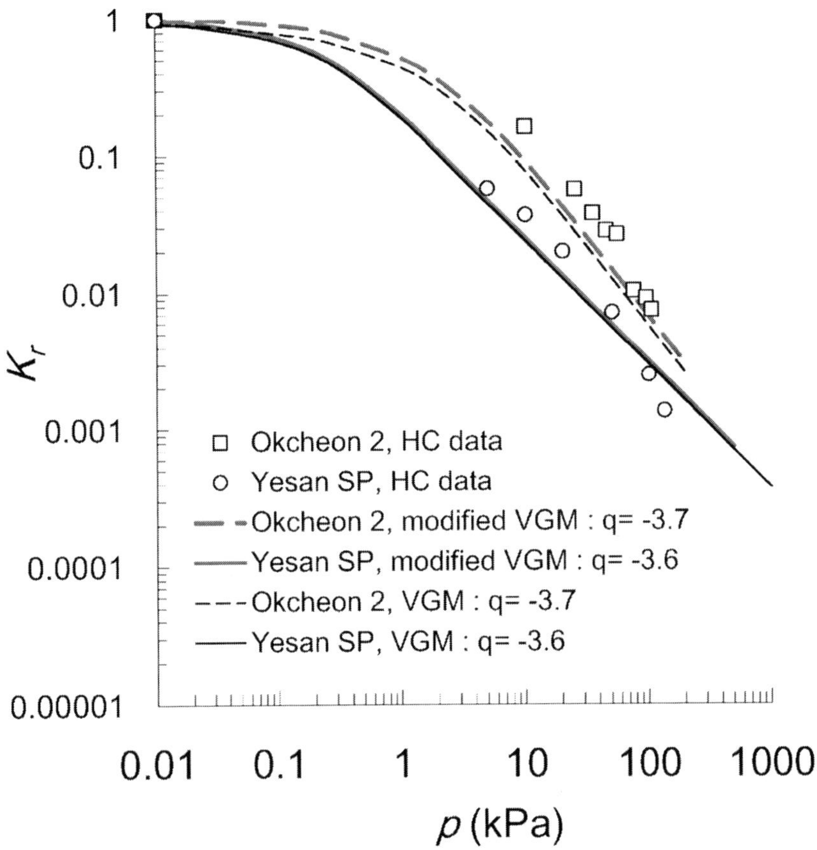

Figure 10. Comparison of the measured hydraulic conductivities with the modified model in Group B.

In group B (n ≥ 1.5) in Figure 10, the modified VGM model predicts the measured Kr reasonably well for Okcheon 2 (R2 = 0.92) and Yesan SP samples. The VGM model could predict similarly to the modified model (R2 = 0.88–0.97 for group B) because n > 1.5, and Kr ~ 1.0 in the nearly saturated state. As shown in Equation (4), Kr is multiplied by Θq, where Θ is deduced from the SWRC of each group. When q is negative, Θq > 1.0 and increases with increasing n. Group B showed the strongest effect of q (−3.6 in average) on HC than Groups A-1 and A-2, because n > 1.5.

6. CONCLUSIONS

The van Genuchten's SWRC was modified near saturation and an improved model of the hydraulic conductivity was proposed by the analytical solutions based on Mualem theory. The modified model preserved both the accuracy of the SWRC fit to the measurement and the stability of the numerical solutions in infiltration analysis using the unsaturated HCs. The modified HC model included the same parameters as van Genuchten's SWRCs (pb and n) and Mualem's HC functions (q), but an arbitrary suction, pb', was also estimated to be pb/50 to determine the tangential point in van Genuchten's SWRC.

The experimental evidence of the SWRC and HC for a total of eight samples was examined and categorized by three groups to verify the HC model. In the SWRCs, both the VG model and the modified model fitted the measured data quite well on all samples of each group because the modified VG SWRC fitted the VG curve differently just in the region of low matric suction.

Near saturation, the calculated hydraulic conductivity in groups A-1 and A-2 when $n < 1.5$, was reduced abruptly in the VGM model but maintained the saturated value in the modified VGM model. The modified model could prevent the numerical instability induced by the VGM model near saturation.

The measured data of the unsaturated HC were predicted well using the modified VGM model. The tortuosity parameter q was evaluated directly using the experimental data for Korean residual soils.

ACKNOWLEDGMENTS

This study was supported by the Yeungnam University Research Grant (214A380204), which is greatly appreciated.

AUTHOR CONTRIBUTIONS

Seboong Oh participated in design of the study, analysis of the experimental data, and writing the manuscript. Yun Ki Kim participated in the experiments and their analysis. Jun-Woo Kim contributed to the analysis of the experimental data and preparing the manuscript.

REFERENCES

1. Campbell, J.D. Pore Pressures and Volume Changes in Unsaturated Soils. Ph.D. Thesis, University of Illinois at Urbana-Champaign, Urbana, IL, USA, 1973.

2. Mualem, Y. A new model for predicting the hydraulic conductivity of unsaturated porous media. Water Resour. Res. 1976, 12, 513–522.

3. Van Genuchten, M.T. A Closed-form Equation for Predicting the Hydraulic Conductivity of Unsaturated Soils1. Soil Sci. Soc. Am. J. 1980, 44, 892–898.

4. Fredlund, D.G.; Xing, A.; Huang, S. Predicting the permeability function for unsaturated soils using the soil-water characteristic curve. Can. Geotech. J. 1994, 31, 533–546.

5. Dourado Neto, D.; de Jong van Lier, Q.; van Genuchten, M.T.; Reichardt, K.; Metselaar, K.; Nielsen, D.R. Alternative Analytical Expressions for the General van Genuchten-Mualem and van Genuchten-Burdine Hydraulic Conductivity Models. Vadose Zone J. 2011, 10, 618–623.

6. Vogel, T.; Cislerova, M. On the reliability of unsaturated hydraulic conductivity calculated from the moisture retention curve. Transp. Porous Med. 1988, 3, 1–15.

7. Vogel, T.; van Genuchten, M.T.; Cislerova, M. Effect of the shape of the soil hydraulic functions near saturation on variably-saturated flow predictions. Adv. Water Resour. 2001, 24, 133–144.

8. GEO-SLOPE International. GeoStudio 2007; Calgary: Alberta, AB, Canada, 2007.

9. Lu, N.; Wayllace, A.; Oh, S. Infiltration-induced seasonally reactivated instability of a highway embankment near the Eisenhower Tunnel, Colorado, USA. Eng. Geol. 2013, 162, 22–32.

10. Oh, S.; Kim, Y.K. A Modification of van Genuchten-Mualem Model on the Hydraulic Conductivity to Account for Numerical Stability and Empirical Connectivity in Korean Residual Soils. In Unsaturated Soils: RESEARCH & Applications; CRC Press: London, UK, 2014; pp. 1173–1177.

11. Oh, S.; Lu, N. Slope stability analysis under unsaturated conditions: Case studies of rainfall-induced failure of cut slopes. Eng. Geol. 2015, 184, 96–103.

12. Ippisch, O.; Vogel, H.-J.; Bastian, P. Validity limits for the van Genuchten-Mualem model and implications for parameter estimation and numerical simulation. Water Resour. Res. 2006, 29, 1780–1789.

13. Schaap, M.G.; van Genuchten, M.T. A Modified Mualem-van Genuchten Formulation. Vadose Zone J. 2006, 5, 27–34.

14. Durner, W. Hydraulic conductivity estimation for soils with heterogeneous pore structure. Water Resour. Res. 1994, 30, 211–223.

15. Huang, S.; Fredlund, D.G.; Barbour, S.L. Measurement of the coefficient of permeability for a deformable unsaturated soil using a triaxial permeameter. Can. Geotech. J. 1998, 35, 426–432.

16. Kim, Y.K. Soil Slope Design and Stability Evaluation Methodology Considering Hydraulic Conductivity and Rainfall Characteristics. Ph.D. Thesis, Korea Advanced Institute of Science and Technology, Daejeon, Korea, 2009.

17. Kim, Y.K.; Choi, K.L.; Lee, S.J.; Lee, S.R.; Kwon, H.S. Estimation of Unsaturated Permeability Function from Water Retention Characteristics for Korean Weathered Soils. J. Korean Geotech. Soc. 2010, 26, 449–460. (In Korean). [Google Scholar]

18. ASTM International. Standard Test Methods for Measurement of Hydraulic Conductivity of Saturated Porous Materials Using a Flexible Wall Permeameter, D 5084. In Annual Book of ASTM Standards; ASTM International: West Conshohocken, PA, USA, 2002.

19. Van Genuchten, M.T.; Leij, F.J.; Yates, S.R. The RETC Code for Quantifying the Hydraulic Functions of Unsaturated Soils; EPA 600/2–91/065; United State Environment Protection Agency (EPA): Washington, WA, USA, 1991.

20. Peters, A. Simple consistent models for water retention and hydraulic conductivity in the complete moisture rang. Water Resour. Res. 2013, 49, 6765–6780.

21. Iden, S.C.; Durner, W. Comment on "Simple consistent models for water retention and hydraulic conductivity in the complete

moisture range" by A. Peters. Water Resour. Res. 2014, 50, 7530–7534.

22. Schaap, M.G.; Leij, F.J. Improved prediction of unsaturated hydraulic conductivity with the Mualem-van Genuchten model. Soil Sci. Soc. Am. J. 2000, 64, 843–851.

23. Peters, A.; Durner, W. A simple model for describing hydraulic conductivity in unsaturated porous media accounting for film and capillary flow. Water Resour. Res. 2008, 44.

24. Salas, J.D.; Govindaraju, R.; Anderson, M.; Arabi, M.; Frances, F.; Suarez, W.; Lavado, W.; Green, T.R. Introduction to Hydrology. In Handbook of Environmental Engineering, Volume 15: Modern Water Resources Engineering; Wang, L.K., Yang, C.T., Eds.; Humana Press-Springer Science: New York, NY, USA, 2014.

25. Peters, A.; Durner, W.; Wessolek, G. Consistent parameter constraints for soil hydraulic functions. Adv. Water Resour. 2011, 34, 1352–1365.

11

CHAPTER

DETECTION OF PERMEABLE BODIES: FROM LABORATORY MEASUREMENTS TO SEISMIC MEASUREMENTS

Jean Luc Mari[1] and Béatrice Yven[2]

[1] IFPEN, France

[2] ANDRA, France

1. INTRODUCTION

The seismic reflection method has the advantage of providing a picture of the subsurface in three dimensions (3D) with a regular grid. In high resolution seismic surveys, the size of the grid cell is about tens of meters for horizontal distances, and of several meters for vertical distances. The classical approach of seismic processing leads to obtain migrated seismic sections in 2D or migrated seismic blocks in 3D. It is indispensable to have a good velocity model to carry out the migration process. The migrated sections can then be transformed into acoustic impedance sections if well data, mainly acoustic and density logs, are available. The procedures used to obtain acoustic impedance sections are often referred to as Model-based seismic inversions which require an a priori impedance model (obtained from well data) which is iteratively refined so as to give a synthetic seismic section to match the seismic section to be inverted. The impedance model is the fruit of a processing sequence which is usually done in time. 3D seismic impedance blocks and logging data are often used for geological model building in time. The geological model must be then converted from time to depth, thanks to the computation of a time to depth conversion model.

The final impedance model can be converted into porosity by using an empirical relationship between porosity and acoustic impedance established at well locations. To model porosities, an other option is to use porosity at wells location and interpolate between the wells by means of kriging. Partly due to the small number of wells, this outcome is really smooth and usually does not seem geologically consistent. More dense information can be integrated in order to improve the estimation of porosity. As porosity is linked to acoustic impedance, it is relevant to use dense seismic acoustic impedance information. So a collocated cokriging of porosity integrating the seismic information is performed by using the normalized acoustic impedance as the secondary variable [1]. 3D cube makes possible to provide 3D imaging of the connectivity of the porous bodies [2]. Core analysis is usually carried out to establish porosity vs. permeability laws [3].

It has been shown that it is possible to extract new attributes from seismic sections, leading to a better understanding of the distribution of the porous and permeable bodies [4]. The attributes are also used to detect the impermeable layers. The methodology is based on laboratory experiments which have shown that a formation permeability indicator can be obtained via the computation of four input data: P-wave frequency, attenuation, porosity and specific surface. The procedure has been firstly conducted in acoustic logging to estimate permeability of porous layers and to detect water inflows [5]. In seismics, the processing is performed in order to measure these parameters. The analytic signal is used to compute the instantaneous frequency and attenuation (Q factor). The porosity and specific surface are computed from seismic impedances obtained by acoustic inversion of the migrated seismic sections. The input parameters are used to compute a new index named Ik-Seis factor. (Indicator (I) of permeability (k) from acoustic or seismic (Seis) data).The chapter is written mainly from two journal articles: "Detection of porous and permeable formations: from laboratory measurements to seismic measurements" [4] and "Characterization of geological formations by physical parameters obtained through full waveform acoustic logging" [5].

In a first step we show how the methodology has been transposed from laboratory measurements to geophysical data. Then we illustrate the potential of the proposed procedure via field examples: acoustic logging, 2D seismic line with acoustic impedance inversion after migration, 3D seismic section with elastic impedance inversion after migration and depth conversion.

2. FROM LABORATORY EXPERIMENTS TO GEOPHYSICAL DATA

We present a short review of the laboratory measurements conducted by P. Morlier and J.P. Sarda [6]. The possibilities of using laboratory results for field geophysical applications are then discussed.

2.1. Laboratory measurements

Laboratory experiments [6] have shown that the attenuation of a clean formation can be expressed in terms of three structural parameters: porosity, permeability and specific surface. Both theoretical and experimental studies have identified the relationship between acoustic attenuation and petrophysical parameters:

$$\delta=(C.S/\varphi).(2\pi.k.f.\rho_f/\mu)^{1/3} \tag{1}$$

with:

δ : attenuation (dB/cm), f : frequency (Hz), ρf : fluid density, μ : fluid viscosity (centipoise)

φ : porosity, S : Specific surface (cm^2/cm^3), C : calibration coefficient, k : permeability (mD).

Figure 1 is an example of laboratory measurements on sandstone core plugs. The upper part of the figure shows the results obtained on cores with constant specific surface, the lower part on cores with variable specific surface, the specific surface being estimated on the basis of the average pore radius measurement. It is necessary for computing the permeability from eq 1, to measure the attenuation of the formation and calculate the effective specific surface of the formation.

Fabricius et al. [7] have found that the specific surface with respect to grain volume (Sg) is apparently independent from porosity. In an attempt to remove the porosity effect on Vp/Vs and mimic a reflected φ vs log (Sg) trend, they propose to use the following relationship between porosity φ, Vp/Vs and Sg

$$\log(Sg.m)=a.\varphi+b.(Vp/Vs)+c \text{ with } Sg=S/(1-\varphi) \tag{2}$$

where it should be observed that Sg is multiplied by m to make Sg dimensionless. To establish eq 2, Fabricius et al. [7] have looked at ultra sonic data, porosity, and permeability of 114 carbonate core plugs.

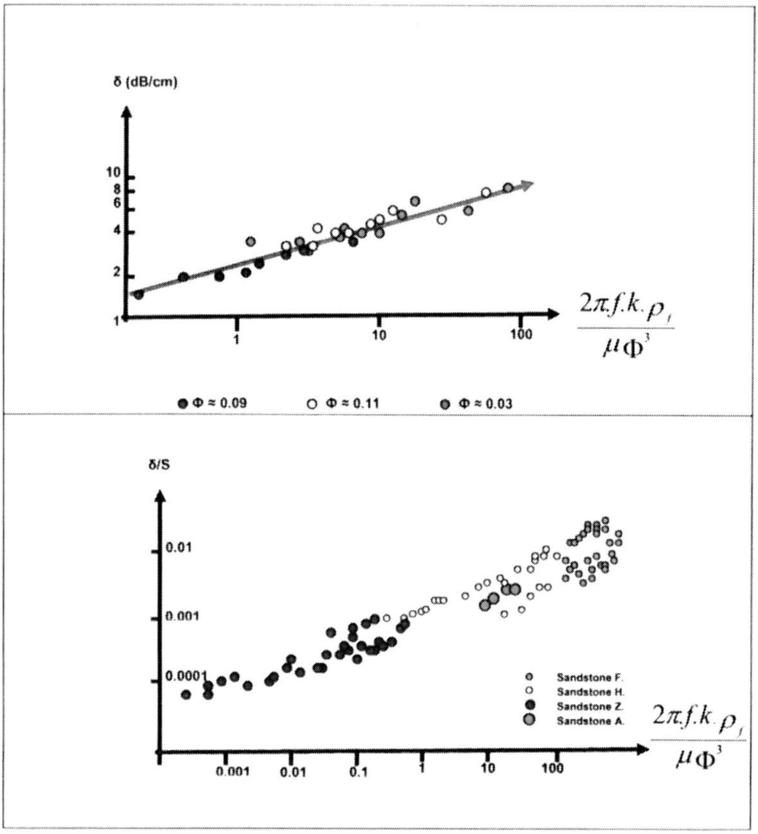

Figure 1. Relationship between attenuation and petro-physical parameters [6]. Laboratory measurements on cores with constant specific surface (top) and with variable specific surface (bottom).

An ideal porous medium made up of spheres (porosity 30%) of any size from cm (pisolites), mm (oolites) to a fraction of mm has an inverse trending specific surface that is the surface of solid exposed to the intergranular void. Fine grained well rounded sands or sandstones have a higher specific surface because the number of grains per unit volume is high. Clay content either as a coating of grains or as lumps or lenses enhances the specific surface up to a limit of effectiveness as long as pore connections prevail. The mineralogical nature of the grains, either silica (quartz or any siliceous rock debris) or carbonate is of little effect on the specific surface as long as secondary changes for instance diagenesis (for instance dolomitization), dissolution (karstification) or fracturation did not occur. Any process that simplifies the mineral surfaces exposed to the pore space tends to decrease the specific surface, whereas fractures and fissures, often occurring as networks or swarms, create new specific surfaces but very little added porosity.

Such processes remain minimal in tectonically quiet basins (Paris basin) where and when connate waters remain in equilibrium with the reservoir minerals. In subsurface exploration when no exposures more existing wells (logs, core and cutting information's) are available, it is difficult to assess the degree of transformation that a granular sedimentary reservoir formation has undergone through the elapsed geologic time.

2.2. To geophysical measurements

In practice, the parameter Ik-Seis (Indicator (I) of permeability (k) from acoustic or seismic (Seis) data) computed from equation 1 is proportional to permeability k.

$$\text{Ik-Seis} = (\varphi.\delta/S)^3/f = (\varphi/SQ)^3/f \tag{2}$$

with f : P-wave frequency, Q quality factor, δ : attenuation,

S : specific surface, φ : porosity

The natural rocks are never perfectly elastic. The viscoelastic media always exhibit a wave amplitude decline as a function of time, independently of geometrical effects. From a wave propagation point of view, this results in attenuation on the one hand, and on the other hand in a dispersion of the propagation velocity, i.e. velocity depends on wave frequency. In order to conduct calculations and measurements on these parameters, it is necessary to build a model, the best known and used being the constant Q model. Q, called quality factor, is inversely proportional to the attenuation often designated by δ or α. In the case where Q is constant with frequency, which is true in the domain of frequencies used with seismic, we have: $\alpha = \pi f/Q$, f being the frequency.

Wave propagation modelling in a saturated porous medium reveals the existence of three propagation waves: two P-waves and one S-wave. The two P-waves (P_1-wave and P_2-wave) have two different propagation velocities V_{p1} and V_{p2} ($V_{p1} > V_{p2}$)and two different characteristic particle motions corresponding to a movement in which overall and fluid displacements are in phase for the P_1-wave (fast P-wave) and out of phase for the P_2-wave (slow P-wave) [8]. The laboratory experiments conducted by Plona [9] have confirmed the Biot 's theory [8]. Frequency is very important. Two major domains, separated by a critical frequency must be distinguished. Above the critical frequency, it is possible to estimate a permeability knowing that the calculated permeability is only an approach of hydraulic permeability. As shown in figure 2, above 2 kHz, permeability has an influence on velocities and attenuations [10]. The attenuation may reach a maximum for frequency in the order of 10 kHz, which is the domain of full waveform acoustic logging. It is the reason why several authors [2, 5-7, 10-12] attempted to predict permeability from acoustic data. The historical focus has been on predicting permeability from P-wave velocity and attenuation.

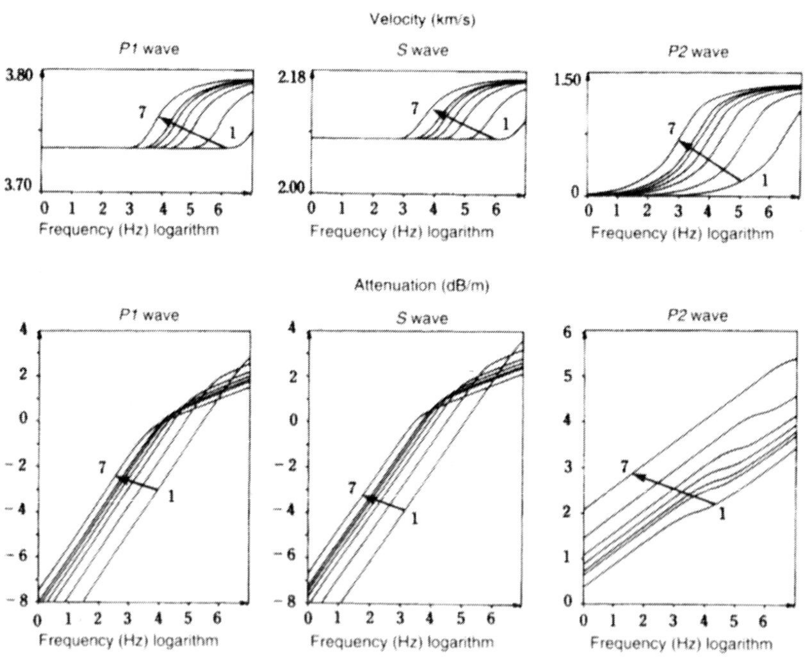

Figure 2. Theoretical sensitivities of velocity and attenuation as a function of permeability, for a 19%-porosity sandstone (Schmitt, 1986). Curves 1 to 7 were calculated for a permeability of 2, 32, 200, 500, 1000, 1500 and 5000 milliDarcys. P_1-wave and P_2-wave are respectively the fast and the slow P-waves.

The transmission of an acoustic wave through geological formations is used for formation characterisation, in the acoustic frequency domain (ranging between 1 and 25 kHz). Acoustic logging allows the measurement of the propagation velocities and frequencies of the different waves (P-wave, S-wave and Stoneley wave) which are recorded by an acoustic tool. The analysis of the acoustic waves recorded simultaneously on both receivers of the acoustic tool is used to compute additional logs, defined as acoustic attributes, useful for the characterization of the formation, such as amplitude, shape index, wavelength and attenuation logs. Singular value decomposition (SVD) filtering method can be used to attenuate the noise, to measure the attenuation and to extract the acoustic wavelets [4].

Above the critical frequency, in the domain of acoustic frequencies, the Ik-Seis factor can be seen as a pseudo- permeability log after calibration on core data or hydraulic tests. Below the critical frequency (low frequency approximation), in the domain of seismic frequencies, the Ik-Seis factor can only be seen as a relative indicator which varies from 0 for less porous and permeable bodies to 1 for more porous and permeable bodies. In the same way, the specific surface is a pseudo specific surface which varies from 0 for less shaly bodies to 1 for more shaly bodies.

The seismic data must be inverted in order to obtain seismic impedance sections. If an elastic inversion is done, it is possible to obtain the elastic impedances Ip and Is. At well location, it is usually possible to obtain cross plots between acoustic impedance and porosity φ and to define a law between the two. Usually a linear or polynomial law can be extracted. The Ip, Is and φ quantities are used to compute the seismic specific surface (equation 2). If an acoustic inversion is done, well logs must be used to define an experimental law between Ip and Is. The analytic signal [13] is computed in order to extract, from the migrated seismic section, the variation of the seismic frequency and the Q factor versus time. The instantaneous frequency gives the frequency variation versus time and the envelop decrease leads to an estimation of the Q-factor. However a high signal to noise ratio is required. For that purpose, a singular value decomposition (SVD) filtering method [13] is used to enhance the coherent reflections and to attenuate the noise. Whatever the geophysical method (acoustic logging or reflection seismic surveying), equation 3 is used to obtain the Ik-Seis factor. The Ik-Seis factor can be used to detect in acoustic logging or on seismic sections permeable and impermeable bodies. For that purpose, we need to compute four quantities : P-wave frequency f, Q factor or attenuation δ, specific surface S and porosity φ. More information concerning the data processing and analysis is given in [4] and in the field examples.

3. FIELDS EXAMPLES

2D and 3D seismic data were recorded in France at the boundary of the Meuse and Haute-Marne departments in the vicinity of the Andra Center (National radioactive waste management Agency). The acoustic data were recorded in well "EST 431" located on the Ribeaucourt township, in the national forest of Montiers-sur-Saulx, 8 km North-North-West of the Andra Center. Figure 3 indicates the location of 2D and 3D seismic surveys and of the selected seismic lines: line "07EST09" from the 2D survey and the in line 405 from the 3D survey. Figure 3 also indicates the location of the well " EST 431".

In a first step, we present and discuss the results obtained in acoustic logging. In a second step we describe those obtained with seismic data (2D then 3D).

3.1. Geological context

One of the drilling platforms, located in the centre of the studied zone, was used to study formations ranging from Oxfordian to Trias. The analysis presented here concerns borehole "EST 431" and covers the Oxfordian formation. The objective of this borehole is to complement the geological and hydrogeological knowledge of this formation. This formation consists essentially of limestone deposited in a vast sedimentary platform. The limestone facies, which vary from one borehole to another, are generally bio-detritic with reef constructions. The base of the Kimmeridgian shale was observed at –258.3 m (100 m ASL) and the base of the Oxfordian limestone at - 544.3 m (-186 m ASL).In this formation, porosity ranges between 5 and 20% and "porous horizons" of kilometric extension have been identified. As far as hydrogeology is concerned the observed water inflows are usually located in high porosity zones [14]. During the drilling, water inflows were detected at – 368 m and - 440 m. At the end of the drilling, the well was left in its natural water.

3.2. Acoustic logging

The acoustic tool used for the field experiment described in this paper is a flexible monopole tool with two pairs of receivers: a pair of near receivers (1 and 1.25 m offsets) and a pair of far receivers (3 and 3.25 m offsets). The data have been recorded through the far offset configuration. The sampling depth interval is 10 cm. The sampling time interval is 5 microseconds. The length of recording is 5 ms.

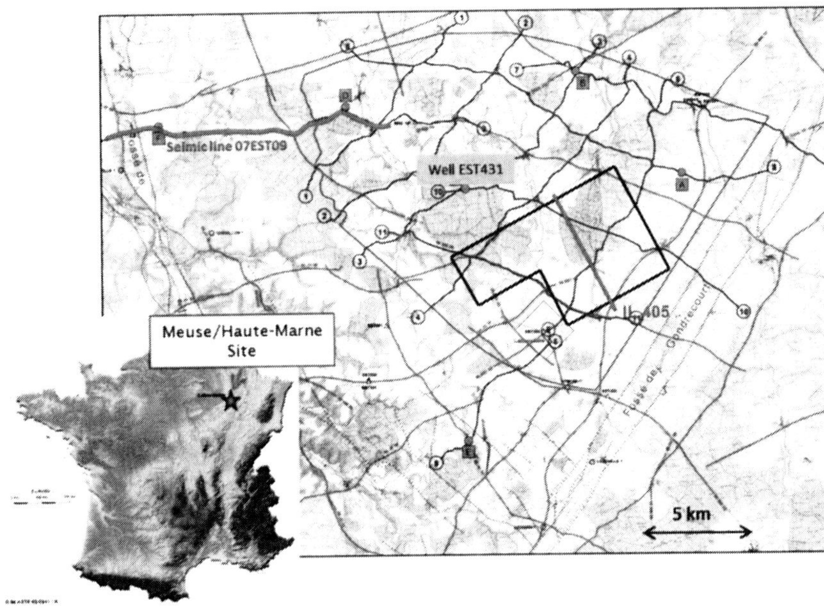

Figure 3. Location map Well EST431 and seismic surveys. The selected seismic lines are the line 07EST09 from the 2D survey and the in line 405 from the 3D survey.

The acoustic log has been run in the Oxfordian carbonate formation, in the 333 – 510 m depth interval. Figure 4 (top left) shows the 3 m constant offset section, opposite the geological description. On the acoustic section, the refracted P-waves appear in the 0.6 – 1.2 ms time interval, the converted refracted shear waves in the 1.2 – 2 ms time interval, and the Stoneley wave in the 2 – 2.4 ms time interval.

On the acoustic section, we can differentiate: an event at 345 m showing a very strong attenuation of all the waves; an interval showing a very strong slowing down of the P and S waves (363 – 375 m); a relatively homogeneous mid-level interval (375 – 400 m); a level which stands out because of its strong variations in P, S and Stoneley velocities (400 – 455 m); a very homogeneous zone below 455 m with easily identifiable P and S waves, and an image of alteration between 501 and 507 m.

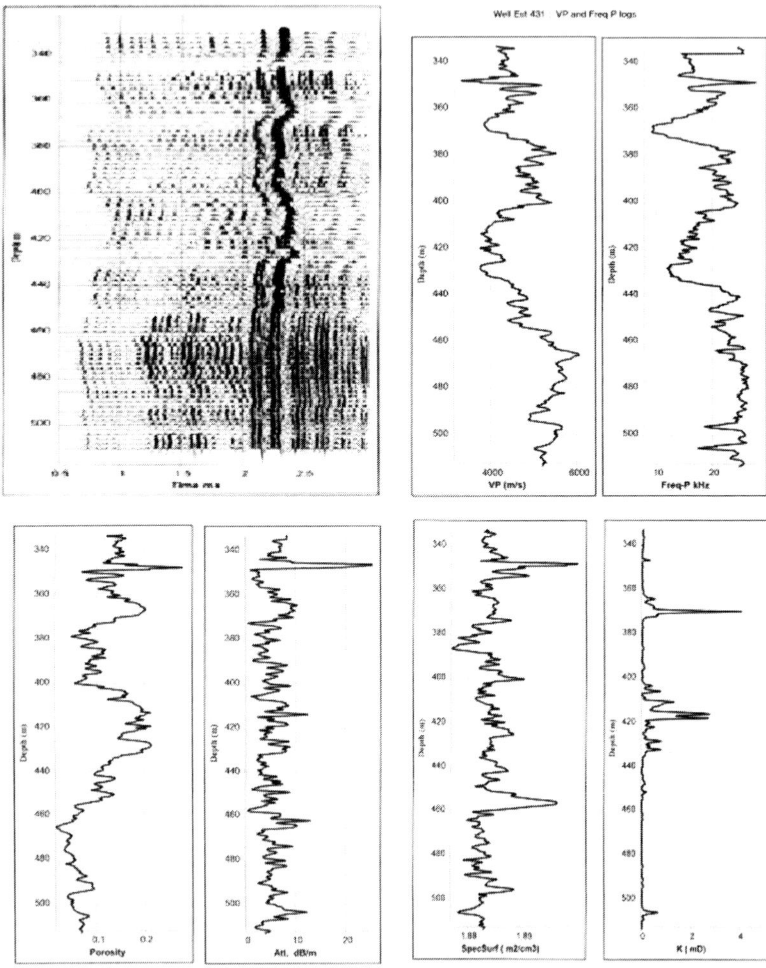

Figure 4. Permeability estimation from acoustic logs [5] Top : acoustic section and P-velocity and frequency logs Bottom from left to right: porosity and attenuation logs; specific surface and predicted permeability (Ik-Seis) logs.

The processing of the acoustic data has been described in detail in [5]. Figure 4 is a display of acoustic logs: P-wave velocity (VP), P-wave frequency, acoustic porosity, P-wave attenuation, Sg specific surface and acoustic permeability. For the carbonate formation, a "V_P - V_S " cross plot led us to define a linear relationship between the two logs and to compute a shear velocity model. The experimental linear relationship computed as a regression line between "V_P and V_S" is $V_S = 0.37\ V_P + 879$. The correlation coefficient between the V_S measurements and the V_S values given by the linear equation is equal to 0.93. For a clean formation, if the matrix and fluid velocities are known, an acoustic porosity log can be computed from the acoustic V_p velocities by using the formula given by Wyllie [15] expressed in velocities. As the maximum P-wave velocity is 6150 m/s at a depth of 464.5 m, the matrix velocity value has been chosen at 6300 m/s, and the fluid velocity at 1500 m/s, since the formation fluid is water. The acoustic porosity log, valid only in the clean part of the formation, shows a strong correlation (correlation coefficient : 0.86) with a NMR porosity log (not displayed here) recorded in the well.

The attenuation (expressed in dB/m) of the formation is computed from the first eigensection (obtained by SVD) of the refracted P-wave acoustic signal recorded by the two adjacent receivers of the acoustic tool. The results obtained by the SVD processing procedure are shown in figure 5.

Figure 4 (bottom right) shows the *Sg* specific surface log and the acoustic permeability log (Ik-Seis) calculated from eq. 2 and 3. The fluid viscosity μ and density ρ_f have been assumed to be constant (μ = 1 centipoise, ρ_f = 1 g/cm^3). The Ik-Seis log detects three permeable zones at 368 m, between 400 and 440 m, and 506 m. The permeable zone located at 506 m corresponds to a high value of conductivity and is characterized by a low porosity (6 %), a 10 dB/m attenuation, but a significant decrease of the P-wave frequency and of the specific surface. During drilling, water inflows have been detected at – 368 m and - 440 m. At the end of the drilling, the well was left in its natural water. The hydraulic tests and conductivity measurements conducted later on did not confirm the inflow at 368 m seen during the drilling, but have validated the 400 – 440 m and 506 m permeable zones detected by the acoustic logging.

A short pumping test was conducted between the 4th and 7th of March, 2008. This test, associated with four geochemical logs, highlighted five productive zones between - 297 and - 507 m. The overall productivity is weak with a 15 L/min flow below 23.5 m drawdown. In general, there is a good correlation between the zones identified through the analysis of the geochemical logs, the natural gamma-ray log and the NMR porosity log. All the identified production zones correspond to low clay content zones.

The mean overall transmissivity of the Oxfordian formation below the EST431 drilling pad is 7.2 10^{-06} m^2/s, ranging between 5 10^{-06} and 1 10^{-05} m^2/s. A total of five inflows have been identified between – 297.5 and -506 m. The most productive inflow (- 506 m) ranges between 2.5 10^{-06} and 5 10^{-06} m^2/s. These inflows are associated with pore porosity and correspond in part to the porous horizons (HP) described in the Oxfordian formation below the Underground Laboratory:

- -297.5 to -301 m inflow, alternation of bioclastic and oolites packstone/mudstone, which corresponds to HP7,
- -328.5 m inflow, coral reefs packstone/grainstone, which corresponds to HP6,
- -413 m inflow, carbonated sand with oolites and oncolites, and numerous coral polyps, which corresponds to HP4, -439 m inflow, oolites and coral polyps grainstone-packstone interface, which corresponds to HP3,
- -506 m inflow, oolites and coral polyps grainstone-packstone interface, no correspondence with the porous horizons. The acoustic porosity and free fluid NMR porosity do not exceed 6%. The inflow, detected by the acoustic permeability log, is characterized by a 10 dB/m attenuation, a significant decrease of the P-wave frequency and of the specific surface.

The logs reveal a strong acoustic discontinuity at a depth of 345 m, clearly visible on the attenuation. The acoustic discontinuity is also revealed by a strong increase of the specific surface, a significant decrease of the acoustic velocities. The acoustic discontinuity is due to the presence of a thin shaly layer in the Oxfordian carbonate formation. It is confirmed by a change in the borehole diameter and a high value of the gamma ray log (not displayed here). Since at that depth, the acoustic porosity log has high values (larger than 25%, figure 4 bottom left), the shaly layer is probably water saturated.

The analysis of the acoustic waves recorded simultaneously on both receivers of the acoustic tool is used to compute additional logs defined as acoustic attributes useful for the characterization of the formation, such as amplitude, shape index and attenuation logs. The results obtained are optimum if the studied wave is extracted from the records and if the signal to noise ratio is high. We show the benefit of using Singular Value Decomposition (SVD) for that purpose [13]. The SVD processing is done on the 2 constant offset sections independently, in a 5 traces (N=5) depth running window. After flattening of each constant offset section with the picked times of the refracted wave, the refracted wave signal space is given by the first eigensection obtained by SVD:

$$\underline{\underline{r}}^{sig} = \lambda_1 \underline{u}_1 \underline{v}_1^T \quad (4)$$

v_1 is the first singular vector giving the time dependence, hence named normalized wavelet, u_1 is the first singular vector giving the amplitude in depth, therefore called propagation vector and λ_1 the associated eigenvalue. The amplitude variation of the refracted wavelet over the depth interval is $\lambda_1 u_1$.

Figure 5 (top) shows the normalized wavelet (v_1) and the associated amplitude ($\lambda_1 u_1$) log versus depth, for the two constant offset sections associated with the two receivers (R1 and R2) of the acoustic tool. Figure 5 (bottom left) also shows the refracted wave signal space versus depth for the two receivers. The amplitude logs have been used to compute the attenuation log (figure 4, bottom left) expressed in dB/m. The correlation coefficient (figure 5, bottom right) between the two normalized wavelets has been computed at each depth. We can notice some anomalies at local depth (358, 390, 460, 492 and 503 m) and a significant decrease of the correlation coefficient in the 400 – 440 m depth interval. The interval corresponds to the porous and permeable zone detected by the Ik-Seis factor (figure 5, bottom right). It is therefore suggested that changes in phase or distortion of the acoustic signal is linked to propagation through a porous and permeable zone. The distortions can be measured by a shape index attribute. To measure the shape variation, an acoustic attribute, named Ic, independent of the energy of the source, has been introduced [10]. The Ic parameter is given by the following equation:

$$Ic = \left(\left(A_2 + A_3 \right) / A_1 \right)^n \qquad (5)$$

where A_1, A_2 and A_3 are the amplitudes of the first three arches, respectively, of the studied signal and n an exponent.

The shape index is computed from formula 5 with an exponent value of 3 (n = 3). Figure 5 (top) allows the comparison between the two shape index logs (Ic-R1 and Ic-R2).

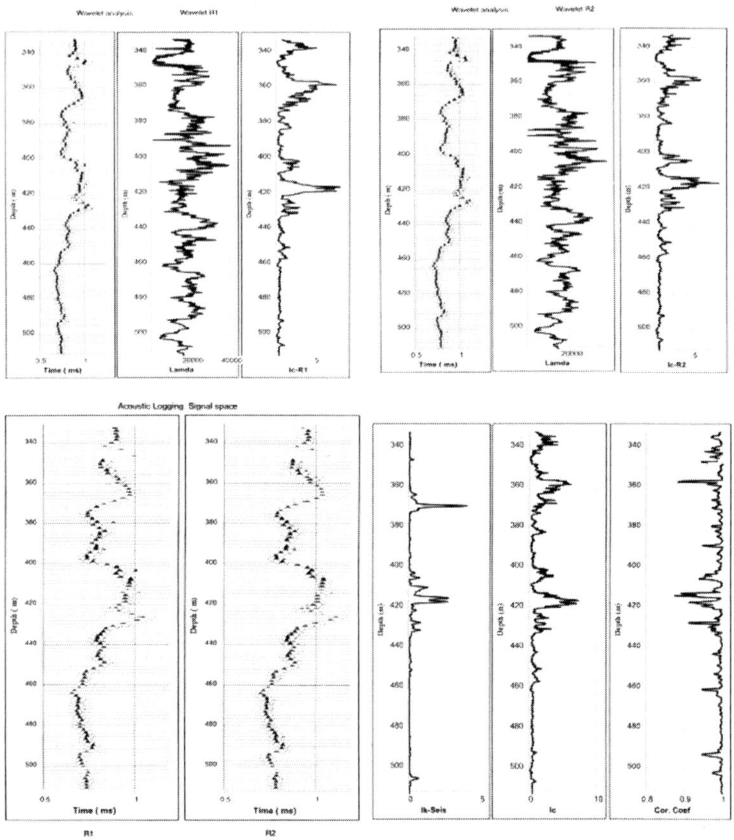

Figure 5. Analysis of refracted waves in acoustic logging by SVD [4]. Top : SVD analysis (Wavelet, amplitude, shape index) on receivers R1 and R2. Bottom from left to right: signal space (R1, R2); predicted permeability (Ik-Seis), shape index and correlation coefficient logs.

The shape index logs highlight anomalic zones, in the 350 – 380 m depth interval with a maximum value at 358 m, and in the 400 – 440 m depth interval with a peak at 415 m which corresponds to the porous layer HP4. In order to reduce the noise to extract the common component of the two shape index logs (Ic-R1 and Ic-R2), the geometric mean of the two has been computed. The resulting shape index log Ic is compared with the Ik-Seis log and the correlation coefficient log (Figure 5, bottom right). The comparison shows a good coherence between the Ic log and the correlation coefficient log. The permeable and porous layers in the 400-440 m depth interval and the inflow at 506 m are seen both by the shape index log and by the Ik-Seis log. Shape index and correlation coefficient logs can be used as a quick look method to detect permeable bodies and inflows which must be confirmed by the Ik-Seis log.

3.3. Seismic surveying: 2D seismic line with acoustic impedance inversion after migration

The 2D seismic line was recorded in 2007 (see location map, figure 3). The 2D design is a split dip spread composed of 240 traces. The distance between 2 traces is 25 m. The source is a vibroseis source generating a signal in the 14 - 140 Hz frequency bandwidth. The bin size is 12.5 m. The nominal fold is 120.

3.3.1. Seismic procedure

A conventional seismic sequence was applied to the data set. It includes amplitude recovery, deconvolution and wave separation, static corrections, velocity analysis, CMP-stacking and time migration.

However, the amplitude recovery has been done in two main steps: spherical divergence compensation by using the Newman's law [16], and attenuation compensation. Besides amplitude changes at each interface, the signal undergoes a general decay as a function of the propagation distance from the source (decaying as 1/r) and with its transmission through the interface. The application of a gain scalar that is a function of t can compensate for these effects. The commonly used gain function is:

$$G_s(t) = t.\, V^2_{rms}(t) / V_1 \tag{6}$$

where t is propagation time, V_1 is the propagation velocity of sound in the first layer and $V_{rms}(t)$ is the rms velocity at time t.

In order to be able to apply the Newman's law, the velocity model given by the rms velocity model must be estimated by velocity analysis. An a priori gain scalar t.V is applied to each shot point. The deconvolution is then performed by spectrum equalization in the 10-130 Hz frequency band. The wave separation by frequency – wavenumber filter (f-k filter) is then done to cancel the direct, refracted and surface waves and to enhance the reflected waves. The data are sorted in CMP gathers and the velocity analysis is performed. The knowledge of the velocity model allows the computation of the gain function $G_s(t)$. The initial gain function is retrieved and replaced by the function $G_s(t)$. After such a processing, a residual decay of the amplitude of the stacked traces has been noticed. The residual decay observed on the envelope of the stacked trace has been used to extract a residual compensation law $G_r(t)$, after a strong smoothing in time of the envelope. The amplitude recovery residual law has been then approximated locally by an exponential law $e^{\alpha t}$, where t is propagation time and α attenuation factor. Figure 6 (top, left) shows the migrated trace at CMP 600 after amplitude recovery by Newman's law, the amplitude recovery function which is used to compensate the attenuation, and the seismic trace after compensation of attenuation.

The migrated section has been filtered by SVD in order to enhance the signal to noise ratio before computing seismic attributes (instantaneous frequency and envelope). Before SVD filtering, the migrated section has been shifted in time to flatten a reference seismic horizon, noted S1. In that case, the SVD processing is done on the migrated section, in a 5 traces (N=5) CMP running window and the signal space is composed of the two first eigensections in order to take into account the local dips which can been present on the time migrated section. Figure 6 (top, right) shows the amplitude recovery function $G_r(t)$, the instantaneous frequency trace f(t) and the associated Q function. The Q factor at time t is obtained from the following equation:

$$Q(t) = \pi f(t)/\alpha \quad \text{with } \alpha = \mathrm{Ln}\left(G_r(t)\right)/t \tag{7}$$

After migration, a model-based stratigraphic inversion (a priori impedance model obtained from well data) provides a 2D impedance model section. The 2D impedance model section has also been shifted in time to flatten the seismic horizon S1. At well locations, the logs of velocity V_p, density ρ and porosity φ have been used to define laws between porosity and acoustic impedance (φ vs Ip) and between V_p and I_p. The porosity vs impedance cross-plot, displayed in figure 6 (bottom right) was used to define a linear law between the two. The porosity law obtained in the Oxfordian limestone and the relation between V_s and V_p ($V_s = 0.37\ V_p + 879$) defined at well EST 431 allow the computation of both the pseudo specific surface and Ik-Seis functions, thanks to equations 2 and 3 (figure 6, bottom left). The processing is done on each migrated CMP independently in order to build the Ik-Seis section. The results are shown in the vicinity of CMP 600 in figure 7 and for the total line in figure 8.

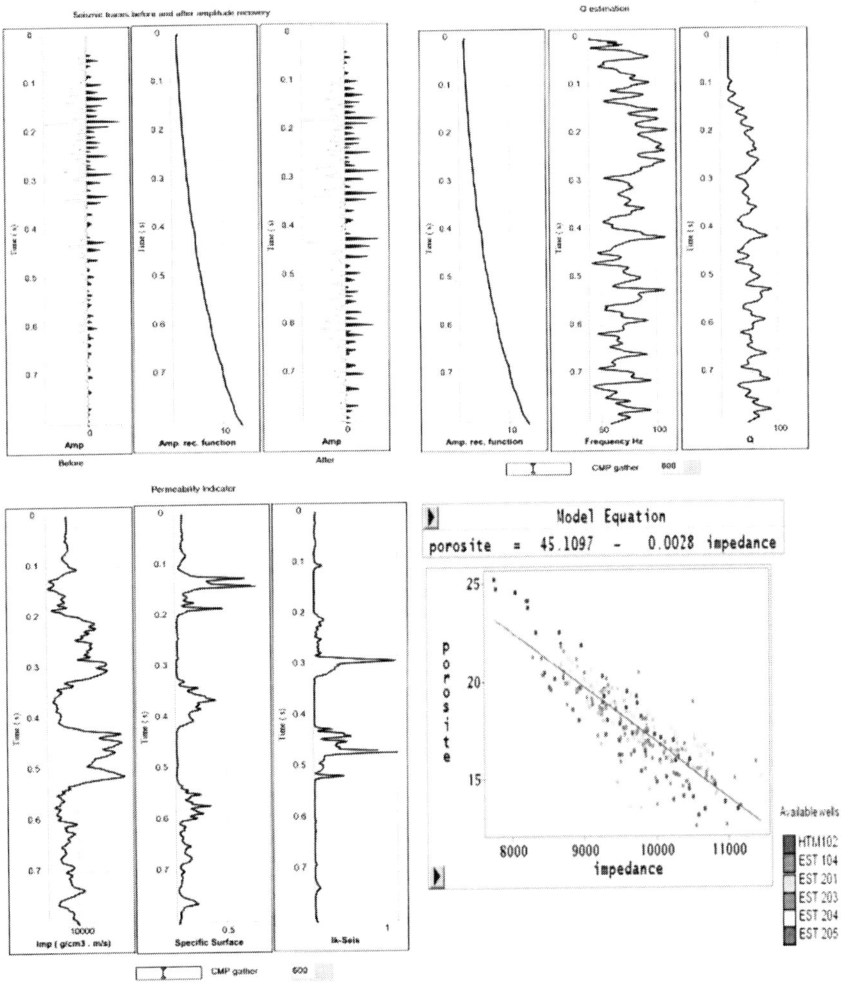

Figure 6. Seismic analysis at CMP 600. [4] Top from left to right: seismic trace before and after amplitude recovery, amplitude recovery function, instantaneous frequency trace and Q seismic trace. Bottom from left to right: acoustic impedance trace, Specific surface and Ik-Seis traces, porosity – acoustic impedance relationship.

3.3.2. Seismic analysis at CMP 600

Figure 6 (bottom left) displays, at location of CMP 600, the acoustic impedance trace, the pseudo specific surface seismic trace with its associated Ik-Seis trace. The specific surface seismic trace clearly shows the main geological units: the Kimmeridgian marls (between 0.1 and 0.2 s), the Oxfordian carbonates (between 0.2 and 0.33 s), the Callovo Oxfordian claystone (between 0.33 and 0.42 s), the Dogger carbonates (between 0.42 and 0.53 s) and the Toarcian claystone after 0.53 s. The shaly units have a high pseudo specific surface and a low value of the Ik-Seis factor. Figure 7 shows the migrated section and the associated Ik-Seis section in the vicinity of the CMP 600. As far as hydrogeology is concerned the observed water inflows are usually located in high porosity zones located in the lower part of the Oxfordian limestone, as it can be seen on the Ik-Seis section between 0.28 and 0.33 s (Figure 7, left). The Ik-Seis section also shows the distribution of the permeable bodies in the Dogger formation (Figure 7, right).

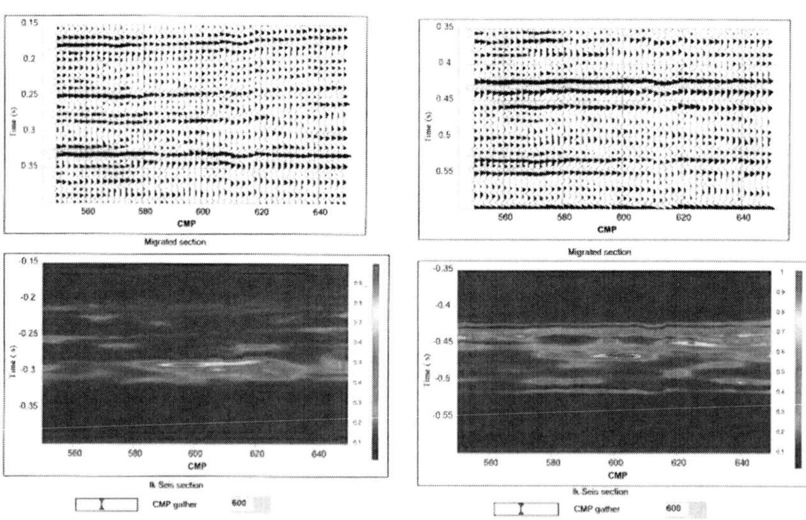

Figure 7. Migrated and Ik-Seis sections at the vicinity of the CMP 600. [4] Left: 150 – 400 ms time interval (Oxfordian carbonate formation), Right: 350 – 600 ms time interval (Dogger carbonate formation).

3.3.3. Seismic line analysis

Figure 8 shows the distribution of the permeable bodies in the carbonate formations via the Ik-seis factor.

Several carbonated shelves have been emplaced during the middle and upper Jurassic, according to second order stratigraphic sequences. The Callovo-Oxfordian claystone is deposited over the Bathonian platform and is in turn overlaid by the Oxfordian platform.

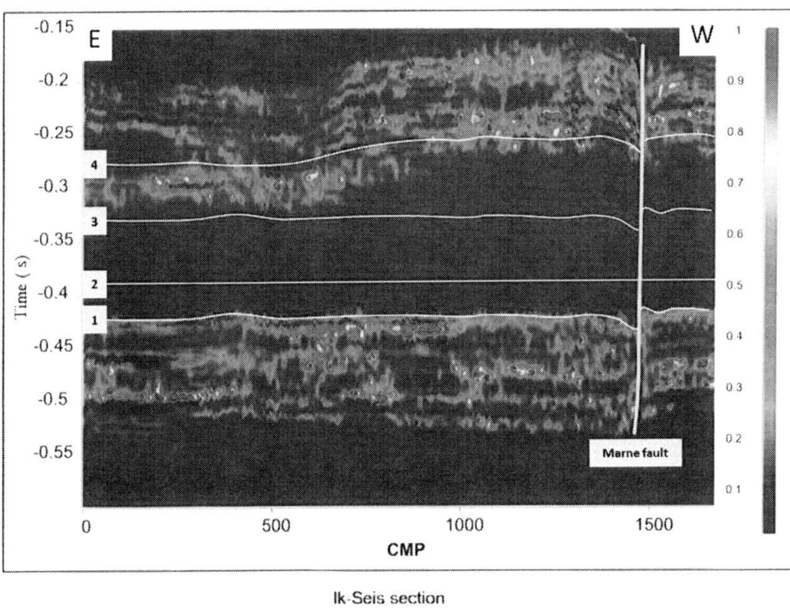

Figure 8. Ik-Seis section [4] 1: base of the Callovo-Oxfordian claystone; 2: carbonated layer used for flattening (S1); 3: top of the Callovo-Oxfordian claystone; Interval 3-4: Oxfordian limestone

The contact between the Dogger formation and the Callovo-Oxfordian claystone is sharp, associated to retrogradation condensed facies. The Dogger carbonate formation displays discontinuous thin porous layers in the Figure 7, but the correlation with the two known thin porous layers cannot be assumed due to the seismic vertical resolution.

The Callovo-Oxfordian claystone displays very low values of the Ik-Seis factor, with no significant lateral variation in good agreement with the known lithology. A carbonated layer, corresponding to a sequence boundary, has been used for flattening. This layer is a known marker-bed in the Paris basin (RIO: repère inférieur oolithique).

The value of the factor in the lower part of the Oxfordian limestone (around 0.30 s) is low in the interval CMP 850 - 1600 according to the transition from high energy inner ramp carbonated facies toward outer ramp marly facies in the west direction. The porosity decreases drastically in the western facies.

3.4. Seismic surveying: 3D seismic section with elastic impedance inversion after migration and depth conversion

The 3D seismic survey was recorded in 2010. The 3D design is a cross spread. The active spread is composed of 12 receiver lines with 120 stations each. The source lines are perpendicular to the receiver lines. The receiver and source line spacing's are respectively 80 m and 120 m. The receiver and source point spacing's are 20 m. The source is a vibroseis source generating a signal in the 14 - 140 Hz frequency bandwidth. The bin size is 10 x 10 m^2. The nominal fold is 60.

A conventional seismic sequence was applied to the data set. It includes amplitude recovery, deconvolution and wave separation, static corrections, velocity analysis and pre-stack time migration. The pre-stack time migration allows a model-based inversion (a priori impedance model obtained from well data) which provides a 3D elastic impedance block. No well being located in the 3D area, additional 2D lines have been recorded to calibrate 3D data on 3 wells situated outside of the 3D area. After migration, the seismic block is depth converted using consistent velocity model thanks to Bayesian kriging [17]. The velocity model is also used for estimating a Ik-Seis model in depth.

3.4.1. Time to depth conversion

Geostatistical approach for time-to-depth conversion of seismic horizons is often used in many geo-modelling projects. From a geostatistical point of view, the time-to-depth conversion of seimic horizons is a classical estimation problem involving one or more secondary variables.The converted depth and associated uncertainty can be estimated using a kriging method which can be constrained by the well markers, velocity model and interpreted horizons. For the multilayer case, the kriging estimator should take into account all the relationships between horizons determined by the velocity model associated to each layer.

The more appropriate kriging method for this problem is the Bayesian Kriging (BK) [17 -19]. Bayesian approach provides an excellent estimator which is more general than the traditional kriging with external drift(s) and fits very well to the needs for time-to-depth conversion of seismic horizons. The advantage of BK as estimator comparing to the others consists in the fact that we can manage simultaneously the uncertainty on the trend velocity model and the local uncertainty defined by the uncertainty of interpreted time maps and local fluctuations of interval velocities.

The input information's for BK are:

- Two-way-time (TWT) maps for interpreted horizons
- Well markers for each horizon
- Prior velocity model and associated uncertainty for each layer
- Local uncertainty definition for each time map (picking uncertainty, and spatial variogram definition)
- Local uncertainty definition of interval velocity for each layer (local velocity fluctuations around the trend model, and spatial variogram definition)

As any Kriging based estimator, the Bayesian Kriging provides as results:

- The estimated variable (estimated depth for each horizon)
- Variance of estimation (associated uncertainty of estimated depth).

The use of BK in depth conversion has the advantage to combine the prior knowledge of the velocity model with a certain degree of uncertainty and the well data. All sources of uncertainty (velocity and time) are integrated in a consistent way in a unique probabilistic model used for estimation or simulation.

For each selected horizon, the Bayesian Kriging provides its estimated depth Z associated with its time t. The " Z versus t " data set is interpolated in the whole space (3D block) at the time sampling rate (1 ms) in order to obtain a time to depth conversion model.

The time to depth conversion procedure is illustrated via the In line 405 extracted from the 3D seismic block (see location map, figure 3).

10 seismic horizons numbered from 1 to 10 have been picked in time and depth converted. The 10 seismic horizons are:

1. Top of Kimmeridgian White Limestones
2. Top of Porous Horizon HP4
3. Top of Lower Oxfordian (Top of target interval)
4. Top of Upper Callovian (RIO)
5. Top of Carbonated Dogger (Base of target interval)
6. Base of Argillaceous limestone and Longwy marls
7. Base of Carbonated Dogger
8. Top of Domerian
9. Base of Lias (base of Gryphees limestone)
10. Top of Beaumont dolomite

Figure 9 shows the acoustic impedance section (I_p) in time and the picked times of the 10 seismic horizons, the time- to – depth conversion model and the depth conversion of the 10 horizons. The time-to-depth conversion model can be used to convert in depth any type of seismic sections (amplitude, velocity, acoustic impedance, Ik-Seis factor).

3.4.2. Ik-Seis section in depth

The Ik-Seis section has been computed in time following the procedure described for the 2D line (figure 8) and then depth converted using the time to depth conversion model shown in figure 9.

Figure 10 shows the Ik-Seis sections in time (a) and in depth (b). The Ik-Seis sections allow the identification of a porous and permeable horizon (HP4) in the Oxfordian limestone (depth : 500 m, CMP : 100). They show the distribution of porous layers in the Dogger formation (between 800 m and 1000 m depth). The contact between the Dogger carbo nate formation and Callovo-Oxfordian argillite is clearly marked. The Callovo-Oxfordian claystone displays very low values of the Ik-Seis factor. The results obtained confirm the observations done on the 2D line (figure 8).

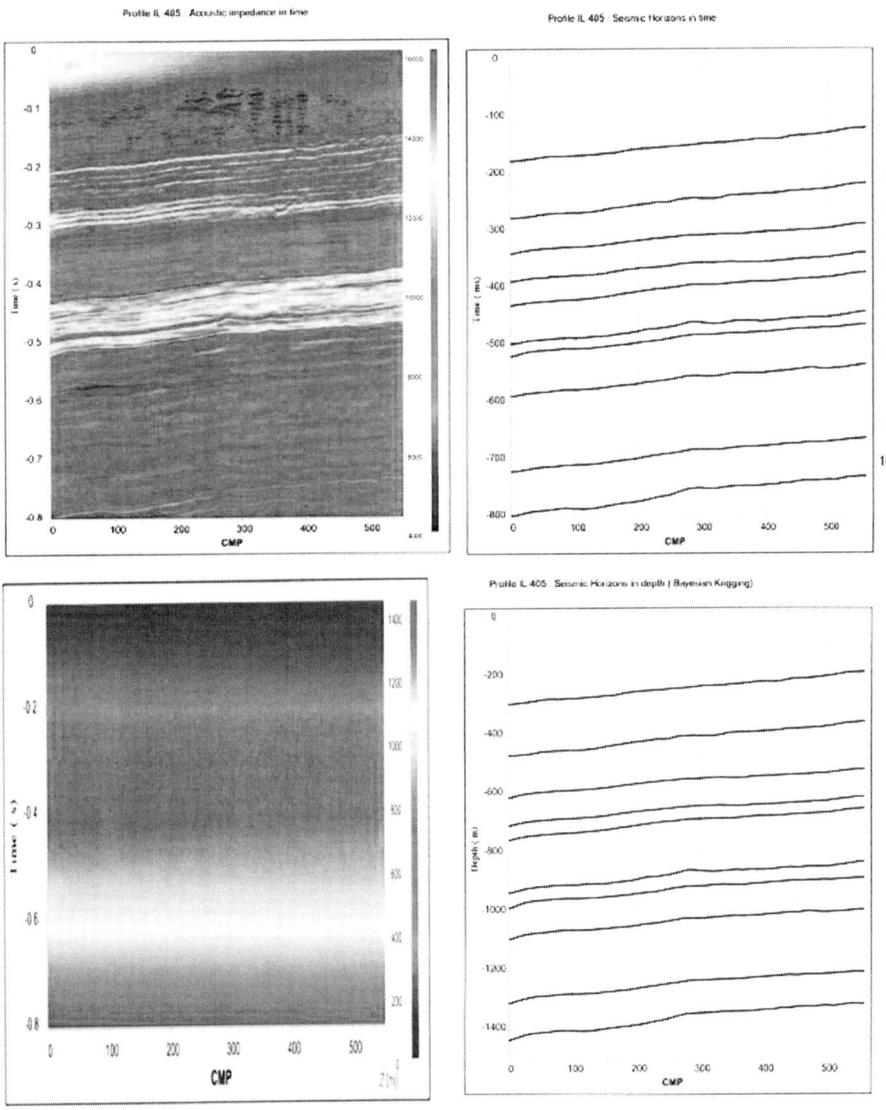

Figure 9. Time to depth conversion. Top : acoustic impedance section in time and the picked times of the 10 seismic horizons, Bottom: time- to – depth conversion model and the depth conversion of the 10 horizons.

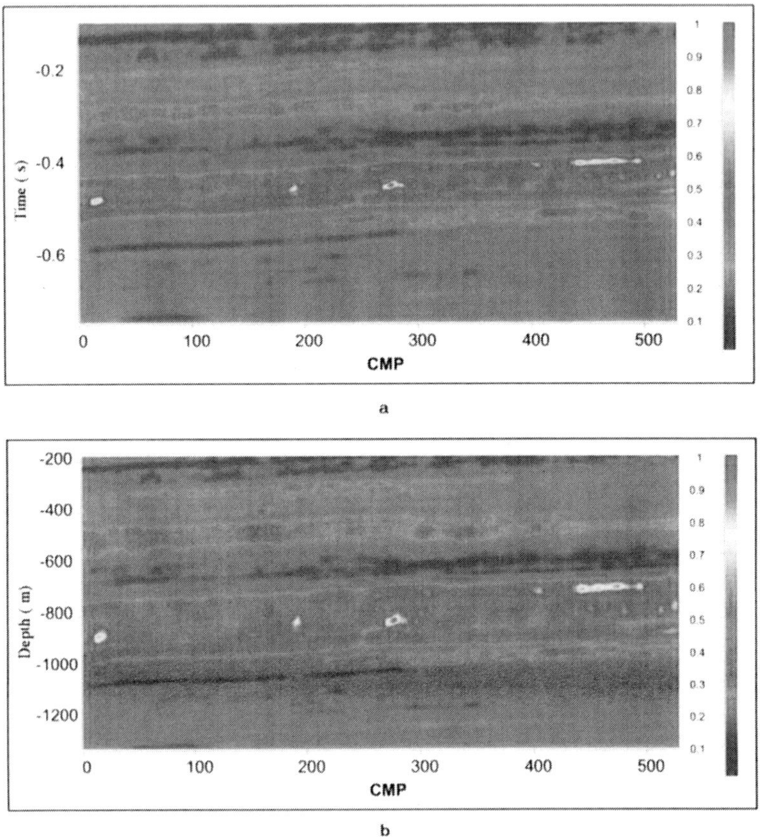

Figure 10. Ik-Seis sections a : in time, b: in depth.

4. CONCLUSION

Knowledge about porosity and permeability is essential to evaluate fluid content and to detect fluid flow. We have presented a procedure which allows to use geophysical data (i.e. full waveform acoustic data and reflection seismic data) for a better understanding of the distribution of the porous and permeable bodies. The methodology is based on laboratory experiments which have shown that a formation permeability indicator, named Ik-Seis factor, can be obtained via the computation of 4 quantities: P-wave frequency, attenuation, porosity and specific surface.

The ability to use laboratory results for field geophysical applications has been discussed. The frequency content is important. Regarding frequencies above 2 kHz, permeability has an influence on velocities and attenuations. The attenuation may reach a maximum for frequency in the order of 10 kHz, this being the domain of full waveform acoustic logs.

Consequently, the procedure has been firstly conducted in acoustic logging to estimate the permeability of porous layers and to detect water inflows. Full waveform acoustic data were recorded in an Oxfordian carbonate formation. The Ik-Seis factor computed in the acoustic frequency domain (ranging between 10 and 25 kHz) has detected permeable zones, both associated with high porosity (20 %) but also with low porosity (6 %). The hydraulic tests and conductivity measurements conducted later on have validated the permeable zones detected by acoustic logging. The benefit of using the SVD method to evaluate the signal to noise ratio, to compute the attenuation log and to extract the acoustic wavelets from the acoustic data has been shown. It has also been observed that the correlation coefficient computed between acoustic wavelets recorded by two adjacent receivers of an acoustic tool significantly decreases in porous and permeable zones. It is therefore suggested that changes in phase or distortion of the acoustic signal is linked to propagation through a porous and permeable zone. The distortions can be measured by a shape index attribute. After calibration on core data or hydraulic tests, the Ik-Seis could be seen as a pseudo acoustic permeability log.

In seismic, after signal to noise ratio enhancement by the SVD method, processing is carried out in order to measure the needed parameters (frequency, attenuation, impedance) to compute the Ik-Seis factor. The analytic signal is used to compute the instantaneous frequency and attenuation (Q factor). The porosity and specific surface are computed from seismic impedances obtained by acoustic inversion of the migrated seismic sections. The Ik-Seis factor should only be used as a relative indicator which varies from 0 for less porous and permeable bodies to 1 for more porous and permeable bodies.

Our results suggest that it is possible to extract a significant Ik-Seis factor from seismic sections. This factor leads to a better understanding of the distribution of the porous and permeable bodies. The potential of the proposed procedure has been demonstrated via a 2D seismic profile and a seismic in-line extracted from a 3D block. We have demonstrated the benefit of combining time–to-depth conversion of seismic horizons by Bayesian kriging, consistent seismic velocity model and acoustic impedance in time for building a 3D geological model in depth. The field data example illustrates the potential of the proposed depth conversion procedure for estimating a Ik-Seis model in depth.

5. ACKNOWLEDGEMENTS

We thank Andra for permission to use the data presented in the field examples. We thank Jean Dellenbach, Jean Paul Sarda and Daniel Guillemot for their valuable help and advices. We thank Arben Shtuka (Seisquare) for very useful discussions on various occasions, specifically for his experience in the use of geostatistical methods for time to depth conversion.

REFERENCES

1. Bourges M, Mari J.L., Jeannée N., A practical review of geostatistical processing applied to geophysical data: methods and applications, Geophysical Prospecting, 2012, 60, 400-412, DOI: 10.1111/j.1365-2478.2011.00992.x
2. Mari J.L., Delay F., Contribution of seismic and acoustic methods to reservoir model building, in Lakshmanan Elango (ed.) Hydraulic Conductivity - Issues, Determination and Applications, InTech, 2011; 329–354, ISBN 978-953-307-288-3. Available from http://www.intechopen.com/articles/show/title/contribution-of-seismic-and-acoustic-methods-to-reservoir-model-building.
3. Zinszner B., Pellerin F.M., A geoscientist's guide to petrophysics, ISBN 978-2-7108-0899-2, Paris, Editions Technip, 2007
4. Mari J.L, Guillemot D., Detection of porous and permeable formations: from laboratory measurements to seismic measurements, Oil & Gas Science an Technology _Rev. IFP Energies nouvelles, 2012, DOI: 10.2516/ogst/2012209.
5. Mari J.L., Gaudiani P., Delay J., Characterization of geological formations by physical parameters obtained through full waveform acoustic logging., J. Phys. Chem. Earth, 2011; DOI:10.1016/jpce.2011.07.11.
6. Morlier P., Sarda J.P., Atténuation des ondes élastiques dans les roches poreuses saturées, Revue de l'Institut Français du Pétrole, 1971 ;26 (9), p. 731-755.
7. Fabricius I. L., Baechle G., Eberli G.P., Weger R., Estimating permeability of carbonate rocks from porosity and Vp/Vs, Geophysics, 2007; 72(5), 185-191, DOI: 10.1190/1.2756081.

8. Biot, M.A.,Theory of propagation of elastic waves in a fluid-saturated porous solid, The Journal of the Acoustical Society of America, 1956; 28(2), 168-191.

9. Plona T.J., Observation of a second bulk compressionnal wave in a porous medium at ultrasonic frequencies, Appl. Phys. Lett., 1980; 36, 259-261.

10. Lebreton F., Morlier P, A permeability acoustic logging, Bulletin of the International Association of Engineering Geology, 1983; 1, 101-105.

11. Klimentos T., McCann C., Relationships among compressional wave attenuation, porosity, clay content, and permeability in sandstones, Geophysics, 1990; 55(8), 998-1014.

12. Yamamoto T., Imaging permeability structure within the highly permeable carbonate earth: inverse theory and experiment, Geophysics, 2003; 68(4), 1189-1201, DOI:10.1190/1.1598103.

13. Glangeaud F., Mari, J.L., Signal Processing in Geosciences, 2000. [CD-ROM] Paris: Editions Technip, ISBN 2-7108-0768-8; 2000.

14. Delay J, Rebours H, Vinsot A, Robin P., Scientific investigation in deep wells for Nuclear waste Disposal studies at the Meuse/Haute-Marne underground research laboratory, North Eastern France, J. Physics and Chemistry of the Earth, 2007; 32,42-57.

15. Wyllie M.R., Gregory R.J., Gardner H.F., Elastic wave velocities in heterogeneous and porous media, Geophysics,1956; 21(1),41-70.

16. Newman P., Divergence effects in a layered earth, Geophysics, 1973; 38, 377-406.

17. Sandjivy L., and Shtuka A., Depth conversion and associated uncertainties using consistent velocity model: a probabilistic unified model based on Bayesian approach, proceedings of the 11th International Congress of the Brazilian Geophysical Society, August 24-28 2009, Salvador, Brazil.

18. Abrahamsen P., Bayesian Kriging for Seismic Depth conversion of Multi-layer Reservoir, in A. Soares (ed.) Geostatistics Troia 92, 1993; 385-398.

19. Omre, H. and Halvorsen, K. B., 'The Bayesian bridge between simple and universal kriging'. Math. Geol., 1989; 21(7), 767-786.

12
CHAPTER

UNSATURATED HYDRAULIC CONDUCTIVITY OF FRACTAL-TEXTURED SOILS

Yongfu Xu[1]

[1] *Department of Civil Engineering, Shanghai Jiao Tong University, Shanghai, China*

1. INTRODUCTION

The increasing concern with groundwater pollution and contamination of soils has stimulated the development of numerous mathematical models of pollutant transport in soils. The most important approaches to model transient water and solute transport in the vadose zone are based on the Richards equation. To solve this equation, the knowledge of the soil hydraulic properties, namely, the soil-water characteristic curve (SWCC) and the unsaturated hydraulic conductivity is required. The laboratory measurements show that the value of unsaturated hydraulic conductivity varies considerably from soil to soil with different water content (Khaleel and Relyea, 1995). Indeed, it is found that the unsaturated hydraulic conductivity decreases by one to three orders of magnitude across a small pressure head range even near saturation (0~10cm pressure head), due to the effects of structural macropores (Jarvis and Messing, 1995). Of all hydraulic properties, the unsaturated hydraulic conductivity is most difficult to measure. Therefore the use of indirect methods has become more and more common to estimate the unsaturated hydraulic conductivity from more easily measured soil properties (van Genuchten et al., 1992).

Modern hydrological models require information on hydraulic conductivity and soil-water retention characteristics. All hydraulic properties, the soil-water characteristics, hydraulic conductivity and soil-water diffusivity (SWD) are closely related to the geometry of a porous media (Brooks and Corey, 1966; Burdine, 1953). Measurements of hydraulic properties are expensive, time-consuming and highly variable (Dirksen, 1991). Prediction of these properties is a viable alternative, especially when the predictive model contains a few parameters sensitive to structural conditions. Porous media (e.g. soils, rocks, etc.) are heterogeneous systems composed of numerous, different and interacting components (van Damme, 1995). The complex nature of these porous media complicates any prediction of their hydraulic properties.

A potentially powerful method results from the pore surface models, in which the soil–water characteristic curve of unsaturated soils is interpreted as statistical measure of its equivalent pore-size distribution (PSD) (van Genuchten, 1980; Corey, 1992). The frequency of different pore radii is related to matric suction by the soil-water characteristic curve (SWCC) and the relation between matric suction and pore radius is described by the Young-Laplace equation. The relative hydraulic conductivity (RHC) of unsaturated soils can be deduced from the soil-water characteristic curve (SWCC) through simplifying assumptions on pore topology and using the Hagen-Poiseuille law as an approximation for water flow. An additional empirical parameter is introduced, which includes all uncertainties. This empirical parameter is often referred to as "tortuosity", but its physical meaning is unclear (Vogel and Roth, 1998). The choice of the analytical model for the soil-water characteristic curve (SWCC) can significantly affect the predicted function of the unsaturated hydraulic conductivity.

Fractals describe hierarchical systems and are suitable to model the heterogeneous soil structure with tortuous pore space (Rieu and Sposito,1991; Xu and Sun, 2002). Toledo et al. (1990) modeled the soil-water characteristic curve (SWCC) and unsaturated hydraulic conductivity using fractal geometry and thin-film physics. Tyler and Wheatcraft (1990) gave the unsaturated hydraulic conductivity functions based on the fractal model for the soil-water characteristic curve (SWCC) and the relative conductive models developed by Mualem (1976) and Burdine (1953). Crawford (1994) studied the influence of heterogeneity of both the solid matrix and the pore space, as well as the shape of the pore boundary, on the saturated and unsaturated hydraulic conductivity. Fuentes et al. (1996) derived an expression for unsaturated hydraulic conductivity using the fractal dimension obtained from the soil-water characteristic curve (SWCC). Hunt and Gee (2002) applied critical path analysis from percolation theory to calculate the unsaturated hydraulic conductivity of soils with fractal pore space. Xu (2004), Xu and Dong (2004) Xu et al. (2004) derived the unsaturated hydraulic conductivity using the fractal model for the pore surface and gave a simple method to determine the fractal dimension. Models of the unsaturated hydraulic conductivity incorporate fractal dimension characterizing scaling of different properties including parameters representing connectivity and tortuosity. Thus, it is encouraged to derive the functions of the soil–water characteristic curve and unsaturated hydraulic conductivity from the fractal model for the pore surface. In this chapter, the soil-water characteristic curve (SWCC) and relative hydraulic conductivity (RHC) function were derived and expressed by the effective degree of saturation based on the fractal model for the pore surface. The proposed soil–water characteristic curve and unsaturated hydraulic conductivity function were examined in detail by the published experimental data. Comparisons between the prediction using both the fractal model and the van Genuchten-Mualem (G-M) model and the measurement of the relative hydraulic conductivity (RHC) were conducted. Using the fractal model for the soil-water characteristic curve (SWCC) and relatively hydraulic conductivity (RHC), one-dimensional rainfall infiltration and slope stability due to rainfall infiltration were analysized in chapter.

2. FRACTALS

Some of the most common and useful examples of fractals are: the Koch curve, the Sierpinski gasket and carpet, and the Menger sponge. All of these fractals enjoy the property of self-similarity. Roughly speaking, a subset of $\Re^n$ is said to be self-similar if it is a union of a number of smaller similar copies of itself. Some of the essential notions of the theory of fractals as it applies to self-similar sets should be briefly reviewed. A previous acquaintance with at least one of the sets mentioned above is useful but not necessary. Given an arbitrary subset *B* of $\Re^n$, a *cover* of *B* is a family *U* of sets $U_0 \subset \Re^n$ such that:

$$B \subset \bigcup_{n} U_{\alpha} \tag{1}$$

If the family *U* is countable (or finite) the cover is said to be countable (or finite). It is customary in that case to indicate the members of the family with a Latin subscript, namely, U_i, where i ranges over the natural numbers.

The diameter of a subset of $\Re^n$ is the least upper bound (i.e., the supremum) of the distance between pairs of points in the subset. By convention, the diameter of the empty set is zero. A δ-cover of $B \subset \Re^n$ is a (countable) cover such that, for every i, diam(U_i)≤δ, where δ is a positive real number and where diam(.) is the diameter function on subsets of $\Re^n$. The members of a δ-cover are indicated by U_i^δ. The s-dimensional Hausdorff measure $H_s(B)$ of B is defined as:

$$H^s(B) = \lim_{s \to 0} \inf \sum_{i=1}^{n} [\mathrm{diam}(U_i^s]^s \tag{2}$$

where the inf extends over all δ-covers. It can be shown that this definition indeed provides an outer measure for all non-negative values of s and for any subset of $\Re^n$. The Hausdorff dimension of B is defined as:

$$\dim_H B = \inf\{s \geq 0 : H^s(B) = 0\} \tag{3}$$

The Hausdorff dimension of a subset of $\Re^n$ cannot exceed n. If $\dim_H B > 0$, then it can be shown that for all values of s strictly smaller (respectively, larger) than the dimension, the s-dimensional Hausdorff measure of B is infinite (respectively, zero). In this sense, the Hausdorff dimension represents a critical value of discontinuity of the s-dimensional Hausdorff measure, thus making Eq. (3) meaningful. Of particular interest is the value of the s-dimensional Hausdorff measure for s equal to the Hausdorff dimension of the set. This is usually called the *Hausdorff measure* of the set. The Hausdorff measure of a set may turn out to have any value, including zero or infinite.

An important feature of the s-dimensional Hausdorff measure is the *scaling property*, namely, the way it changes under similarity transformations of $\Re^n$. Recall that a transformation S: $\Re^n \to \Re^n$ is called a similarity if there exists a real scale factor λ>0 such that, for all x, $y \in \Re^n$, the following equation is satisfied:

$$|S(y) - S(x)| = \lambda |y - x| \tag{4}$$

where |.| denotes the Euclidean distance in $\Re^n$. It is not difficult to prove that for all subsets $B \subset \Re^n$ and for all values of *s*, under a similarity transformation *S* with scale factor λ the *s*-dimensional Hausdorff measure transforms according to the formula:

$$H^s(S(B)) = \lambda^s H^s(B) \tag{5}$$

Another important property of the *s*-dimensional Hausdorff measure is that it is preserved under Euclidean isometries (translations, rotations, reflections). In other words, congruent sets have the same *s*-dimensional Hausdorff measure. The behavior of the Hausdorff measure under general affine transformations, on the other hand, cannot be captured under the umbrella of a simple formula.

The properties just described of the *s*-dimensional Hausdorff measure can be used to obtain a straightforward evaluation of the Hausdorff dimension of self-similar fractals. Indeed, let *B* be the union of *m* copies of λ-scaled mutually congruent copies of itself (with λ<1). If these copies are disjoint Borel sets, we have:

$$H^s(B) = m\lambda^s H^s(B) \tag{6}$$

by virtue of the scaling property (5). Assume now that the Hausdorff measure of *B* (namely, the value of $H_s(B)$ for $s=\dim_H(B)$) is finite and positive. In that case, it follows Eq. (6) that:

$$\dim_H B = -\frac{\log m}{\log \lambda} \tag{7}$$

The fractal dimension for a pore surface can be defined in the following way. Imagine the pores being enclosed by a set of spheres of radius *r*, the number of spheres *N* necessary to do this is clearly a function of radius of the spheres. The definition of the fractal dimension *D* is:

$$D = \lim_{r \to 0} \left[\frac{\ln(N(A,r))}{\ln(1/r)} \right] \tag{8}$$

Equation (8) is often used to determine the fractal dimension by experiments. A typical fractal set, the middle third Cantor set may be constructed from a unit interval by a sequence of deletion operations (Fig. 1). Let E_0 be the interval [0,1], E_1 is the set obtained by deleting the middle third of E_0 so E_1 consists of the two intervals $\left[0, \frac{1}{3}\right], \left[\frac{2}{3}, 1\right]$. Deleting the middle third of E_1 gives E_2, thus E_2 comprises the four intervals $\left[0, \frac{1}{9}\right], \left[\frac{2}{9}, \frac{3}{9}\right], \left[\frac{6}{9}, \frac{7}{9}\right], \left[\frac{8}{9}, \frac{9}{9}\right]$. Proceeding in this like manner, E_i is obtained by deleting the middle third of each interval in E_{i-1} and E_i consists of 2^i intervals with length of 3^{-i}. Using Eq. (8), the fractal dimension of the middle-third Cantor set is $D = \lim_{i \to \infty} \left\{ \frac{\log(2^i)}{\log(1/(3^{-i}))} \right\} = 0.63.$ The Cantor set is often used to model the dust distribution.

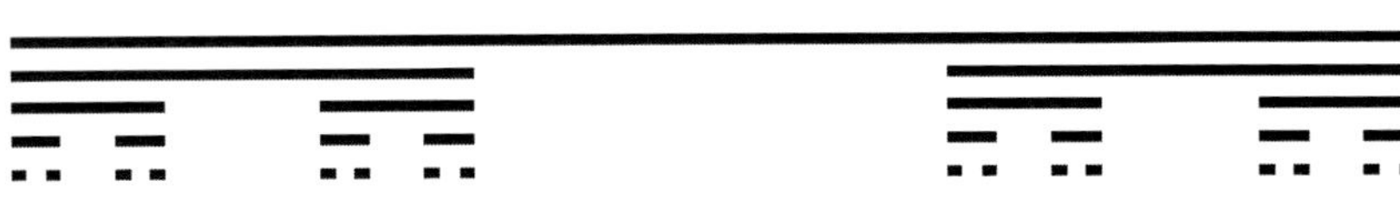

Figure 1. Middle-third Cantor set with fractal dimension D=0.63

3. FRACTAL MODEL FOR THE PORE SURFACE

Many research results show that the soil pore surface is fractal (Avnir and Jaroniec 1989; Xu and Sun 2002). To investigate the impacts of fractal scaling upon hydraulic properties of porous media, a fractal representation of a porous media is developed. A cube of size 1 by 1 is formed by initially subdividing the cube into $1/a^3$ subcubes, each of size a by a. From the original cube, N subcubes are removed and represent a pore of size a by a. The remaining $1/a^3$-N subcubes are then each divided into $1/a^3$ subcubes and N subcubes are removed from each of the original subcubes. Such a recursion algorithm results in a cube everywhere filled with pores of all sizes, with a predominance of small pores. The number of size of a^3 needed to cover the pores equal to or larger than a^3 is given by N. The Menger sponge (Fig. 2) is often used to model porous materials, such as soils. For the Menger sponge, a=1/3, N=7, and the fractal dimension is given by

$$D = \lim_{i \to \infty} \left\{ \frac{\log(20^i)}{\log(1/(3^{-i}))} \right\} = 2.73.$$

Figure 2. The Menger sponge with fractal dimension D=2.73

When we covered a soil pore surface using balls with the same radius, the soil pore surface is covered by N_1 balls with radius r_1, the same surface is covered by N_i balls with radius r_i, and it is covered by N_j balls with radius r_j, and usually $r_1 > r_i > r_j$, it is found that the surface area of the soil pore surface contents has the following relationship: $N_0 r_0^2 < N_i r_i^2 < N_j r_j^2$. If the soil pore surface is a self-similar surface, i.e., the pore surface is a fractal surface, the relationship between the number of covered balls and its radius can be written as (Mandelbrot 1982)

$$N(r) = Cr^{-D} \tag{9}$$

where C is a constant, r is the size corresponding to the pore radius, D is the fractal dimension of the pore surface. All materials have a surface fractal dimension in the whole range of physically meaningful values, i.e., from 2 to 3. The limit D=2 would correspond to a perfectly regular, smooth (Euclidean) surface, whereas D=3 would correspond to a self-similar surface so intricate that it would be space filling.

The fractal dimension of the pore surface can be determined using mercury intrusion porosimetry and adsorption isotherm. These two methods are introduced in detail as follows.

3.1. Mercury intrusion porosimetry

The d-measure is obtained from Eq. (9), and is expressed as follows:

$$M(r) = N(r)r^{d} = Cr^{d-D} \tag{10}$$

For the fractal pore surface, the relationship between the surface area with the radius less than r and the pore radius r can be obtained from Eq. (10). The surface area can written as

$$\begin{aligned} A_p(\le r) &= Cr^{2-D} && a) \\ V_p(\le r) &= Cr^{3-D} && b) \end{aligned} \tag{11}$$

where in 11 a) $A_p(\le r)$ is the surface area of the soil pores with the radius smaller than r. Similarly, the total volume of the pores with the radius less than r is given by 11b), where in $V_p(\le r)$ is the cumulative volume of the soil pores with the radius smaller than r.

Neimark (1992) gave a method to measure the surface fractal dimension of the soil pores using mercury intrusion porosimetry. The soil pore surface can be approximated by the equilibrium interface between mercury and soil particles in the close vicinity of the pore surface. According to the

Young–Laplace equation, the mean radius of the equilibrium interface is expressed as follows:

$$r(p) = \frac{2\sigma \cos\alpha}{p} \tag{12}$$

where $r(p)$ is the mean radius of the equilibrium interface under pressure p, σ is surface tension. From the thermodynamic viewpoint, the equilibrium interface area can be calculated from the balance between the work of formation of the equilibrium interface and the work of mercury intrusion (Neimark 1992), i.e.,

$$A_{\mathrm{p}}(p) = \int_0^{V(p)} \frac{V(p)}{T} \mathrm{d}p \tag{13}$$

where $A_{\mathrm{p}}(p)$ is the equilibrium interface area under pressure p. The surface fractal dimension can be determined theoretically from Eq. (11a). If the slope of the straight line is λ in log $r(p)$ versus log $A_{\mathrm{p}}(p)$, the surface fractal dimension of soil pores is given by

$$D = 2 - \lambda \tag{14}$$

The pore-size distribution (PSD) is usually obtained using mercury intrusion porosimetry. According to Eq.(11b), the fractal dimension of the pore surface can be obtained from the slope of the regressed linear relation in the plane of log$r$ vs. logV_{p}. If the slope of the fitting line in the log$r$-logV_{p} plane is κ, the fractal dimension of the pore surface is written as

$$D = 3 - \kappa \tag{15}$$

The value of fractal dimension spans a large range from 1.0 to 3.0 (Gimenez et al., 1997). Larger fractal dimension is associated with clayey soils (van Damme, 1995).

The accumulative volume of the pores can be measured by many methods. Mercury intrusion porosimetry provides a useful and potentially valuable measurement of the pore volume for the porous medium. The ranges of equivalent pore diameter explored cover almost five orders of magnitude, from several hundred microns down to approximately 16Å (Watabe et al., 2000).

Figure 3 shows the soil pore-size distribution (PSD) obtained from mercury intrusion porosimetry (Watabe et al., 2000). Symbols S-02, S-03 and S-04 are the serial numbers of soil samples, and V_v is the total pore volume in Fig. 3. It is seen from Fig. 3 that the surface of the soil pores can be described by fractal model, and the relationship between $V_p(\leq r)$ and r can be expressed by a linear function in log-log plot. The slopes of the regressed linear relation in the plane of log$r$ vs. $\log(V_p(\leq r)/V_v)$ are 0.34, 0.37 and 0.49, and therefore the fractal dimensions are 2.66, 2.63 and 2.51 for specimens of S-02, S-03 and S-04, respectively. The maximum radius is the radius at which the pore volume reaches the maximum value, and is defined as the intersection between the fitting line and the line of $V_p(\leq r)/V_v$=100%. The maximum radii of soil pores are 0.0125, 0.06 and 0.006 mm for specimens of S-02, S-03 and S-04, respectively. The relationship between the maximum radius R and the air-entry value ψ_e can be expressed by the Young-Laplace equation, i.e. $\psi_e=2\sigma\cos\alpha/R$. The parameters α=0 and σ=0.075kPa mm (Watabe et al., 2000), and the air-entry values are 12kPa, 2.5kPa and 25kPa corresponding to the maximum pore radius R of 0.0125, 0.06 and 0.006 mm for specimens of S-02, S-03 and S-04, respectively. The parameters obtained from the fractal model of the pore surface are listed in Table 1. In Table 1, parameters R and κ are obtained from the fractal model of the pore surface in Fig. 3.

Table 1. Parameters obtained from the fractal model of PSD

Soil type	κ	D	δ	R (mm)	ψ_e (kPa)	Data source
S-02	0.34	2.66	-0.34	0.0125	12	Watabe et al., 2000
S-03	0.37	2.63	-0.37	0.06	2.5	
S-04	0.49	2.51	-0.49	0.006	25	
A horizon	0.22	2.78	-0.22	0.2	0.75	Vogel and Roth, 1998
B horizon	0.13	2.87	-0.13	0.25	0.6	
Toyoura sand	1.33	1.67	-1.33	0.09	1.67	Uno et al., 1998

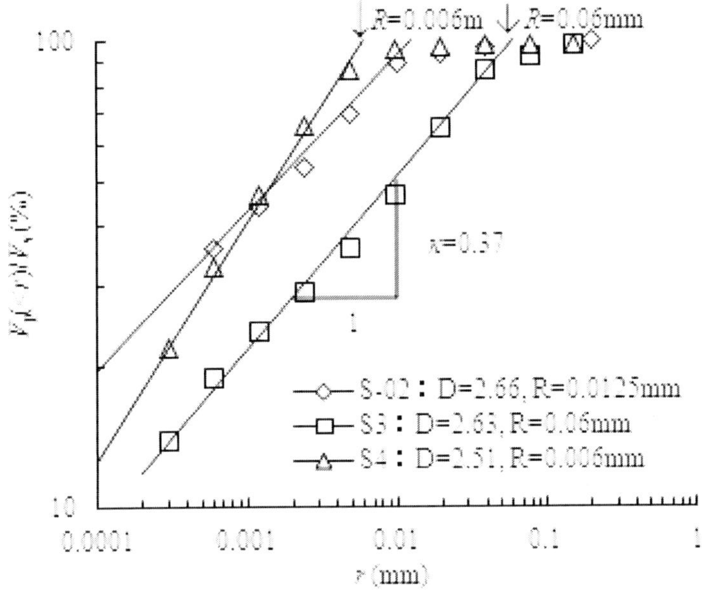

Figure 3. Fractal dimension and the maximum pore radius obtained from PSD for glacial tills (Data from Watabe et al., 2000)

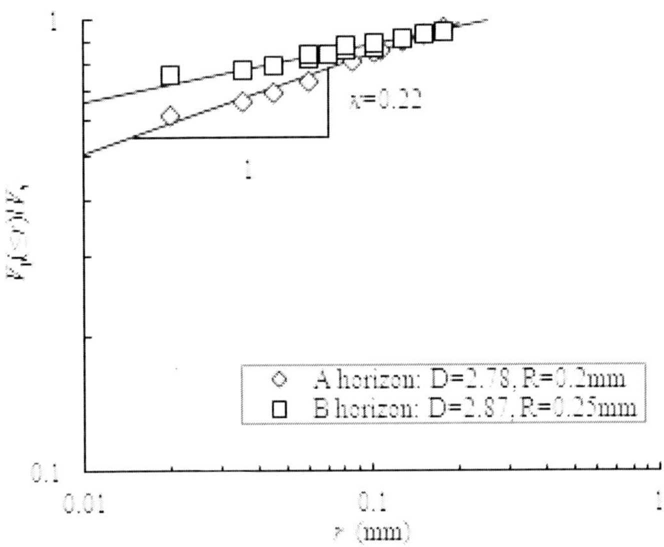

Figure 4. Fig. 4 Fractal dimension and the maximum radius of the PSD of two silty soils (Data from Vogel and Roth, 1998)

Experimental data of the pore-size distribution (PSD) for agricultural silty soils from two horizons, A and B were also taken from Vogel and Roth (1998). The pore-size distribution (PSD) was determined using the three-dimensional reconstructions of the pore space. The pore space was eroded by a spherical structuring element with a given diameter 2r and subsequently dilated by the same structuring element (Vogel and Roth, 1998). In the resulting image all pores smaller than $2r$ were removed. The cumulative volume $V_P(>r)/V_T$ was obtained using the application of successively larger structuring elements by Vogel and Roth (1998). The following relationship is obtained between $V_P(>r)/V_T$ and $V_P(\leq r)/V_T$.

$$\frac{V(>r)}{V_T}+\frac{V(\leq r)}{V_T}=\frac{V_v}{V_T}=\theta_s \tag{16}$$

where V_T is the total volume of soil, V_v is the total volume of void, θ_s is the saturated volumetric water content. The saturated volumetric water contents are 0.502 and 0.448 for the soils at A horizon and B horizon, respectively. Hence the cumulative volume $V(\leq r)/V_T$ can be obtained from the cumulative volume $V(>r)/V_T$. The cumulative volume of $V(\leq r)/V_T$ can be translated into $V(\leq r)/V_v$ according to Eq.(16). The cumulative volume of $V(\leq r)/V_v$ is shown in Fig. 4, where the dots denote the results transformed from the experimental data given by Vogel and Roth (1998) and the solid lines denote the results of the fractal model. The relationship between the pore volume and the pore radius satisfies an approximately linear expression in log-log plot. The slopes (κ) of the regression line are 0.24 and 0.13 for the soils at A and B horizons, respectively. According to Eq. (15), the fractal dimensions of the pore surface are 2.76 and 2.87 for A and B horizons, respectively. The maximum radius is defined as the value at which the cumulative volume of pores reaches the maximum, and $V(\leq r)/V_v$=100%. The maximum radii of the soil pores at the A and B horizons are 0.2cm and 0.25cm, respectively. The air-entry values can be calculated using the Young-Laplace equation, and they are 0.75kPa and 0.6kPa for the A and B horizons, respectively. The fractal dimension of the soil pore and the air-entry value of the soils at the A and B horizons are listed in Table 1. It can be seen from Figs. 3-4 that the pore surface can be modeled by fractal theory, and the pore-size distribution (PSD) satisfies with Eq. (11b). The soil structure can be described by the fractal dimension. The larger the fractal dimension, the more tortuous is the pore space. The fractal dimension is varied with the soil structure, and is different for different soils.

Uno et al. (1998) studied the relationship between the pore-size distribution (PSD) and unsaturated hydraulic properties of Toyoura sand. The pore-size distribution (PSD) of Toyoura sand was measured by the air intrusion method, and is shown in Fig. 5. The relationship between the pore volume and pore radius is approximated by a linear function in log-log plot, and satisfies with Eq. (11b). Hence the pore-size distribution (PSD) of Toyoura sand can be described by fractal model, and its fractal dimension is 1.67 obtained from Fig. 5 not increase with radius, i.e., V_p/V_v=1. The maximum pore radius of Toyoura sand is nearly 0.09 mm in Fig. 5. The surface tension (T) of water is 0.075kPa mm. The air-entry value can be calculated from Young-Laplace equation, and is 1.67kPa on the assumption that α=0. The fractal dimension of the soil pore and the air-entry value of Toyoura sand are listed in Table 1.

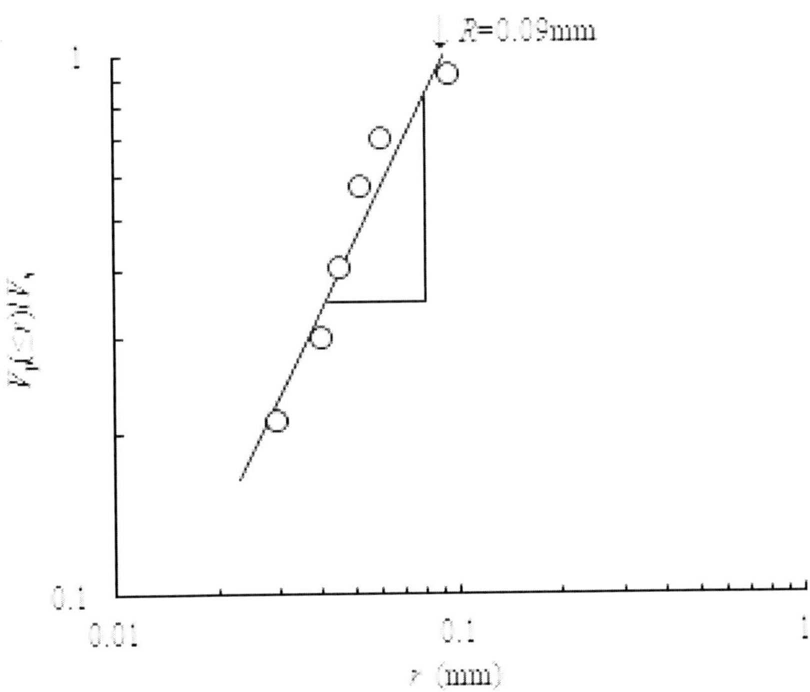

Figure 5. The fractal dimension and the maximum radius evaluated from the PSD of Toyoura sand. (Data from Uno et al., 1998.)

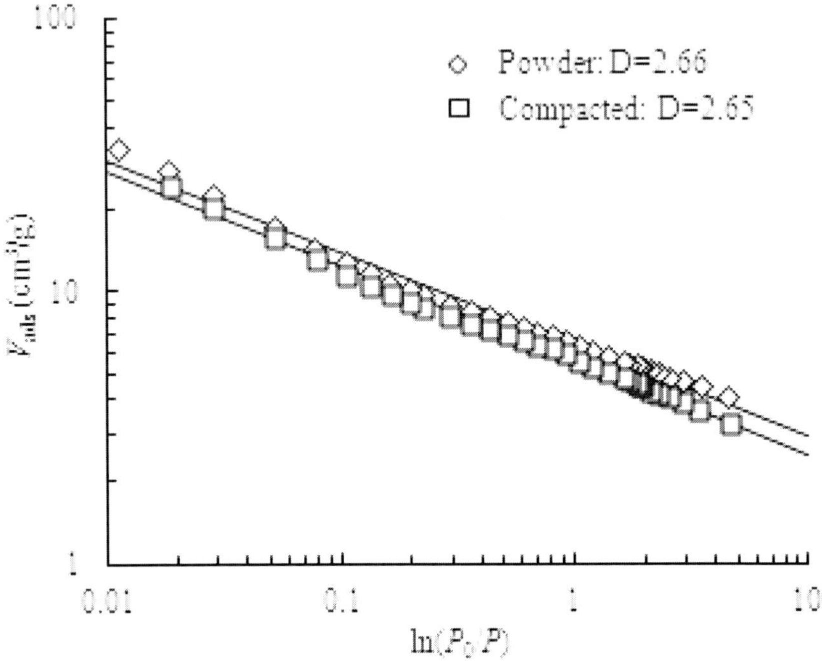

Figure 6. The surface fractal dimension of Tsukinuno bentonite from nitrogen adsorption

3.2. Adsorption isotherm

The fractal dimension of the pore surface serves to characterize the pore surface roughness or irregularities. The magnitude of the fractal dimensions is relevant to many important physicochemical processes namely adsorption, surface diffusion and catalysis. Avnir and Jaroniec (1989) proposed a convenient method to determine the fractal dimension from the nitrogen adsorption. Similarly with the adsorption isotherm equation, the correlation of the water volume absorbed by clay to the vapour pressure is given by

$$\frac{V_{\mathrm{w}}}{V_{\mathrm{m}}} = k\left[\ln\left(\frac{P_0}{P}\right)\right]^{D-3} \tag{17}$$

where V_w is the water volume absorbed by clay, V_m is the volume of montmorillonite, P is the partial water vapour pressure in equilibrium with clay at some water content w and some temperature T and P_0 is the equilibrium water vapour pressure of pure water at temperature T. The surface fractal dimension of Tsukinuno bentonite obtained from nitrogen adsorption is shown in Fig. 6. The surface fractal dimensions of bentonite powder and compacted bentonite are 2.66 and 2.65, respectively.

The adsorption isotherm also allows the calculation of swelling pressure of clay as a function of the water content (Kahr et al., 1990). Satisfactory agreement is found between the calculated and the experimental data for water content bigger than 10%.

$$p_s = -\frac{\bar{R}T}{M_w v_w} \ln\left(\frac{P_0}{P}\right) \tag{18}$$

where p_s is swelling pressure, the maximum axial pressure which is needed to maintain the original sample height, $\bar{R}$ is the molar gas constant, T is the Kelvin temperature, M_w is molecular mass of water and v_w is partial specific volume of water. Combining Eq. (17) with (18), the relationship between normalized water volume by the clay volume and swelling pressure is written as

$$\frac{V_w}{V_m} \propto p_s^{D-3} \tag{19}$$

The surface fractal dimension of clay can be obtained from Eq. (19). The experimental data of the swelling pressure and swelling deformation tests are compiled in Fig.7. It is seen that the relationship between the normalized water content and swelling pressure or vertical overburden pressure can be described by the same linear function which is expressed in Eq. (19). Therefore, Eq. (19) represents a general relationship between the water volume absorbed by clay per unit volume and vertical pressure supplied to specimen in swelling process. The surface fractal dimension of

Tsukinuno bentonite is 2.63, which nearly equals to that obtained from Fig. 6. In Fig. 6, C_b is the bentonite content.

Swelling deformation tests of Wyoming bentonite were also conducted by Millins et al. (1996) and Studds et al. (1998). The relationship between the clay void ratio (e_c) and vertical pressure is shown in Fig. 8. Relationship between the clay void ratio (e_c) and normalized water volume is given by

$$e_c = \frac{1}{C_m} \frac{V_w}{V_m} \tag{20}$$

where C_m is the montmorillonite content, and C_m=75%. It is seen that the slope of loge_c-log$p$ equals to that of log(V_w/V_m)-log$p$ from Eq. (20) for C_m being constant. Thus, the surface fractal dimension of Wyoming bentonite can be obtained from the linear relationship of loge_c-log$p$. The surface fractal dimensions of Wyoming bentonite are 2.64, which obtained from the swelling deformation tests conducted by different authors.

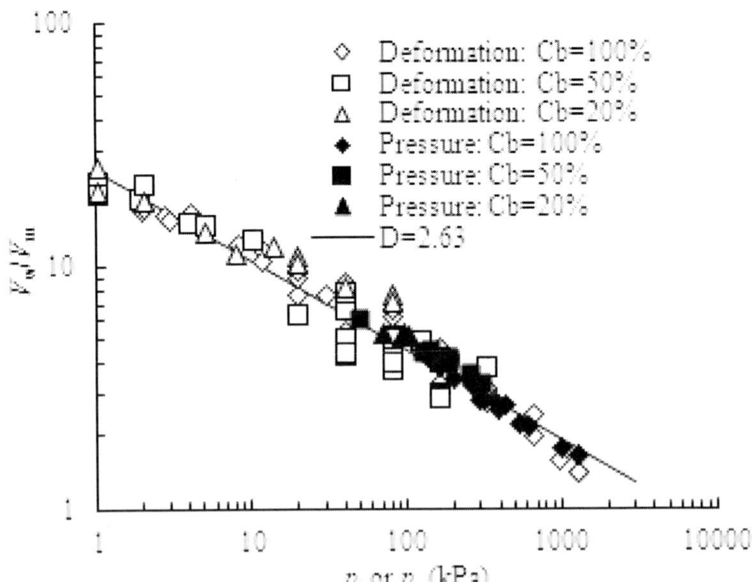

Figure 7. Relationship between absobed water volume and swelling pressure or vertical overburden pressure of Tsukinuno bentonite

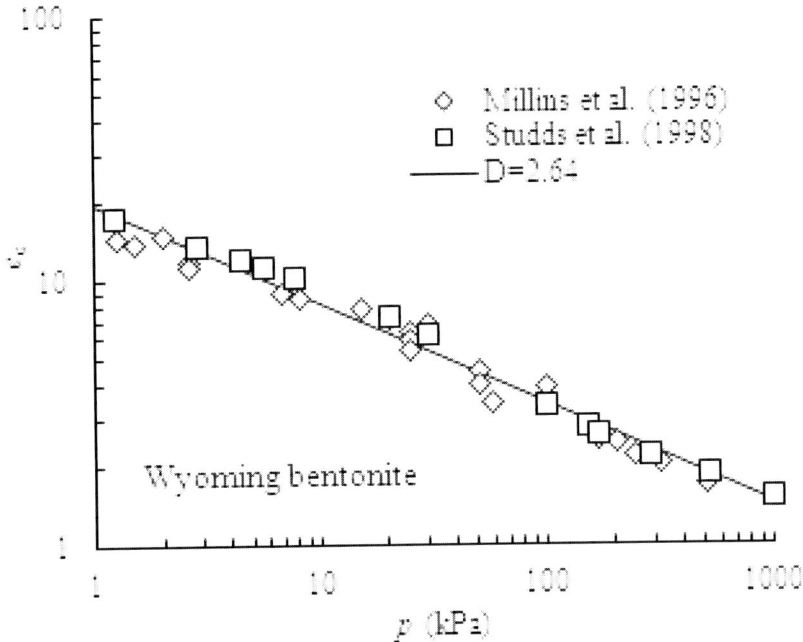

Figure 8. Relationship between clay void ratio and vertical overburden pressure of Wyoming bentonite

3.3. Correlation between the fractal dimension of PSD and that of the GSD

The pore-size distribution (PSD) is not often and easily measured experimentally. The grain-size distribution (GSD) is easily measured and the fractal dimension of the pore surface can be conveniently obtained from the grain-size distribution (GSD). The relationship between the pore-size distribution (PSD) and the grain-size distribution (GSD) can be constructed for the soils with a homogeneous fabric. Consider a pore limited by a number of grains N_g, the pore radius r is related to the radius r_g of the smallest grain as follows (Watabe et al., 2000)

$$r = f_p r_g \tag{21}$$

where f_p is a constant less than 1.0. If the particles are randomly arranged, the probability of having a grain of radius less than r_g is dP_g, the probability of having the N_g-1 grains with a radius greater than r_g is $(1-P_g)^{N_g-1}$. The probability P of having a pore with a radius less than r is given by (Watabe et al., 2000)

$$dP = N_g\left(1-P_g\right)^{N_g-1} dP_g \tag{22}$$

It is assumed that the grain-size distribution (GSD) satisfies the fractal model, and is given by

$$P_g = \left(r_g/R_g\right)^{3-D_g} \tag{23}$$

where D_g is the fractal dimension of the grain-size distribution (GSD), R_g is the maximum radius of soil grains. The probability P is given by (Watabe et al., 2000)

$$P = 1-\left[1-\left(\frac{r}{f_p R_g}\right)^{3-D_g}\right]^{N_g} \tag{24}$$

If $r/(f_g R_g)<<1$, Eq.(24) can be written as

$$P = N_g\left(\frac{r}{f_p R_g}\right)^{3-D_g} \tag{25}$$

It is seen that the pore surface has the same fractal dimension as that obtained form the grain-size distribution (GSD) from Eqs. (23) and (24).

Comparisons between the pore-size distribution (PSD) and the grain-size distribution (GSD) are shown in Fig. 9, where $P=V(\leq r)/V_v$ and $P_g=M(\leq r)/M_T$, V and M are the pore volume and grain mass, respectively. It is seen from Fig. 9 that both the pore-size distribution (PSD) and the grain-size distribution (GSD) can be expressed by fractal model, and they have the nearly same fractal dimension.

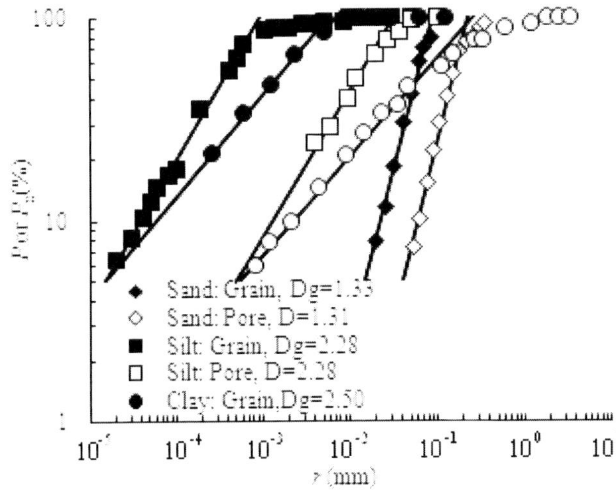

Figure 9. Correlation of the pore-size distribution (PSD) to the grain-size distribution (GSD).

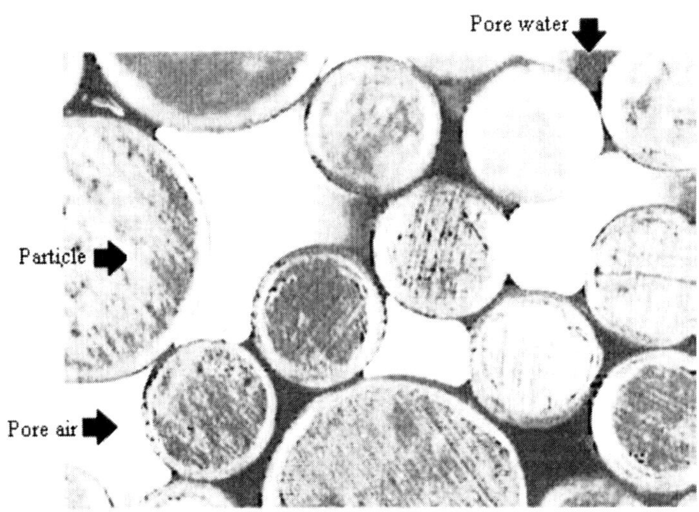

Figure 10. Model test for simulating the water distribution

4. SOIL-WATER CHARACTERISTIC CURVES (SWCC)

The soil-water characteristic curve defines the relationship between water content (gravimetric water content w, volumetric water content θ) or degree of saturation (S) and matric suction (ψ). The soil water characteristic curve (SWCC) has been used extensively in the study of unsaturated soils and relates the water content and soil matric suction. The relationship between the matric suction and the radius of the incurvated surface between the pore-air and pore-water is expressed by the Young-Laplace equation. The hydraulic and mechanical properties of unsaturated soils are correlated to the microstructures of soils through the Young-Laplace equation.

For unsaturated soils, the pore water is usually localized in small micropores, and the micropores are usually fully filled with water. The water content in macropores is very little, sometimes, there is no water in macropores in highly unsaturated soils (Fig. 10). The water distribution (black domain) of model test shown in Fig. 10 proved the above assumptions. In the model test, the soil grains of different sizes were simulated by the aluminum rods with different radius (Matsuoka, 1999). The water mainly distributed in the small pore and no water exists in the large pore for unsaturated soils in the model test.

Three air-water distribution states were identified as saturated funicular state, complete pendular state, and partial pendular state, respectively. All the voids are filled with liquid in the saturated state. Air bubbles are present in the voids in the funicular state, and the liquid phase is continuous. There is no continuity in the water phase in the pendular state. Pore water within unsaturated soils can be divided into three forms (Wheeler and Karube, 1996) (Fig. 11): bulk water within those void spaces that are completely flooded, meniscus water surrounding all inter-particle contact points that are not covered by bulk water and absorbed water (which is tightly bounded to the soil particles and acts as parts of the soil skeleton). The relationship between the soil-water characteristic curve (SWCC) and the pore-water forms is shown in Fig. 11. When the volumetric water content is less than its residual value, the pore-water is tightly absorbed by soil particle and cannot move freely. This absorbed water is seen as a part of soil particles. Thus, the contribution of the filled pores with radius $r \rightarrow dr$ to the water content is given by

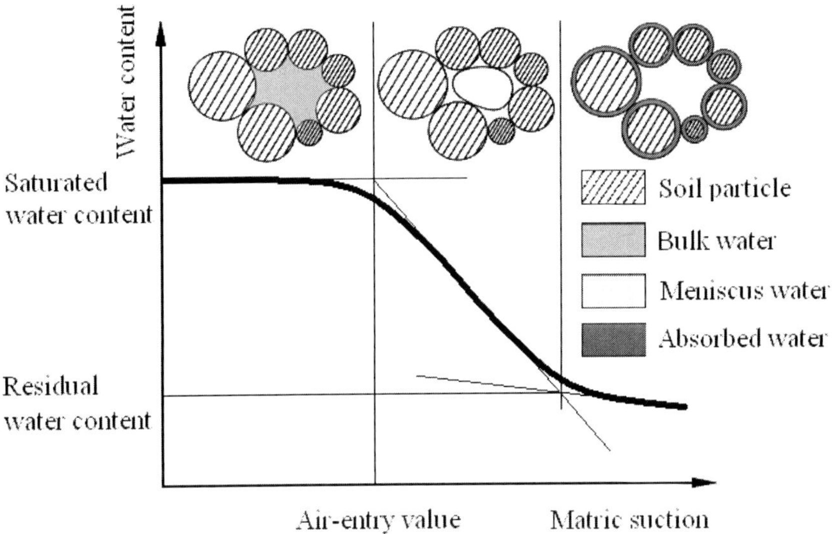

Figure 11. Relationship between the SWCCs and the pore-water forms of unsaturated soil (Matsuoka, 1999)

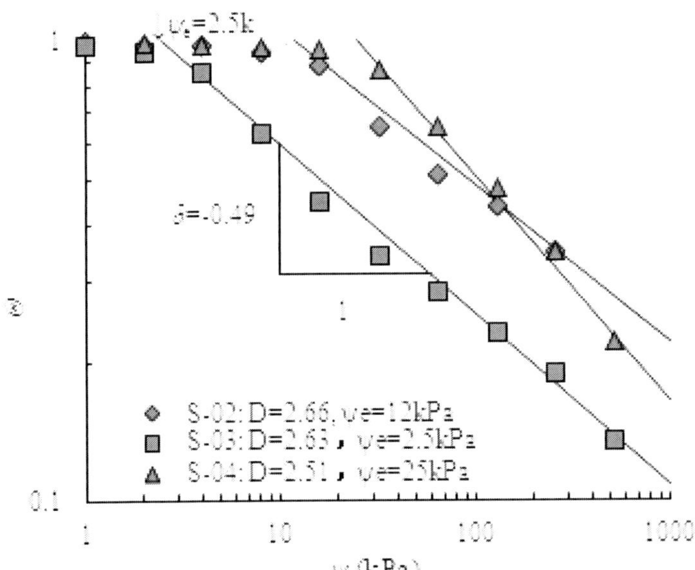

Figure 12. Comparison between the prediction and experiments of the SWCCs (Data from Watabe et al., 2000)

$$d\Lambda = \frac{N4\pi r^2 dr}{V_T}$$

(26)

where Λ is the relative volumetric water content, and $\Lambda=\theta-\theta_r$, θ and θ_r are the actual and residual volumetric water content, respectively. The residual volumetric water content is the volumetric water content at which the effectiveness of matric suction to cause further removal of water requires vapour migration. Substituting Eq. (9) into Eq. (26), Λ is given by

$$\Lambda = Br^{3-D}$$

(27)

where $B=4\pi C/[V_T(3-D)]$. Similarly with Eq. (27), the relatively volumetric water content at saturation is written as follows

$$\Lambda_s = BR^{3-D}$$

(28)

where R is the maximum radius of soil pores.

The relationship between matric suction ψ and the pore radius is obtained from the Young-Laplace equation. Thus, the soil-water characteristic curve (SWCC) is obtained as follows

$$\Theta = \left(\frac{\psi}{\psi_e}\right)^{\delta}$$

(29)

where $\delta=D-3$, Θ is the normollized volumetric water content, and is written as follows:

$$\Theta = \frac{\theta - \theta_r}{\theta_s - \theta_r} = \frac{S - S_r}{100 - S_r} \quad (30)$$

where θ, θ_r and θ_s are the volumetric water content, residual volumetric water content and saturated volumetric water content, S and S_r are the degree of saturation and residual degree of saturation. The residual volumetric water content is not always made routinely, in which case it has to be estimated by extrapolating available soil-water characteristics data towards lower water content, such as shown in Fig. 11. van Genuchten (1980) defined the residual volumetric water content as the water content for which the gradient $d\theta/d\psi$ becomes zero at high matric suction. From a practical point of view it seemed sufficient to define θ_r as the water content at some large matric suction, e.g. at the permanent wilting point ψ_e=1500kPa.

Equation (29) is the soil-water characteristic curve (SWCC) derived from the fractal model for the pore surface. The normalized volumetric water content is equal to unity at values of matric suction up to the air-entry value and equal to zero after residual saturation. The only two parameters ψ_e and D, which have obvious physical meaning, are used in Eq. (29) to express the soil-water characteristic curve (SWCC). The fractal dimension of the pore surface and the maximum pore radius can be evaluated from the pore-size distribution (PSD) measured using the mercury intrusion tests. The air-entry value can be calculated from the Young-Laplace equation using the maximum pore radius. Thus, the soil-water characteristic curve (SWCC) can be calculated from Eq. (29) using the fractal dimension and the air-entry value obtained from the mercury intrusion tests.

The soil-water characteristic curves (SWCC) were calculated from Eq. (29) using the fractal dimension of the pore surface and the air-entry value for specimens of S-02, S-03 and S-04, which listed in Table 1. The value of the normalized volumetric water content is calculated from Eq. (30). Here S_r=7.5% (Watabe et al., 2000). Experimental data show in good accord with the calculation of the soil-water characteristic curve (SWCC) in Fig. 12.

The experimental data of the soil-water characteristic curve (SWCC) measured in a multi-step outflow experiment were given by Vogel and Roth (1996) for silty soils. The saturated volumetric water contents are 0.502 and 0.448 for silty soils at the A and B horizons, respectively. The residual volumetric water content is equal 0 for both A and B horizons. Comparisons between the calculations of Eq. (29) and experimental data of the soil-water characteristic curves (SWCC) were shown in Fig. 13. The calculation of the soil-water characteristic curves (SWCCs) for silty soils was obtained from Eq. (29). The fractal dimensions of the soil pore and the air-entry values were listed in Table 1.

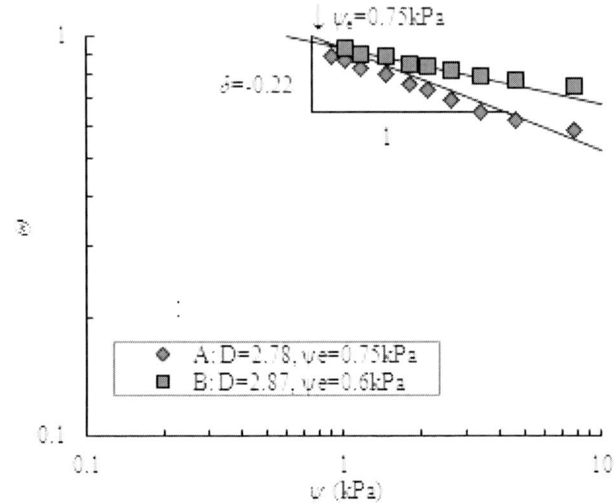

Figure 13. Comparison between prediction and experiments of SWCCs (Data from Vogel and Roth, 1996)

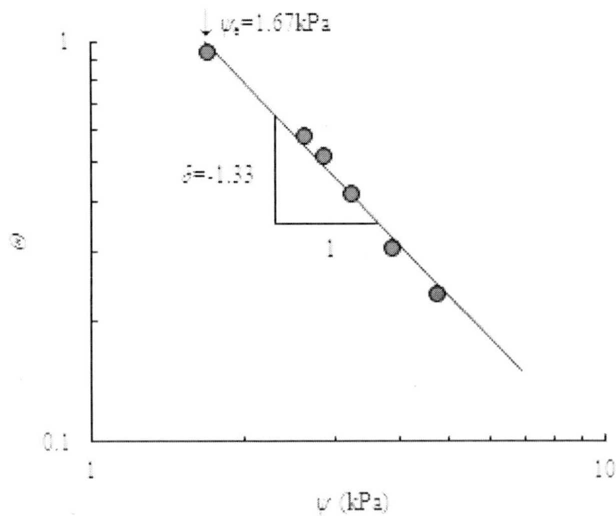

Figure 14. Comparison of the SWCCs for Toyoura sand (Data from Uno et al., 1998)

Uno et al. (1998) gave the experimental data of the soil water characteristic curves for Toyoura sand. Using the fractal dimension and the air-entry value obtained from the pore-size distribution (PSD), the soil-water characteristics can be simulated from Eq. (29) for Toyoura sand. The parameters obtained from the pore-size distribution (PSD) and used to predict unsaturated hydraulic conductivity are listed in Table 1. θ_s=0.425, θ_r=0. Comparison between the prediction of Eq. (29) and experimental data of the soil-water characteristics is shown in Fig. 14 for Toyoura sand.

Stingaciu et al. (2009) evaluated the feasibility of using nuclear magnetic resonance (NMR) relaxometry measurements to characterize pore size distribution and hydraulic properties in four porous samples with different texture and composition. The sandy samples FH31 and W3 were loaded in the penetrometer and packed at the same packing density as for the NMR measurements. The Mix8 and MZ samples were used as solid conglomerates. From the relaxation time distribution functions, the cumulative pore-size distribution functions were calculated with the average surface relaxivity. The normalized cumulative pore-size distributions functions were displayed in Fig. 15. The fractal dimension of pore surface and the maximum pore radius can be obtained from Fig. 15, and were listed in Table 2.

Table 2. Parameters for the prediction of SWCCs

Sample	Method	Bulk density (g/cm^3)	θ_s	θ_r	D	R (mm)	ψ_e (kPa)
FH31	Pressure plate	1.58	0.32	0.02	0.85	0.065	2.31
W3	MSO	1.42	0.33	0.058	1.51	0.0175	8.6
Mix8	Rosetta	1.45	0.41	0.061	1.52	0.00053	290
MZ	Pressure plate	1.60	0.44	0	2.29	0.00035	429

The soil-water characteristic curves (SWCC) were determined using the standard sand bed, pressure cell or multistep outflow method, and were plotted in Fig. 16. SWCC of FH31 and MZ were based on pressure plate measurements. SWCC of W3 was determined using multistep outflow (MSO) and that of Mix8 using ROSETTA software (Schaap et al., 2001). The matric suction as plotted in the abscissa (Fig. 16) was transformed into pore diameter using the Young-Laplace equation. In addition, the ordinate was the normalized volumetric water content. The predictions of SWCC were conducted using Eq. (29) with the parameters listed in Table 2. The parameters were derived from the pore-size distribution (PSD) shown in

Fig. 15. Comparisons of the predictions and experimental data were shown in Fig. 16. The predictions of Eq. (29) were in good accord with the experimental data.

In general, the soil-water characteristic curves (SWCC) were usually interpreted as cumulative distribution functions in comparison to pore-size distribution (PSD) functions Using the soil-water characteristic curves (SWCC), pore-size distribution can be extracted for a given porous medium on the basis of an empirical law that related the pore suction to the effective pore radius. The soil-water characteristic curves (SWCC) and the pore-size distributions (PSD) are related by correlations as: (1) the absolute values of the slopes of SWCC and PSD are equal; (2) the air-entry value of SWCC is related to the maximum pore radius of PSD through the Young-Laplace equation.

The following conclusions can be obtained from Figs. 12-16: (1) The surface of the soil pore can be described by fractal model, and the surface fractal dimension of the soil pore and the air-entry value can be evaluated from the pore-size distribution (PSD). The soil-water characteristic curves (SWCCs) can be calculated using fractal model of the soil pore. (2) The fractal dimension and air-entry value obtained from the soil pore-size distribution (PSD) are equivalent to those obtained from the soil-water characteristic curve (SWCC). Hence the fractal dimension and air-entry value can be obtained from the fitting of the soil-water characteristic curve (SWCC) if the pore-size distribution (PSD) were not measured.

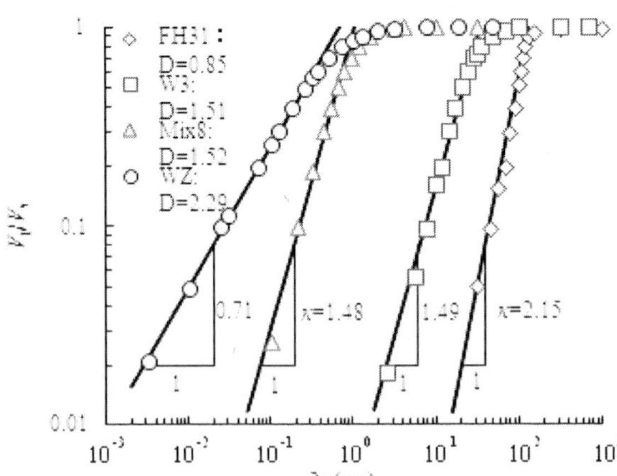

Figure 15. Cumulative pore-size distribution of for the three artificial substrates (FH31, W3, and Mix8) and the natural soil Merzenhausen (MZ), from which the fractal dimension of pore surface and the maximum pore radius were derived (Data from Stingaciu et al. 2009)

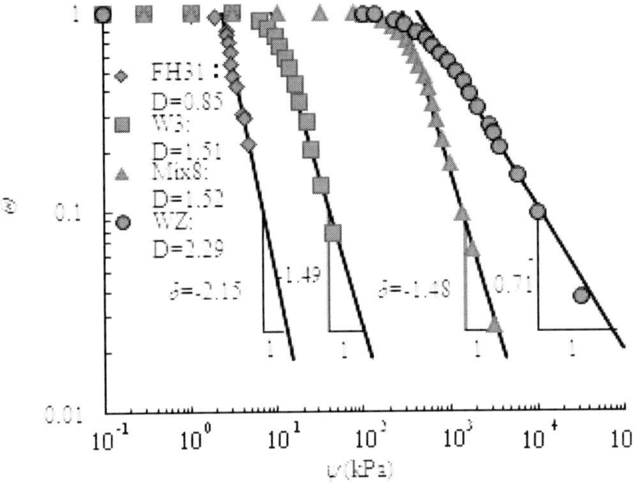

Figure 16. Comparison between the predictions of Eq. (29) and experimental data of SWCCs (Data from Stingaciu et al. 2009)

5. UNSATURATED HYDRAULIC CONDUCTIVITY AND DIFFUSIVITY

5.1. Brief review

Brutsaert (2000) made a brief review of the capillary model, which was cited as follows.

It is very difficult to determine unsaturated hydraulic conductivity through experiments. Hydraulic conductivity for unsaturated soils is a function of soil-water content θ. For this reason, over the years many attempts have been made to represent the hydraulic conductivity by simple parametric equations, in terms of other properties of the soil which are easier to determine. One of these was of the following form (Brutsaert, 2000):

$$k_r = \Theta^{\omega} \tag{31}$$

In the past, the power form of Eq. (31) has been derived on the basis of some widely different conceptual models of the pore geometry. There have been some marked differences in the obtained values of ω, depending on the adopted model of the pore geometry.

Uniform pore size models

A simple approach was that porous medium was characterized by some equivalent uniform pore size and the variability of the pore sizes was not considered. This approach invariably resulted in a constant value of ω. The capillary tube model of Averyanov (Polubarinova-Kochina, 1952) led to ω=3.5, whereas the extension of the hydraulic radius model of Kozeny (Kozeny, 1927) to unsaturated soils by Irmay (1954) produced ω=3.

Parallel models

In parallel models, the pore system was assumed to be equivalent with a bundle of uniform capillary tubes of many different sizes. The pore-size distribution was derived from the soil-water characteristics (SWCC), which can be related to the effective pore radius r by the Yong-Laplace equation. The true mean velocity in each pore was described by a Poiseuille-like equation for creeping flow. Purcell (1949) and Gates and Tempelaar-Lietz (1950) first used this approach. Several subsequent applications of the parallel model can be written in the common form (Brutsaert, 1967):

$$k_r = \frac{\beta \int_0^{\Theta} \psi(x)^{-2} \mathrm{d}x}{\beta_0 \int_0^{1} \psi(x)^{-2} \mathrm{d}x} \tag{32}$$

where x is the dummy variable representing Θ. The variable $\beta=\beta(\Theta)$ is related to the tortuosity, and β_0 is its value at satiation, when Θ=1. The tortuosity concept was originally introduced by Carman (1956) as an improvement on the non-uniform hydraulic radius model of Kozeny (1927) and it can be expressed as $T=(L_e/L)^2$, in which L_e is the actual or microscopic path length of the fluid particles in the pores and L is their apparent or macroscopic path length along the Darcy stream lines. The parameter β was introduced in the parallel model by Wyllie and Spangler (1952), because without it, Eq. (32) yielded values considerably larger than their experimental data. Burdine (1953) proposed on the basis of his experimental data that $\beta/\beta_0=\Theta^2$.

A further development was performed by adopting the following soil-water characteristic curve (SWCC) (Brooks and Corey, 1966):

$$\Theta = 1 \qquad \text{for } \psi \leq \psi_e$$

$$\Theta = \left(\frac{\psi}{\psi_e} \right)^b \qquad \text{for } \psi \leq \psi_e \tag{33}$$

Brooks and Corey (1966) integrated (32) with Burdine's assumption for β (Burdine, 1953) to obtain Eq. (31), with an exponent

$$\omega = 3 - \frac{2}{b} \tag{34}$$

It should be noted that, without Burdine's assumption for the tortuosity, the parallel model Eq. (32) applied with Eq. (33) would have yielded

$$\omega = 1 - \frac{2}{b} \tag{35}$$

Series-parallel models

The theoretical construction of this model also started with a bundle of parallel pores, each with a different but uniform size. However, these pores were then cut normally to the direction of flow with two resulting faces, and finally after some random rearrangement of the tubes the faces are joined again. This way account was taken of the random variations of the pore sizes, not only in the plane normal to the direction of flow, but also along the direction of flow. The discharge rate in each single pore, which consisted of two sections in cases, was assumed to be governed by the section with the smaller diameter. The pore-size distribution was derived form of the soil-water characteristics by means of the Yong-Laplace equation. The true velocity in each pore was obtained by means of a Poiseuille-like equation.

This model was originally proposed by Childs and Collis-George (1950) in a finite-difference scheme to calculate hydraulic conductivity from experimental data of the soil-water characteristics (SWCC). It was subsequently reformulated in integral form by Brutsaert (1968), to allow the derivation of more concise analytical expressions for hydraulic conductivity. The integral form can be written as

$$k_w = \left[\left(\frac{2T}{\gamma} \right)^2 \frac{\theta_0^{\ 2}(1-S_r)^2}{G} \right] \times \left[\int_0^\Theta \int_0^x \psi(x)^{-2} \mathrm{d}y \mathrm{d}x + \int_0^\Theta \psi(x)^{-2} \int_0^\Theta \mathrm{d}y \mathrm{d}x \right] \tag{36}$$

where G is a geometrical constant. In the case of Poiseuille's equation, G=8. It was trivial to show by integration by parts that the first double integral on the right was identical to the second. Hydraulic conductivity can be expressed concisely as (Brutsaert, 2000):

$$k_r = \frac{\int_0^\Theta (\Theta - x)\psi(x)^{-2} \mathrm{d}x}{\int_0^1 (1-x)\psi(x)^{-2} \mathrm{d}x} \tag{37}$$

The capillary model similar with Eq. (37) in form was originally proposed for by Fatt and Dykstra (1951).

5.2. Fractal model for hydraulic conductivity

Soil pore systems are viewed as a collection of interconnected voids. The individual voids are considered to have two equilibrium states: either filled by water or empty. Flow within the soil pores is assumed to be described by (Burdine, 1953; Mualem, 1976)

$$v = -\frac{r^2 g}{c\eta} \frac{\mathrm{d}h}{\mathrm{d}x} \tag{38}$$

where v is the average velocity within the pore; r is the radius of the pore; g is the gravity acceleration; η is the kinematic viscosity; c is a constant depending on the pore geometry; h is the hydraulic head, x is the position axis. Let us consider a porous slab of thickness Δx isolated from a homogenous soil column by two parallel cross-sections normal to the x axis. The areal porosity at each face of the slab is the same and it is equal to the volumetric porosity. The major simplifying assumption is that the actual configuration of slab may be replaced by a set of capillary tubes with different radii, parallel to the x axis. The flux dq flowing through the connections between pores of radius $r_1 \rightarrow r_1 + \mathrm{d}r_1$ on one side of the slab (at

x) and pores of radius $r_2 \rightarrow r_2+\mathrm{d}r_2$ on the other side of the slab (at $x+\mathrm{d}x$) can be expressed by (Mualem, 1976)

$$\mathrm{d}q = \beta r_{\mathrm{e}}^{2}(r_1, r_2, \rho) A_{\mathrm{e}}(r_1, r_2, \rho) \frac{\mathrm{d}h}{\mathrm{d}x} \mathrm{d}r_1 \mathrm{d}r_2 \tag{39}$$

where β is a constant which incorporates the fluid with the matrix properties; $r_e(r_1, r_2, q)$ is the effective radius; $A_e(r_1, r_2, q)$ is the effective area; ρ is the maximum radius of the water filled pores at volumetric water content θ. According to the Darcy equation, the hydraulic conductivity k_w is given by

$$k_{\mathrm{w}} = \frac{q}{\mathrm{d}h/\mathrm{d}x} = \beta \int_0^{\rho} \int_0^{\rho} r_{\mathrm{e}}^{2}(r_1, r_2, \rho) A_{\mathrm{e}}(r_1, r_2, \rho) \mathrm{d}r_1 \mathrm{d}r_2 \tag{40}$$

The relatively hydraulic conductivity (k_r) is the ratio of the hydraulic conductivity at any volumetric water content (k_w) to the hydraulic conductivity at saturation (k_s). The relative hydraulic conductivity (RHC) is related to the radius of the soil pores, and is written as (Mualem, 1976)

$$k_r = \frac{k_w}{k_s} = \frac{\int_0^{r} \int_0^{r} r_e^{2}(r_1, r_2, r) A_e(r_1, r_2, r) dr_1 dr_2}{\int_0^{R} \int_0^{R} r_e^{2}(r_1, r_2, r) A_e(r_1, r_2, r) dr_1 dr_2} \tag{41}$$

where r is the maximum radius of the water filled pores at the volumetric water content θ. It is assumed that the pore distribution at the two cross-section is completely random. The probability of pores of radius $r_1 \rightarrow r_1+\mathrm{d}r_1$ at x to encounter pores of radius $r_2 \rightarrow r_2+\mathrm{d}r_2$ at $x+\mathrm{d}x$ is given by

$$A_e(r_1, r_2) dr_1 dr_2 = f(r_1) f(r_2) dr_1 dr_2 \tag{42}$$

The effective radius is assumed to be equal to the radius *r* reduced by a factor, which is equal to the ratio between the effective area to flow and the actual pore area. The effective radius is given by (Mualem, 1976)

$$r_e = r\frac{\int_0^r \int_0^r A_e(r_1, r_2, r)dr_1 dr_2}{\int_0^R f(r_1)dr_1} = r\Lambda^{1/2} \tag{43}$$

Substituting Young-Laplace equation and Eqs.(42) and (43) into Eq.(41), the relative hydraulic conductivity (RHC) is obtained as

$$k_r = \left(\frac{\psi}{\psi_e}\right)^{\xi} \tag{44}$$

where ξ=3D-11. Substituting Eq. (29) into Eq. (44), the relative hydraulic conductivity (RHC) unction expressed by effective degree of saturation is written as follows

$$k_r = \Theta^{\xi/\delta} \tag{45}$$

Equations (44) and (45) are the relative hydraulic conductivity (RHC) function derived from the fractal model of the pore surface. The parameters in the proposed hydraulic conductivity function have an obvious physical implication, and can be determined from the pore-size distribution (PSD). Eqs. (44) and (45) offer a powerful, physical-based method to calculate the relative hydraulic conductivity (RHC). Eq. (44) is nearly similar with the Brooks and Corey equation (Brooks and Corey, 1966) in form, and the parameters in Eq. (44) have obvious physical implication. Eqs. (44) and (45) offer a powerful, physical-based method to calculate the relative hydraulic conductivity (RHC).

The expression of the soil-water diffusivity (SWD) can be derived from the unsaturated hydraulic conductivity and soil-water characteristics, and written as follows:

$$d = k_{w}\left|\frac{d\psi}{d\theta}\right| \tag{46}$$

where d is the soil-water diffusivity (SWD), k_w is the saturated hydraulic conductivity. Substituting Eqs. (44) and (45) into Eq. (46), the soil-water diffusivity (SWD) is written as

$$d = k_{s}\psi_{e}\left(\frac{\psi}{\psi_{e}}\right)^{\varsigma} \tag{47}$$

$$d = k_{s}\psi_{e}\Theta^{\varsigma/\delta} \tag{48}$$

where $\varsigma=2D-7$.

As given in Eqs. (29), (44), (45), (47) and (48), methods to evaluate the soil-water characteristics (SWCCs), unsaturated hydraulic conductivity (RHC) and soil-water diffusivity (SWD) are proposed using the fractal model for the pore surface. The fractal dimension and the maximum radius of soil pores can be obtained from the pore-size distribution (PSD). The air-entry value can be calculated from the Young-Laplace equation using the maximum radius of soil pores. The soil-water characteristic curves (SWCC), unsaturated hydraulic conductivity (RHC) and soil-water diffusivity (SWD) can be predicted using the fractal dimension and air-entry value, which can be determined from the pore-size distribution (PSD). According to Eq. (29), the fractal dimension of the pore surface and the air-entry value can also be evaluated from the fitted soil-water characteristic curve (SWCC). Thus, the unsaturated hydraulic conductivity (RHC) and soil-water diffusivity (SWD) can also be estimated from Eqs. (44), (45), (47) and (48) using the fractal dimension and air-entry value evaluated from the soil-water characteristic curve (SWCC).

1. PREDICTIONS OF SWCC AND RHC FROM PSD

Vogel and Roth (1998) inferred effective hydraulic properties of unsaturated soil from the structure of the pore surface. The water retention characteristic and the hydraulic conductivity were simulated by network models with 32^3=32768 nodes. The hydraulic conductivity, k_w, is determined at each step of desaturation by imposing a pressure gradient Δp across the ends of the network. Water flow q_{ij} in a bond between the nodes i and j through a horizontal surface A is described by Poiseuille's law. The hydraulic conductivity was calculated as $=qL/(A\Delta p)$, where L is the vertical length of the network. Comparison between the predictions of Eq. (44) and experimental data given by Vogel and Roth (1998) was shown in Fig. 17. The parameters used to do prediction were listed in Table 1, which were obtained from the fractal model for the pore-size distribution (PSD). The predictions were in good accord with the experimental data.

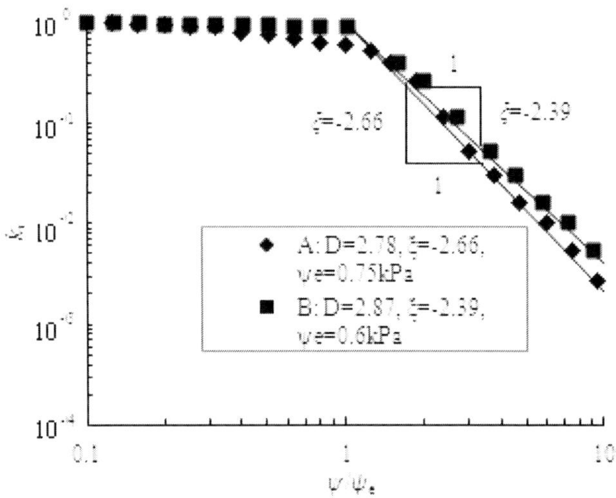

Figure 17. Comparison between the predictions of Eq. (44) and experimental data of the RHC (Data from Vogel and Roth, 1998)

Uno et al. (1998) studied the relationship between the pore-size distribution (PSD) and unsaturated hydraulic properties of Toyoura sand. The relationship between the relative hydraulic conductivity (RHC) and the normollized volumetric water content Θ can be calculated from Eq. (45) using the parameters listed in Table 1. The prediction of Eq. (45) satisfactorily agrees with the experimental data of relative hydraulic conductivity (RHC) for Toyoura sand in Fig. 18. The proposed function of relative hydraulic conductivity (RHC) (Eq. (45)) is verified by the experimental data in Fig. 18.

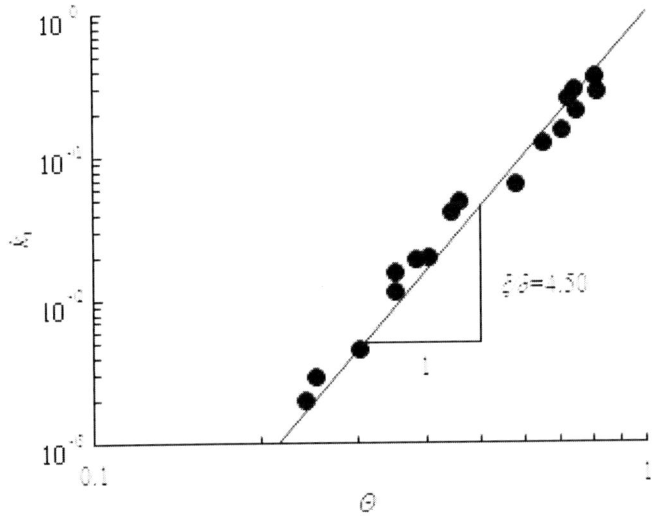

Figure 18. Comparison between predictions of Eq. (45) and experimental data of the RHC for Toyoura sand. (Data from Uno et al., 1998)

2. PREDICTION OF RHC AND SWD FROM SWCC

The characteristics of the pore-size distribution (PSD) were prerequisite to determine the relative hydraulic conductivity and soil-water diffusivity of unsaturated soils in Eqs. (44), (45), (47) and (48). A close relationship was established between the soil-water characteristic curve (SWCC) and the pore-size distribution by fractal model for the pore surface. The pore-size distribution (PSD) can be estimated from the soil-water characteristic curve because many soils show that the soil-water characteristic curve is equivalent to the pore-size distribution (PSD) function (Watabe et al., 2000). Thus, the fractal dimension of the pore surface and the air-entry value can be estimated from the fitting of the soil-water characteristic curve (SWCC). Thus, unsaturated hydraulic properties, such as relative hydraulic conductivity (RHC) and soil-water diffusivity (SWD) can be predicted using the fractal dimension and air-entry value obtained from the soil-water characteristic curve (SWCC).

Smettem and Kirkby (1990) offered the hydraulic properties of haploxeroll loam. Experimental data of the soil-water characteristic curve (SWCC) were shown in Fig. 19. The fractal dimension of the pore surface can be obtained from the fitted soil-water characteristic curve (SWCC). Through fitting the soil-water characteristic curve (SWCC), we obtained that δ=-0.37 for haploxeroll loam from Fig. 19. Fractal dimension of the pore surface can be calculated from Eq. (29) using the fitting value of δ. The fractal dimension (D) of the pore surface is 2.63 for haploxeroll loam. The air-entry value is 0.14kPa obtained from Fig. 19.

The hydraulic properties of unsaturated soils can be predicted using the fractal dimension of the pore-size distribution (PSD), obtained from fitted soil-water characteristic curve (SWCC). The fundamental parameters for predictions of hydraulic properties were tabulated in Table 3. The comparisons between predicted result and experimental data of the relative hydraulic conductivity (RHC) were shown in Fig. 20. The coefficient of hydraulic conductivity at saturation is 140 mmh^{-1}. Fig. 20 depicted the relationship between the relative hydraulic conductivity (RHC) and water potential for haploxeroll loam. The parameter ξ was -3.11 calculated from Eq. (44) using the fractal dimension D=2.63 for haploxeroll loam. The predictions of the relative hydraulic conductivity (RHC) calculated from the proposed function Eq. (44) are in satisfactory agreement with the experimental data.

The relationship between the soil-water diffusivity and water potential for haploxeroll loam was also shown in Fig. 20. The values of k_s and ψ_e are 140mmh^{-1} and 0.14kPa, respectively, used in Eq. (47). The parameters ς was -1.74, calculated from Eq. (47) using the fractal dimension D=2.63. The prediction of the soil-water diffusivity was the line passing through the point of (0.14, 0.64) with the slope of -1.74. The prediction of the soil-water diffusivity nearly agreed with the experimental data. The proposed function for the soil-water diffusivity was verified by the comparison between prediction and measured results. The parameters obtained from the soil-water characteristic curve (SWCC) and used to predict the relative hydraulic conductivity (RHC) and soil-water diffusivity (SWD) for haploxeroll loam were listed in Table 3.

Table 3. Parameters used to predict the unsaturated hydraulic conductivity

Soil type	D	δ	ξ	ξ/δ	ς	ψ_e (kPa)
Haploxeroll loam	2.63	-0.37	-3.11	8.41	-1.74	0.14
Toyoura sand (wetting)	1.60	-1.40	-6.2	4.43	-3.8	/
Toyoura sand (drying)	1.26	-1.74	-7.22	4.15	-4.48	/
Loamy sand	2.68	-0.32	-2.96	9.25	-1.64	2.0
McGee Ranch soil	2.51(GSD)	-0.49	-3.47	7.08	-1.98	4.6

Figure 21 shows the experimental data and the fitting of the soil-water characteristic curve (SWCC) for loamy sand. The experimental data are scaled from Simunek et al. (1999). The parameter δ was -0.32, which is obtained from the fitting of the soil-water characteristic curve (SWCC) for loamy sand from Fig. 26. Hence the fractal dimension of the pore surface was 2.68 calculated from Eq. (29) using the fitting value of δ for loamy sand. The air-entry value was given by the intersection between the line expressed by Eq. (29) and the line of Θ=1 in log-log plot. The air-entry value is 2kPa obtained from Fig. 21.

The parameters used to predict the hydraulic properties of loamy sand were listed in Table 3. Substituting D=2.68 and ψ_e=2kPa in Eq. (44), the predictions of relative hydraulic conductivity (RHC) can be obtained. The comparisons between predictions and experimental data of the relative hydraulic conductivity were shown in Fig. 22. The parameter ξ is -2.96 calculated from Eq. (44) using the fractal dimension D=2.68 for loamy sand. The predictions of the relative hydraulic conductivity were in good accord with the experimental data. The proposed function for the relative hydraulic conductivity was validated by the good agreement between the predictions and measurements.

The parameters of fractal model were first calibrated by matching the measured moisture-suction data. It should be noted that the experimental values of hydraulic conductivity were not matched by adjusting the model parameters. The calibrated SWCC model parameters were, instead, directly used to predict the relative hydraulic conductivity. The soil-water characteristic curves (SWCC) and their calibration of eight soils were shown in Fig. 23. The calibration of SWCC model parameters were listed in Table 4. As seen in these figures, the measured moisture-suction data for these soils were unavailable for the full range (0%-100%) of the degree of saturation. Predicting the suction beyond the available experimental data range was a challenging task since the pattern of variation was unknown (Ravichandran and Krishnapillai, 2011).

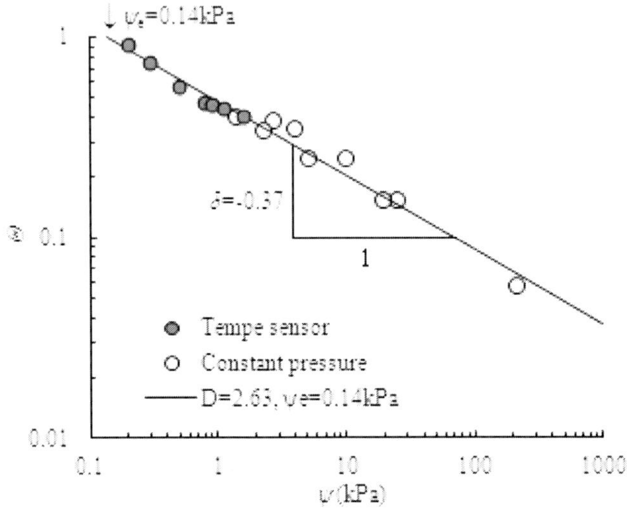

Figure 19. Fractal dimension of the pore surface and air-entry value evaluated from SWCC (Data from Smettem and Kirkby, 1990)

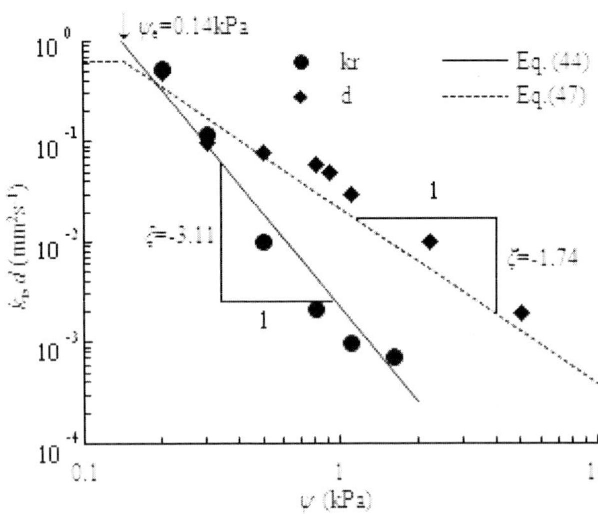

Figure 20.between predictions and experimental data of RHC and SWD for haploxeroll loam (Smettem and Kirkby, 1990)

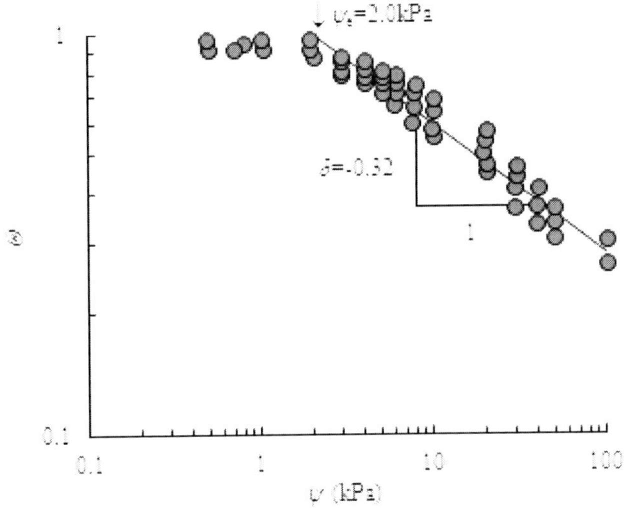

Figure 21. Fractal dimension and the air-entry value obtained from the fitting of SWCC for of loamy sand (Data from Simunek et al., 1999)

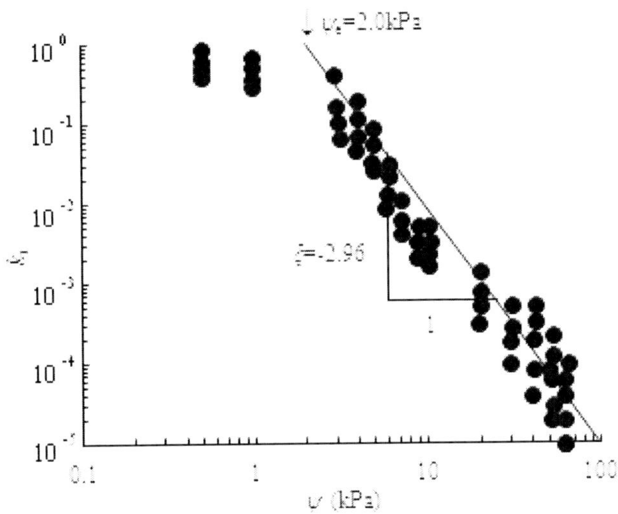

Figure 22. Comparison between measurements of RHC and prediction of Eq. (44) for loamy sand (Data from Simunek et al., 1999)

Table 4. Parameters of the fractal model for eight selected soils

Soil	θ_s	S_r (%)	D	ψ_e (kPa)	Reference
Lakeland	0.375	30-100	1.75	2.5	Elzeftawy and Cartwright, 1981
Superstition	0.5	30-100	2.7	1.2	Richards, 1952
Columbia sandy loam	0.458	50-100	2.29	4.5	Brooks and Corey, 1964
Touchet silt loam	0.43	20-100	2.09	7	Brooks and Corey, 1964
Silt loam	0.396	50-100	2.59	15	Reisenauer, 1963
Guelph loam	0.52	45-100	2.75	3	Elrick and Bowmann, 1964
Yolo light clay	0.375	45-100	2.87	1.5	Moore, 1939
Speswhite kaolin	0.56	55-100	2.85	15	Peroni et al., 2003

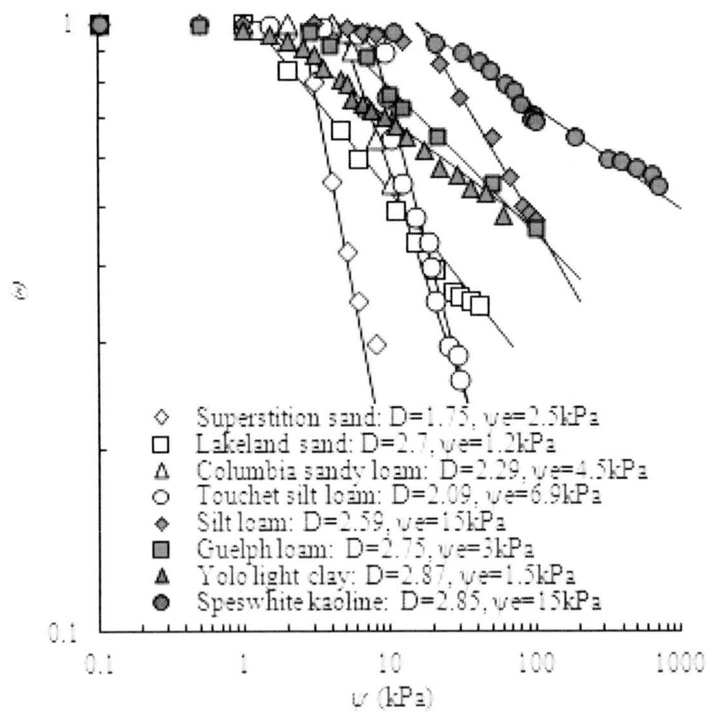

Figure 23. Fractal dimension and air-entry value obtained from the fitting of SWCC

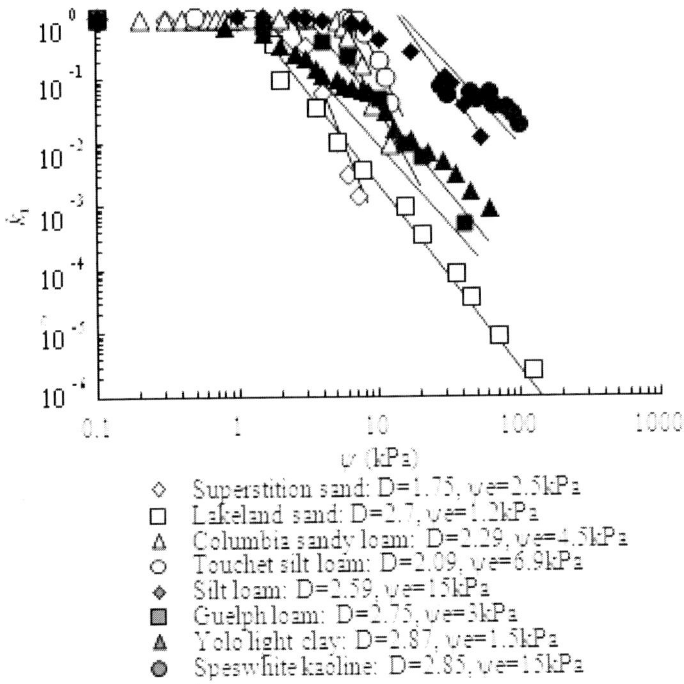

Figure 24. Comparison between predictions of Eq. (44) experiments of RHC

The relative hydraulic conductivity of the above mentioned eight soils were predicted using the fractal model with the same fitting parameters listed in Table 4. The fitting parameters were obtained by matching the experimental SWCC. Fig. 24 illustrated the prediction of Eq. (44) and experimental data of relative hydraulic conductivity. It should be noted that the fractal model parameters were not calibrated or adjusted to match the measured hydraulic conductivity values. The predictions of Eq. (44) were comparable for sand and loam, as shown in Fig. 24. The fractal model showed slight deviations from the measured data of clay. In the case of Yolo light clay and Speswhite kaolin, the difference between the predictions and the experimental data increased as the suction increases (Fig. 24). The fractal model showed better predictions at lower suction range. However, the accuracy of fractal model in the higher suction range could not be verified because the experimental results were available only for the lower suction ranges.

It is seen that it is feasible to estimate the relative hydraulic conductivity of unsaturated soils using the fractal dimension of the soil pore surface from Figs. 19-24. A practical method to express the relative hydraulic conductivity was proposed to use the fractal dimension and the air-entry value, which can be obtained from the fitting of the soil-water characteristic curve (SWCC).

3. PREDICTION OF RHC AND SWD FROM GSD

The feasibility to determine the soil hydraulic properties using the fractal dimension of the grain-size distribution (GSD) is examined by the experimental data of McGee Ranch soil. The grain-size distribution (GSD) was measured by Hunt and Gee (2002) and was shown in Fig. 25. The fractal dimension of the grain-size distribution (GSD) is 2.51, obtained from Fig. 25.

Hunt and Gee (2002) gave the values of θ_s and θ_r were 0.4 and 0.01, respectively, the air-entry value was 4.6kPa. Using the fractal dimension of the grain-size distribution (GSD), the soil-water characteristic curve (SWCC) and the relative hydraulic conductivity (RHC) can be determined. The parameters of the grain-size distribution (GSD) used to determine the hydraulic properties are listed in Table 3. Comparisons between the predictions of Eq. (29) using the fractal dimension of the grain-size distribution (GSD) and the experimental data of soil-water characteristic curve (SWCC) are shown in Fig. 26. It was seen that a good result is obtained to predict the soil-water characteristic curve (SWCC) using the fractal dimension of the grain-size distribution (GSD).

Hunt and Gee (2002) gave the values of saturated hydraulic conductivity was 0.001cm/s. Comparisons between the predictions of Eqs. (44) and (45) and experimental data were shown in Fig. 27. The prediction of Eqs. (44) and (45) nearly satisfied the experimental data of relative hydraulic conductivity in Fig. 27. In Fig.27, the predictions of Eq. (45) deviated from the experimental data near saturation, and extend to equal to the experimental results at low water content. The soil hydraulic conductivity can be approximately determined using the fractal dimension of the grain-size distribution (GSD).

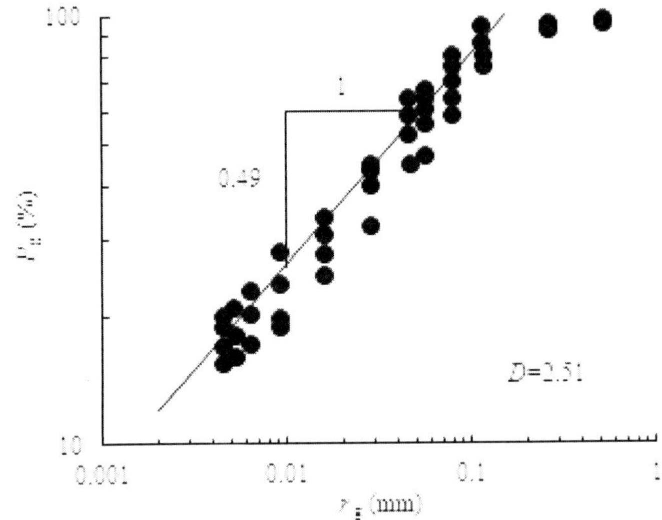

Figure 25. Fractal dimension of GSD for McGee Ranch soil (Data from Hunt and Gee, 2002)

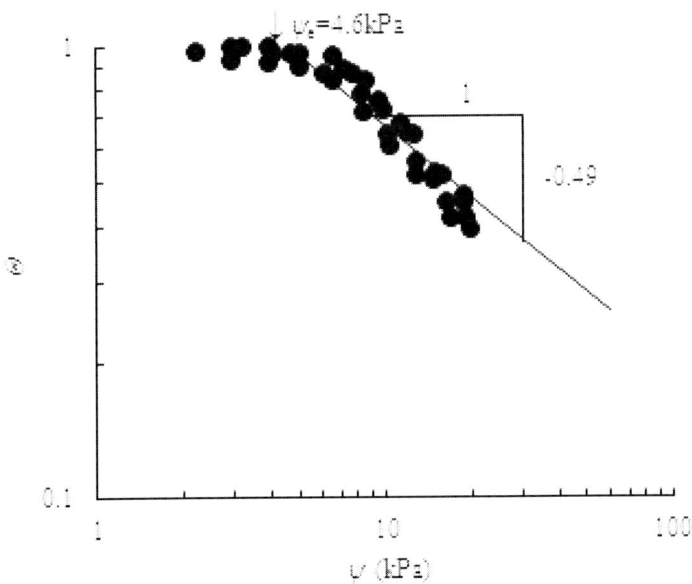

Figure 26. Comparisons between predictions and experiments of SWCC for McGee Ranch soil (Data from Hunt and Gee, 2002)

6. COMPARISON OF THE FRACTAL MODEL WITH THE VAN GENUCHTEN–MUALEM (G-M) MODEL

6.1. van Genuchten–Mualem (G-M) model for relative hydraulic conductivity (RHC)

van Genuchten (1980) derived an empirical relationship to describe the soil-water characteristic curve (SWCC):

$$\Theta = \frac{1}{(1+|\alpha\psi|^{n})^{m}} \tag{49}$$

where α, n and m are the van Genuchten curve-fitting parameters, and m=1-1/n.

Burdine (1953) and Mualem (1976) presented a similar model to estimate the unsaturated hydraulic conductivity from pore size distributions inferred from soil water retention characteristics. The models share a number of similarities, allowing them to be written in a general form as (Hoffmann-Riem et al., 1999)

$$k_{\mathrm{r}} = \frac{k_{\mathrm{w}}}{k_{\mathrm{s}}} = \Theta^{l}\left(\frac{\int_0^{\Theta}\psi^{-\chi}\mathrm{d}\Theta}{\int_0^{1}\psi^{-\chi}\mathrm{d}\Theta}\right)^{\gamma} \tag{50}$$

where k_{w} is the unsaturated hydraulic conductivity at any water content, k_{s} is the saturated hydraulic conductivity, Θ is the normalized water content, ψ is matric suction. Generally accepted parameter values for the parameters (l, χ, γ) are (0.5, 1, 2) and (2, 2, 1) for the Mualem (1976) and the Burdine (1953) models, respectively.

Using the Mualem model, van Genuchten (1980) derived a closed-form to determine the relative hydraulic conductivity (RHC) at a degree of saturation,

$$k_r = \frac{\left[1 - |\alpha\psi|^{n-1}\left(1 + |\alpha\psi|^n\right)^{-m}\right]^2}{\left(1 + |\alpha\psi|^n\right)^{m}} \tag{51}$$

$$k_r = \Theta^{\frac{1}{2}}\left[1 - \left(1 - \Theta^{\frac{1}{m}}\right)^m\right]^2 \tag{52}$$

Equations (51) and (52) are the representations of the G-M model, which is the most widely and popularly used to predict the hydraulic conductivity.

The other models for relative hydraulic conductivity (RHC) were also introduced as follows.

6.2. van Genuchten–Burdine (G-B) model for relative hydraulic conductivity (RHC)

Hydraulic conductivity resulting from the Burdine capillary models (Burdine, 1953) was expressed as

$$k_r = \Theta^2\left[1 - \left(1 - \Theta^{\frac{1}{m}}\right)^m\right] \tag{53}$$

Replacing Eq. (49) in Eq. (53) yielded

$$k_r = \frac{\left[1 - |\alpha\psi|^{n-2}\left(1 + |\alpha\psi|^n\right)^{-m}\right]}{\left(1 + |\alpha\psi|^n\right)^{2m}} \tag{54}$$

where m=1-2/n.

Note that for both Eqs. (52) and (54), because m is less than 1, the derivative dk/dθ is infinite for θ_s, whatever the value of n.

6.3. van Genuchten–Fatt & Dykstra (G-FD) model for relative hydraulic conductivity (RHC)

The capillary model proposed by Fatt and Dykstra (1951) was expressed as

$$k_r = \frac{\int_0^{\Theta} (\Theta - x)\psi(x)^{-2-c}\,dx}{\int_0^1 (1 - x)\psi(x)^{-2-c}\,dx} \tag{55}$$

Evaluating $\psi(x)$ from Eq. (49), the integrals of Eq. (55) with the condition of m>-(2+c)/2n and n>2+c resulted in:

$$k_r = \frac{\Theta B(a_1, b) I_{\Theta^{1/m}}(a_1, b) - B(a_2, b) I_{\Theta^{1/m}}(a_2, b)}{B(a_1, b) - B(a_2, b)} \tag{56}$$

where a_1=m+(2+c)/n, a_2=2m+(2+c)/n, b=1-(2+c)/n, $I_x(a, b)$ and $B(a, b)$ are the Incomplete Beta and Beta functions of positive arguments a and b, respectively, and given by:

$$B(a,b)=\int_0^1 x^{a-1}(1-x)^{b-1}\mathrm{d}x,\quad I_y(a,b)=\frac{\int_0^y x^{a-1}(1-x)^{b-1}\mathrm{d}x}{B(a,b)} \tag{57}$$

The results presented by Touma (2009) were obtained with c=0.5 and a_1=1.

Even though the conductivity predicted by the capillary model of Fatt and Dykstra (1951) gave the best results compared with both the quasi-analytical solution and observations, the resulting expression was not easy to use because it was necessary to evaluate Incomplete Beta and Beta functions. In order to simplify G-FD model, Θ was expressed by Eq. (49) and k_r was according to Eq. (31) with ω=2+2.5/mn. The fitted result in a value of b was close to -mn, and a conductivity curve close to that predicted by Eq. (56).

6.4. Brooks & Corey-Brutsaert model for relative hydraulic conductivity (RHC)

Brutsaert (2000) presented a capillary model expressed as Eq. (37). Combining this capillary model with the soil-water characteristic curve (SWCC) of BC model (Brooks and Corey, 1966), relative hydraulic conductivity was written as Eq. (31) in form, and the parameter ω was given by

$$\omega = 2\text{-}\frac{2}{b} \tag{58}$$

Combining of BC model for the soil-water characteristics with the Mualem and the Burdine capillary models, the relative hydraulic conductivity curves were also expressed as Eq. (31) in form, and the parameter ω was given by

$$\omega = 3+\frac{2}{mn} \tag{59}$$

when BC is applied with the Burdine condition, and

$$\omega = 2.5+\frac{2}{mn} \tag{60}$$

when applied with the Mualem condition.

Here we focused on the G-M model, as it was the most widely used in soil science and hydrology.

van Genuchten (1980) found that the predictions of the G–M model were less accurate for Beit Netofa clay. The fractal dimension and the air-entry value evaluated from the fitted soil-water characteristic curve (SWCC) are 2.89 and 60kPa, respectively (Fig. 28) The fitted parameters of the van Genuchten model, α and n are 0.0015 and 1.17, respectively from Fig.28. The parameters for the determination of the relative hydraulic conductivity (RHC) are listed in Table 5.

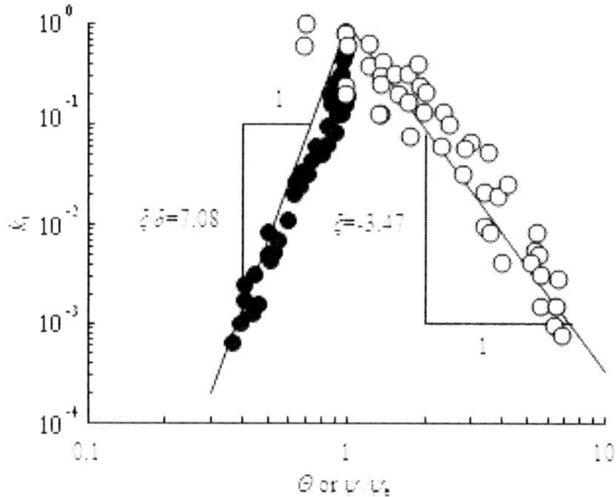

Figure 27. Comparisons between predictions and experimental data of RHC for McGee Ranch soil (Data from Hunt and Gee, 2002)

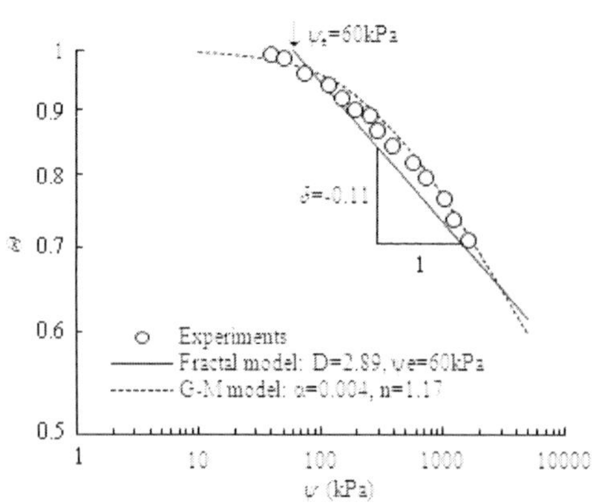

Figure 28. Fractal dimension and the parameters of G-M model obtained from SWCC for Beit Netofa clay (Data from van Genuchten, 1980)

Table 5. Parameters obatined from SWCC for prediction of RHC

Soil type	Fractal model				G-M model	
	δ	D	ξ	ξ/δ	α (kPa^{-1})	n
Beit Netofa clay	-0.11	2.89	-2.33	21.2	0.004	1.17
Hanford glacial soil 0-099	-0.62	2.38	-3.86	6.23	0.015	1.71
Hanford glacial soil 2-1637	-0.82	2.18	-4.46	5.44	0.076	1.89
Grenoble sand	-0.64	2.36	-3.92	6.13	0.4	2.17

The saturated and residual water contents of Beit Netofa clay are 0.446 and 0, respectively. Comparisons between the predictions of both the fractal model (Eq. (44)) and the G–M model (Eq. (51)) and the experimental data of the relative hydraulic conductivity (RHC) for Beit Netofa clay are shown in Fig. 29. It is seen that predictions of the fractal model (Eq. (44)) satisfactorily agree with the experimental data, especially at high matric suction, while the predictions of the G–M model (Eq. (51)) are found to deviate from the experimental data. However, the fractal model shows a sharp corner at the air-entry value point, which is disagree with the measured data that usually show smooth transition after the air-entry value point. This default does not reduce the advantages to predict the relative hydraulic conductivity (RHC) using the fractal dimension obtained from the pore-size distribution (PSD), because the hydraulic conductivity near the point of the air-entry value nearly equal to the saturated conductivity.

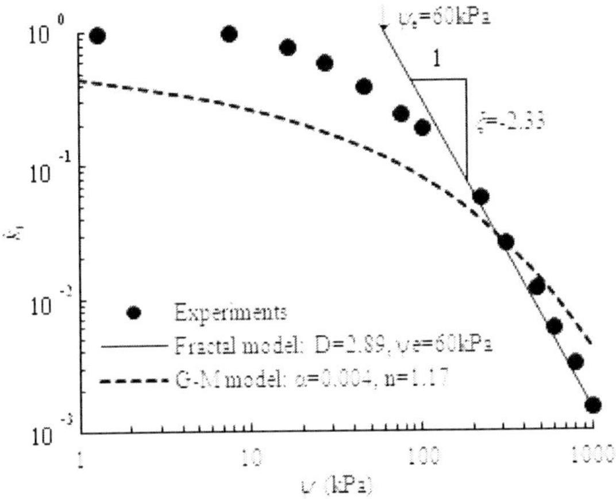

Figure 29. Comparison between the predictions of the RHC using fractal model and G-M model (van Genuchten, 1980)

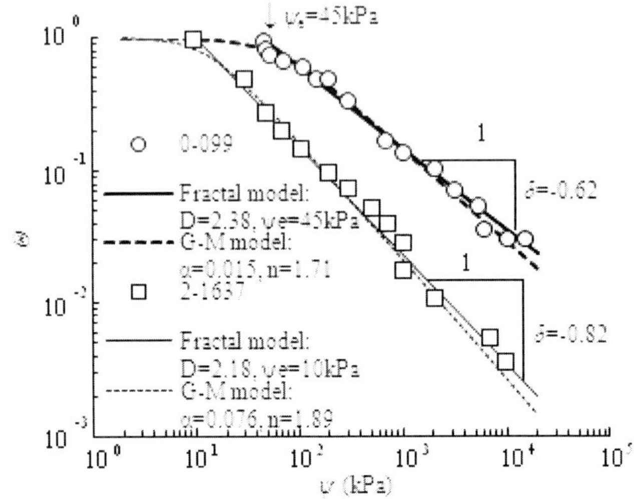

Figure 30. Fractal dimension and G-M parameters obtained from SWCC for Hanford glacial (Khaleel and Relyea, 1995)

The measured soil–water characteristic curves of Hanford glacial soil by Khaleel and Relyea (1995) were shown in Fig. 30. The saturated water contents (θ_s) are 0.338 and 0.303 for the samples 0–099 and 2–1637, respectively. The residual water contents (θ_r) are 0.039 and 0.025 for the samples 0–099 and 2–1637, respectively. From Fig. 30, it is seen that the soil-water characteristic curves (SWCC) of Hanford glacial soil are in good accord with the fractal model. The fractal dimensions (D), according to Eq. (29), are 2.35 and 2.18 obtained from the soil-water characteristic curves for the samples 0–099 and 2–1637 of Hanford glacial soil, respectively.

From the experimental data of the soil-water characteristic curve (SWCC), the parameters for determining relative hydraulic conductivity (RHC) using the fractal dimension are listed in Table 5. The larger the fractal dimension, the larger the complexity of the pore connectivity will be. The larger the fractal dimension, the larger the change in the water content with the same change in matric suction will occur. From the fittings of the soil–water characteristic curve in Fig. 30, the parameters (a and n) in the van Genuchten model (1980) for the soil–water characteristic curve (SWCC) are obtained, and are also listed in Table 5. Comparisons between the predictions of the fractal model (Eq. (45)) and the experimental data of the relative hydraulic conductivity (RHC) for Hanford glacial soil are shown in Fig. 31. The predictions of the fractal model (Eq. (45)) nearly agree with the experimental data of the relative hydraulic conductivity (RHC) for Hanford glacial soil in Fig. 31. The predictions of the G–M model (Eq. (52)) deviate from the experimental data of the relative hydraulic conductivity (RHC) in Fig. 31. It is seen that the predictions of the fractal model are better than those of the G–M model for Hanford glacial soil in Fig. 31.

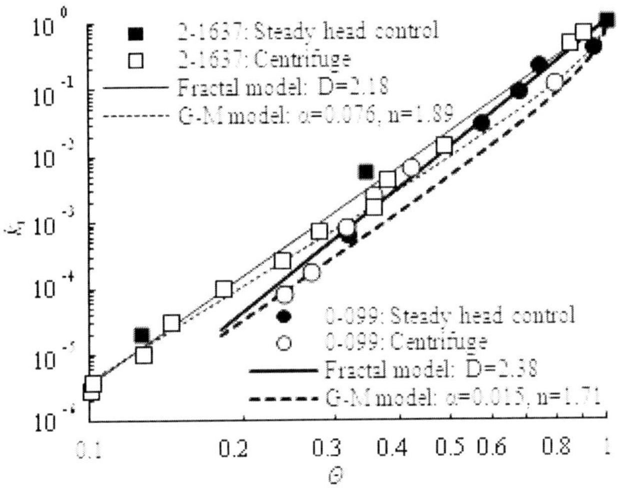

Figure 31. Comparison between the predictions using fractal model and G-M model (Data from Khaleel and Relyea, 1995)

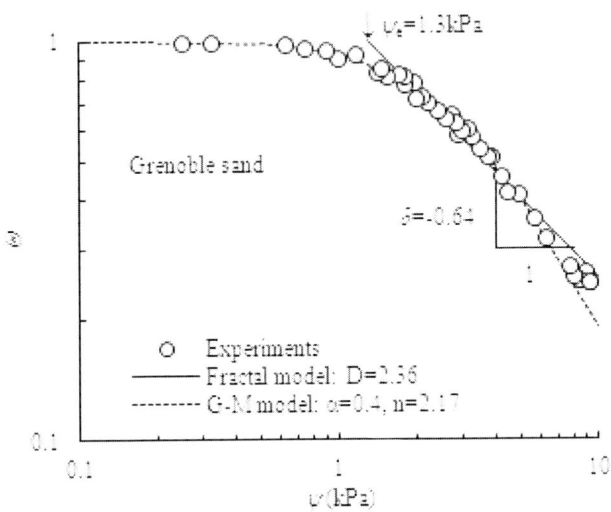

Figure 32. Fitted SWCC curves

The combinations of Eq. (29) with fractal capillary model were tested on Grenoble sand. Data points for the soil-water characteristic curve (SWCC) and hydraulic conductivity were taken from Touma (2009). The infiltration experiment was conducted under a constant head of 2.3cm (Touma & Vauclin, 1986). The solid lines in Figure 37 were the fitted curves for Eq. (29), and the resulting predicted hydraulic conductivity were shown in Fig. 32. Fractal model gave good results for the Grenoble sand. The parameters obtained through the fitting results of soil-water characteristic curves (SWCC) were used to predict hydraulic conductivity curve. Comparison between the fractal model and the G-M model were shown in Figs. 33. It was found that the prediction of hydraulic conductivity curve using Eq. (45) closed to the experiments for Grenoble sand.

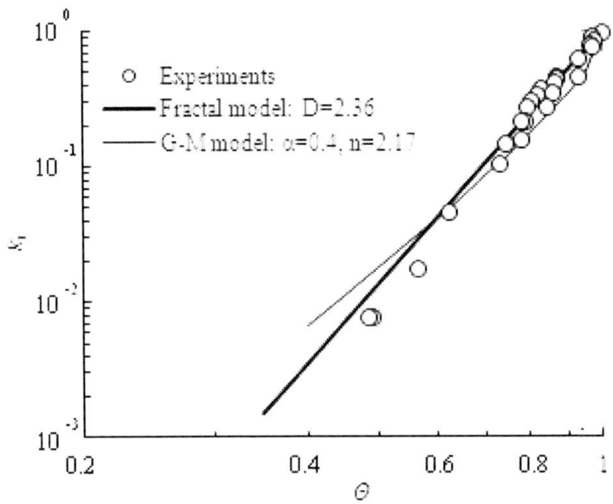

Figure 33. Comparison between fractal model and G-M model

7. RAINFALL INFILTRATION OF UNSATURATED SOILS

Rainfall induced slope failures often occur as relatively shallow failure surfaces orientated parallel to the slope surface and are observed and analyzed by different mechanisms. The effect of seepage on slope stability is introduced in the analyses by calculating the critical depth for an infinite slope with rainfall infiltration, while the effect of both negative and positive pore water pressures on the stabilities of initially unsaturated slopes are explained and coupled with infinite slope analysis and pore-air flow analysis methods in order to present a predictive formulation of slope failures that occurs in rainfall evens and is derived from a fractal model on unsaturated soil. The formulation serves as a baseline analysis method for evaluating potentially unstable slopes.

The pore water pressure pattern that develops in the initially unsaturated soil will occur as a transient process as the infiltration moves downward into the soil profile. Several factors should been taken into account in unsaturated analyses. The shear strength of the soil mass and the development of seepage forces which both depend on the evolution of the pore water pressure profile must been addressed in detail.

Here, an individual soil slice can be treated as a one-dimensional column with vertical infiltration. Suppose that any lateral flow between the adjacent slices will be equal on the up-slope and down-slope boundaries. Then considering vertical infiltration only could meet the flow continuity requirement. These one-dimensional infiltration analyses are addressed by the saturated/unsaturated seepage finite element software SEEP/W (GEO-SLOPE International Ltd.2007) coupled with the air flow software AIR/W (GEO-SLOPE International Ltd.2007).

For a homogenous, isotropic soil, one-dimensional water flow equation (GEO-SLOPE International Ltd.2007) is given by:

$$m_w \gamma_w \frac{\partial H_w}{\partial t} = \frac{\partial}{\partial y}\left(k_w \frac{\partial H_w}{\partial t}\right) + m_w \frac{\partial P_a}{\partial t} + Q_w \tag{61}$$

where H_w is the total water energy potential comprised of both pressure and elevation potentials; P_a is the pore air pressure; k_w is hydraulic conductivity; y is y coordinate; m_w is the slope of SWCC; t is time; and Q_w has units of length per time.

For the air conversation of mass, we can arrive at the pore air general mass balance equation (GEO-SLOPE International Ltd.2007) as like:

$$\left(\rho_w m_w + \frac{\theta_a}{\bar{R}T}\right)\frac{\partial P_a}{\partial t} = \frac{\partial}{\partial y}\left[\frac{\rho_a k_a}{\gamma_{0a}}\frac{\partial P_a}{\partial y} + \frac{\rho_a{}^2 k_a}{\rho_{0a}}\right] + \rho_a \lambda_w m_w \frac{\partial H_w}{\partial t} \tag{62}$$

where k_a is air permeability; θ_a is volumetric air content; ρ_a is air density; γ_{oa} is initial air unit weight; ρ_{oa} is initial air density; $\bar{R}$ is ideal air constant and $\bar{R}$=287J/(kg K) for dry air; T is temperature; H_w is total head and P_a is pore air pressure. These two governing equations can be used to determine total pressure head profile for the water and the pore air pressure, and then suction is attained.

The relationships relating volumetric water content and hydraulic conductivity to suction must be known to solve the Eq. (61). And similarly to calculate the pore air pressure the pore air permeability function is also wanted. To find the effect of different fractal dimensions and air-entry values of different soil types, series of parameter studies have conducted. Different types of Soil-water characteristic curves and hydraulic conductivity curves were shown in Figs. 34 and 35. In the legends of these pictures, Soil 10, 2.1,-5, for example, means a type of soil with air-entry value, ψ_e=10kPa, the fractal dimension, D=2.1 and saturated hydraulic conductivity k_s=1×10^{-5}m/s. It is assumed that θ_r=0. The pore air permeability function curve used in these studies was shown in Fig.36, and it remained a constant (GEO-SLOPE International Ltd.2007).

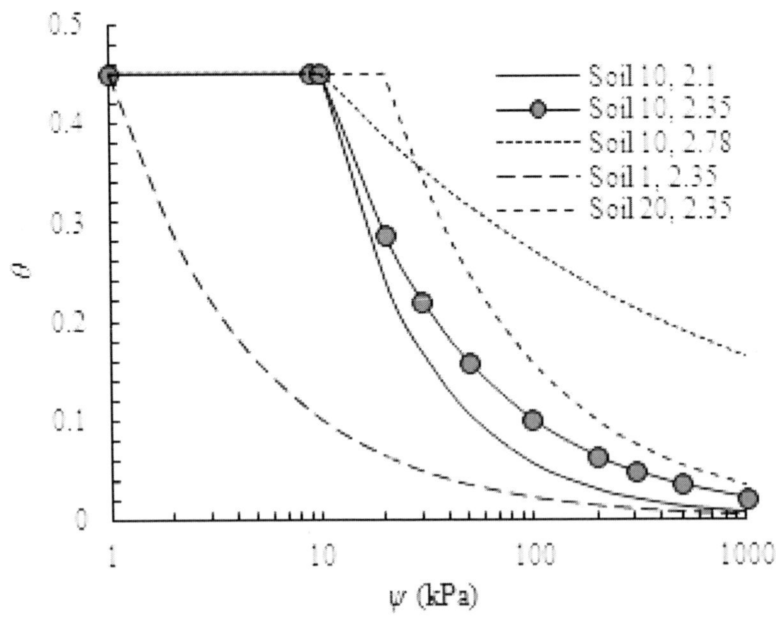

Figure 34. Soil-water characteristic curves for soils

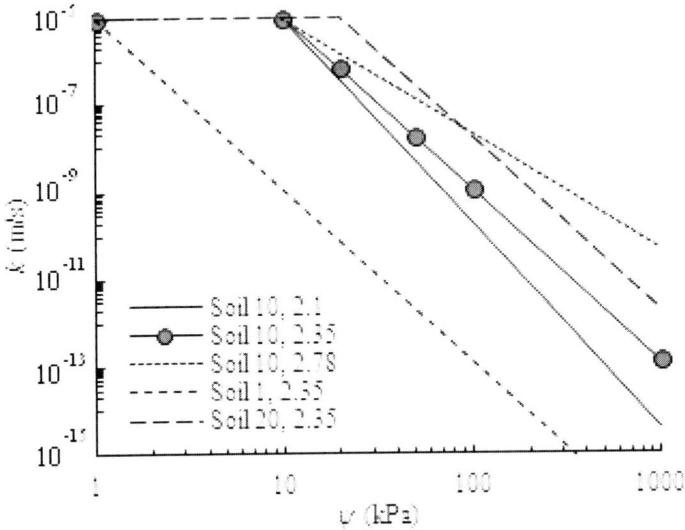

Figure 35. Hydraulic conductivity curves f

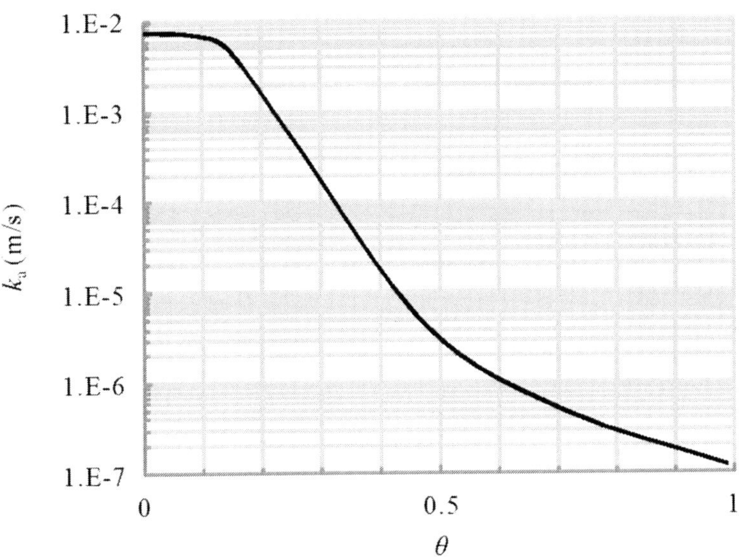

Figure 36. Pore air permeability function used in study

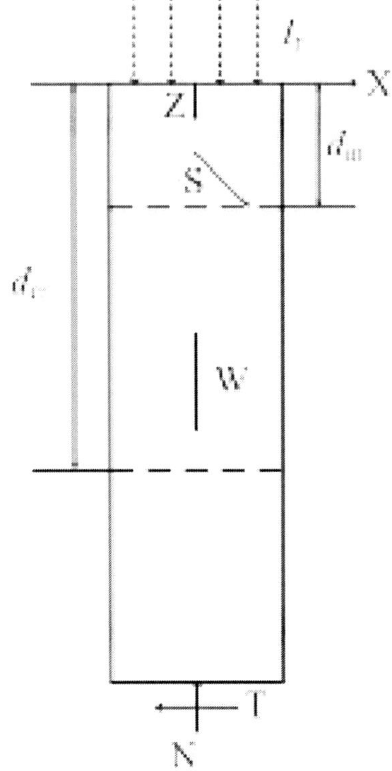

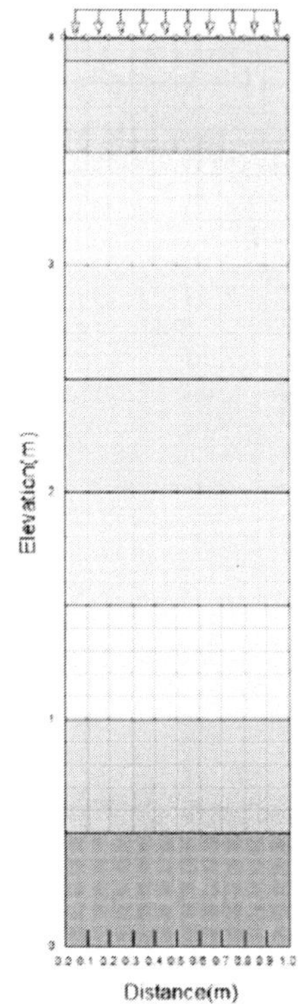

Figure 37. Sketch map of the one-dimensional infiltration model

A 4m-deep soil column was modeled for the one-dimensional analyses, as shown in Fig. 37, and infiltration occurs from the top of the column, while the bottom was set to be a pervious boundary. A linear hydrostatic suction distribution presented for the initial conditions. Analyses for the rainfall infiltration into different types of soils were performed using the unsaturated characteristic curves shown in Figs. 34-36. In applying a top boundary condition to simulate the infiltration of rainfall, it is important to realize that an influx boundary depends on the relationship between the saturated hydraulic conductivity and the rainfall intensity. In the following analyses, if a rainfall intensity greater than or equal to k_s with the non-infiltrating rainfall running off the slope, the top boundary condition is set to a total head equal to 4m; while if field results indicate that the rainfall intensity is less than k_s, then a flux type boundary condition is more appropriate.

Since initial failures often have small depth-to-length ratios and form failure planes parallel to the slope surface, the use of infinite slope analysis in modeling the infiltration process by vertical one-dimensional analysis makes it justified in describing the physical process of failure initiation. However, the methods used in traditional infinite slope analysis must be modified to take into account the variation of the suction profile that results from the infiltration process. There are two distinct failure mechanisms can be initiated by the infiltration process. The failure takes place due to positive pore pressure and it takes place while the suction still exists.

The shear strength of unsaturated soil can be represented by fractal model, and written as follows (Xu, 2004):

$$\tau = c' + (\sigma - u_a)\tan\phi' + \psi_e^{3-D}\psi^{D-2}\tan\phi' \tag{63}$$

where τ is the shear strength; c' is the effective cohesion; $(\sigma\text{-}u_a)$ is the total normal stress; φ' is effective friction angle; ψ is the suction; ψ_e is the air-entry value. The stability envelope can be represented as (Collins and Znidarcic, 2004):

$$d_{cr} = \frac{c' + \gamma_w h_c \tan\phi' - \gamma_w h_p \tan\phi'}{\gamma\cos^2\beta(\tan\beta - \tan\phi')} \tag{64}$$

where d_{cr} is the critical depth; β is slope angle; h_p is the pore passive pressure and h_c is the pore negative pressure.

We can get suction with analyses of both pore water pressure and pore air pressure in the soil slope under rainfall infiltration. Combining Eq. (63) with Eq. (64), the new stability envelope can be derived and written as follows:

$$d_{cr} = \frac{c' + \psi_e^{3-D}\psi^{D-2}\tan\phi' - \gamma_w h_p \tan\phi'}{\gamma\cos^2\beta(\tan\beta - \tan\phi')} \tag{65}$$

The critical depth for infinite slope failure is now a function of the suction, ψ, pressure head h_p, the given material and slope characteristics, c', φ', γ, ψ_e, ξ, γ_w, and β. In this way, the slope stability issues related to the decrease in shear strength from a loss of soil suction and the development of seepage forces from positive pressure head generation can be clearly understood.

Because the infiltration results and the slope stability results were both presented in terms of the pressure head and the suction profile, the two analyses can be coupled to yield a comprehensive method for determining the location and time of failure for a slope if the soil, slope, and rainfall parameters are given. The methodology of the coupled analysis involves plotting a stability envelope as defined by Eq. (65) for specified soil and slope parameters, over a given infiltration profile generated from the particular rainfall and unsaturated characteristic curves. Coulomb failure and the initiation of slope mobilization are defined at the points where the infiltration trace intersects the stability envelope. Intersections of the infiltration trace and the stability envelope indicate points at which the slope is unstable and can be thought of as "critical depths" of failure. It must therefore be assumed that if the slope is initially stable, the point at which the stability envelope inter sects the initial suction distribution line will not contribute any further information about the failure mechanisms from infiltration.

Here, factors controlling the unsaturated soil slope under infiltration are studied and the results which combine infiltration analyses and stability analyses are shown in Figs.43-47. Infiltration analyses were performed by software SEEP/W coupled with AIR/W in order to calculate the suction file and the stability envelope as defined by Eq.(65) for slope parameters, β=40°, γ=19.6kN/m^3, γ_w=10kN/m^3, and shear strength parameters c'=5kPa, φ'=15°. Other parameters are studied in a systematic way.

7.1. Effect of fractal dimension

Soil column was saturated due to rainfall infiltration from the initial hydrostatic suction profile. It was shown that infiltration took place in unsaturated soil much faster for the soil with higher fractal dimension. Three fractal dimensions (D=2.1, 2.35 and 2.78) were calculated respectively and the results were shown in Fig.38. It was seen from Fig. 38 that the higher initial hydraulic conductivity existed on the top of soil column which had a higher fractal dimension. On the other hand, the stability envelope for the soil with higher fractal dimension was steeper. There was an obvious inflection point in the stability envelope of the soil with the low fractal dimension. The results also indicated that it was the relative shape of the unsaturated characteristic curves (Fig.34 and Fig. 35) that has a controlling effect on the suction distribution.

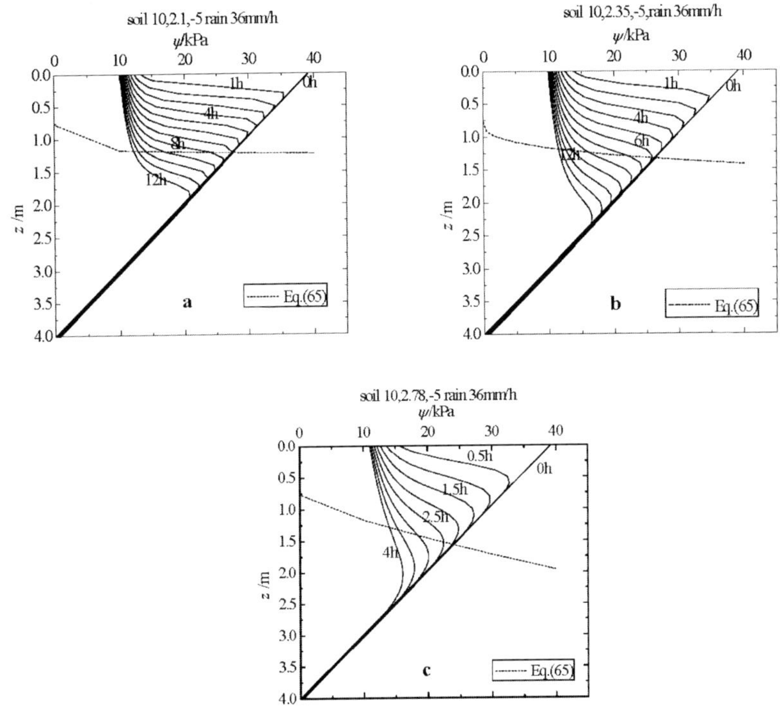

Figure 38. Infiltration results for soils with different fractal dimension and the new stability envelope

7.2. Effect of air-entry value

Three air-entry values (ψ_e=1kPa, 10kPa and 20kPa) are calculated respectively and the results as shown in Figs.39. Compared with the results of three different air-entry values, both the infiltration trace and the stability envelope have changed with the different air-entry values. When the air-entry value is relatively small (ψ_e=1kPa), as shown in Fig.39a, passive pore water pressure was generated on the upper soil column at the time t=32h. Due to the low initial permeability on the top of soil column and a state of lower saturation, fewer pores initially filled with water and there are fewer channels available for fluid transport and consequently the flow of water is hindered. With the commencement of infiltration at the top boundary, water is forced into a soil which is not capable of transporting it efficiently, which results in the development of positive pressure heads as the infiltration front progresses downward. While the air-entry value is relatively larger, as shown in Fig.39c, infiltration moves downward much faster and suction has reached to the air-entry value. Passive pore water pressure occurs at the bottom of the soil column which means that there is little unsaturated zone in the middle of soil column during the rainfall infiltration. Besides, the stability envelope is much smooth when the soil has a relatively low air-entry value, and consequently the critical depth is smaller.

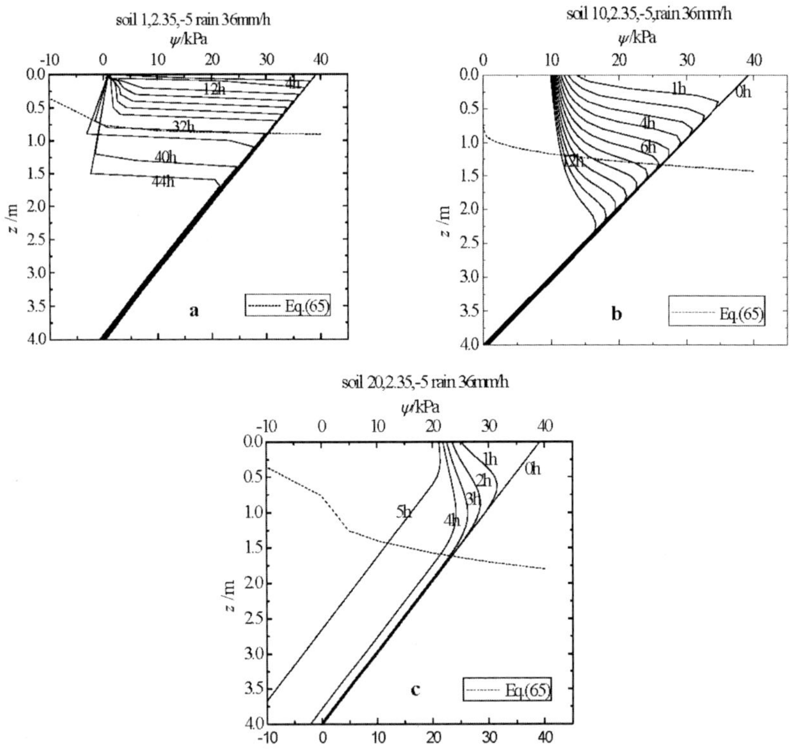

Figure 39. Infiltration results for soils with different air-entry value and the new stability envelope

7.3. Effect of saturated hydraulic conductivity

The dominate factors which control the stability of unsaturated soil slope under rainfall infiltration are the soil saturated hydraulic conductivity and rainfall intensity. While the soil saturated hydraulic conductivity determines the water transportation ability of soil and the infiltration quantity most depends on the rainfall intensity. In this parameter study series, three different saturated hydraulic conductivity values (k_s=1×10-

5m/s, 1×10-6m/s, 1×10-7m/s) are taken into account and the results are shown in Fig.40.

When the rainfall intensity is much larger than the saturated hydraulic conductivity, soil in the top region of the column has been saturated soon after the infiltration begins and its suction almost reaches to zero at the time t=1.5h. It can be seen that the suction at the base is increasing over time as a positive number, which means that the soil is actually de-saturating at its base while wetting up above. This is caused by the air which becomes trapped when the top has saturated and starts to resist water infiltration. As the situation that the rainfall intensity is much smaller than the saturated hydraulic conductivity as shown in Fig.40a, the suction of the top region decreases with the infiltration developing, however, the top region has remained unsaturated and the infiltration depth is also small due to the small rainfall quantity into the soil.

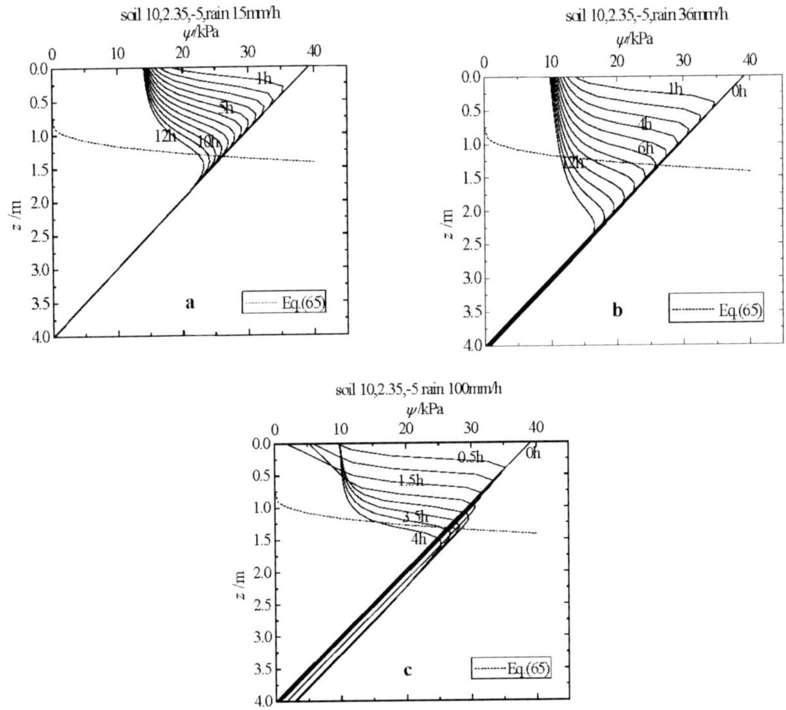

Figure 40. Infiltration results for soils due to different rainfall intensities

7.4. Effect of rainfall intensity

The rainfall intensity for a certain saturated hydraulic conductivity has taken into considered to simulate the infiltration process. The range is changed from 0.3mm/h to 100mm/h and the relationship between the time of failure and rainfall intensity for the three soil types of saturated hydraulic conductivity has plotted in Fig.41. These curves decrease as exponent forms similarly while the magnitude of different curves has changed dramatically. The time at which the soil slope failure occurs can be about ten hours for the soil saturated hydraulic conductivity k_s equals to 1×10^{-5}m/s, and for the situation that k_s equals to 1×10^{-6}m/s, the time can range from 10 to 100 hours. While much more time is needed for the lower saturated hydraulic conductivity soil type.

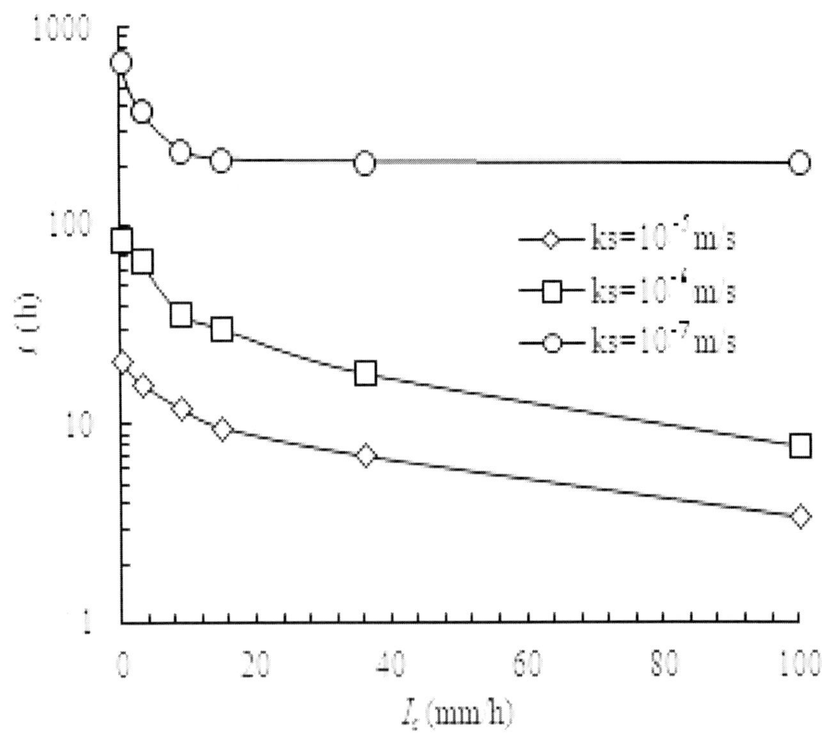

Figure 41. The time of failure vs. rainfall intensity

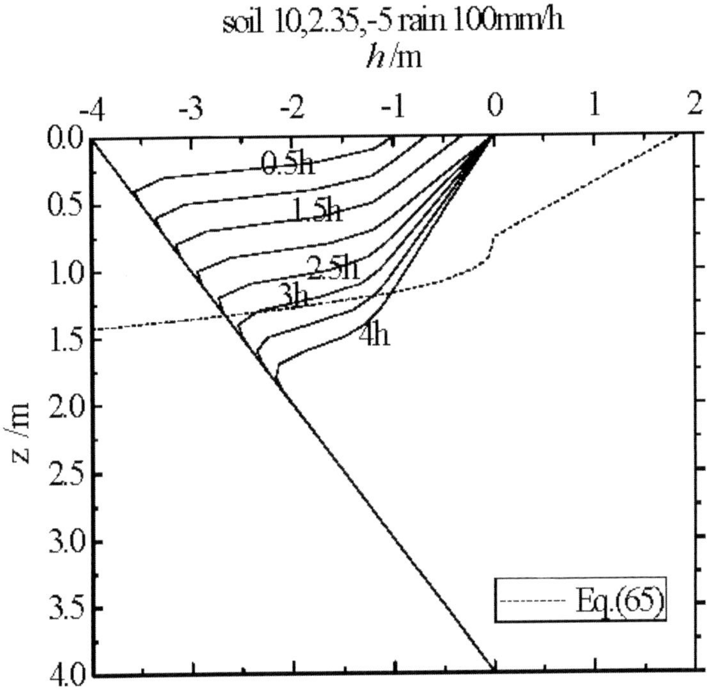

Figure 42. Infiltration results without air phrase consideration

7.5. Infiltration analyses coupled with the air flow analyses

In the infiltration analyses, air flow movement in the soil pore is also conducted. The effect of the pore air flow movement is especially obvious in the situation when the rainfall intensity is much larger than the soil saturated hydraulic conductivity. Compared with the results of infiltration analyses coupled with air flow analyses, the results without air phrase consideration are shown in Fig.42. In this condition, the negative pore water pressure gradually reaches zero at the top of soil column. While the pore water pressure remains unchanged at the bottom of the column. If we take the pore air flow movement into consideration, the results are different as shown in Fig.40c. Besides, the suction could be calculated more precisely when we attain both the pore water pressure and the pore air pressure by simulating infiltration analyses coupled with air flow analyses.

8. SLOPE STABILITY ANALYSES DUE TO RAINFALL INFILTRATION

It was widely recognized that the rainfall infiltration took a great role in causing landslides, while the relative importance of soil properties, rainfall intensity, initial water table depth and slope geometry in inducing instability of a homogenous unsaturated soil slope under different rainfall was investigated through a series of parametric studies (Rahardjo, et al, 2007). Soil properties and rainfall intensity were found to be the primary factors controlling the instability of slopes due to rainfall, while the initial water table depth and slope geometry only played a secondary role.

The factors affecting the stability of a slope were considered to be the soil properties, rainfall intensity, initial depth of the groundwater table, and the slope geometry (i.e., slope angle and slope height). To assess the effects and relative contribution of controlling factors, a series of parametric studies were performed on a typical geometry of a homogeneous soil slope shown in Fig. 43. It had the boundary conditions as follows: ab, bc, cd=q=I_r(rainfall intensity); ah, de, fg=q=0m^3/s (i.e. no flow boundary); and ef, gh=h_t(total head at the side). Four slope heights, H_s,(3m, 4m, 5m and 6m), three slope angles, α,(26.6, 33.7 and 45.0), eight initial depths of groundwater table (GWT), H_w,(0.5m, 1m, 2m, 3m, 5m, 7.5m, 10m and 15m), three fractal dimensions D (2.1, 2.35 and 2.78), four air-entry values ψ_e (1kPa, 10kPa, 20kPa and 50kPa), five values of saturated hydraulic conductivity, k_s (10^{-4}m/s, 10^{-5}m/s, 10^{-6}m/s, 10^{-7}m/s and 10^{-8}m/s) and five rainfall intensities I_r(3mm/h, 9mm/h, 15mm/h, 36mm/h and 100mm/h each for 24h duration) were used in six cases of parametric studies. The shear strength parameters of the soils used in the parametric study were c'=10kPa, effective angle of internal friction, φ'=26°, and unit weight of soil, γ=20kN/m^3.

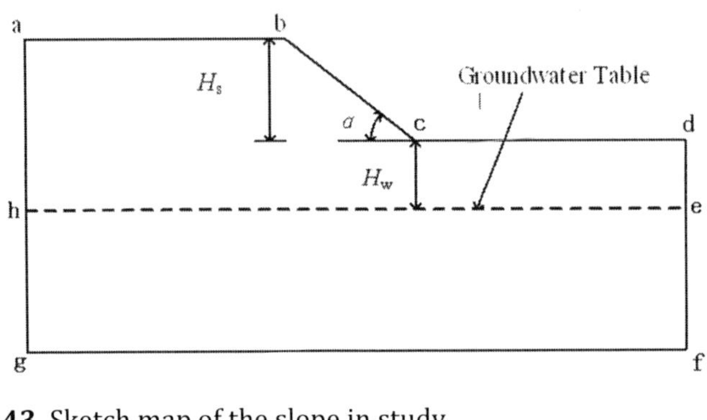

Figure 43. Sketch map of the slope in study

The fractal model of the soil-water characteristic curves for unsaturated soil was written as Eq. (29). The function of relative hydraulic conductivity was used as Eqs. (44) and (45). The derived SWCC and RHC for the all six soils were shown in Figs. 34 and 35. In the legends of these pictures, the symbol S in the soil names represent "soil", the first number means the air-entry value, and the second one means the fractal dimension.

In the seepage analysis, the governing partial differential equation (GEO-SLOPE International Ltd.2007) for a two-dimensional transient water flow used in the finite element seepage model was written as follows:

$$m_w \gamma_w \frac{\partial (H-y)}{\partial t} = \frac{\partial}{\partial y}\left(k_w \frac{\partial H_w}{\partial t} \right) + m_w \frac{\partial P_a}{\partial t} + Q \quad (66)$$

where H is the total head; k_x are k_y ars hydraulic conductivity in x, y direction respectively ; Q is the applied boundary flux; m_w is the slope of the SWCC and γ_w is the unit weight of water. Eq. (66) was solved by SEEP/W software (Geo-Slope 2007). The boundary conditions used in the transient seepage analysis were shown in Fig. 43. And the initial condition for the analyses was a hydrostatic condition with a limiting pore water pressure of -75kPa. Then the SEEP/W software generated the negative pore water pressure as the initial condition. The pore-water pressures obtained from the seepage analysis were then used in the slope stability analyses to calculate the factor of the slope safety, F_s.

The equation for unsaturated shear strength was used as Eq. (63). In the study, we chose c'=10kPa, φ'=26°. The Morgenstern-Price Method was adopted in the slope stability analysis. This method considered both shear and normal inter-slice forces, satisfied both moment and force equilibrium, and allowed for a variety of user-selected inter-slice force function. The slope stability analysis using Morgenstern-Price Method was performed using SLOPE/W software (Geo-Slope 2007). The pore-water pressures, u_w, obtained from the transient seepage analyses using SEEP/W were added to SLOPE/W to be incorporated in the slope stability analyses.

The results were presented with attention on the effects of the factors which impacting on the soil slope stability under rainfall for 24h with a combination of various controlling factors.

8.1. Effect of soil properties

The variation in factor of safety with time for a homogeneous soil slope of constant slope height H_s=6m, initial groundwater table depth H_w=2m, subjected to rainfall intensities of 3, 9 and 15mm/h of respective soil for 24h with a combination of various soil types (different fractal dimensions and air-entry values) were analysized as follows.

1. Effect of fractal dimension

In Figs.44, the fractal dimension, D was 2, 2.35 and 2.78, respectively. The air-entry value, ψ_e remained 10kPa. It is shown that the reduction of slope stability is larger in the condition of larger fractal dimension under the condition of the same rainfall intensities and the saturated hydraulic conductivity. This means that the soil slope is much safe if the fractal dimension is low while the saturated hydraulic conductivity remains constant. The factor of safety decreases in a short time under rainfall infiltration when the fractal dimension is relatively large, as shown in Fig.44c, while the decrease occurs at a longer time for the low fractal dimension soil type, as shown in Fig.44a.

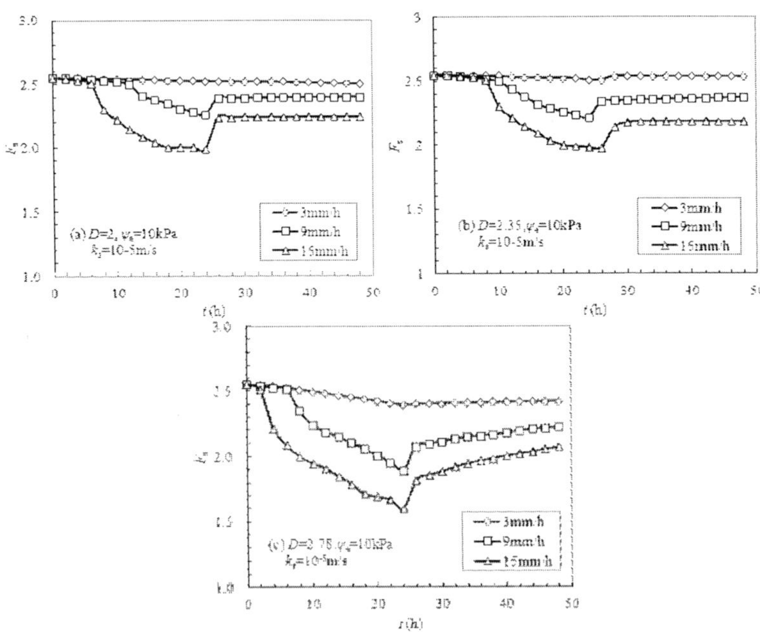

Figure 44. Effect of fractal dimension on slope stability

If the air-entry value and the saturated hydraulic conductivity remain unchanged, the slope of the hydraulic conductivity function is small when the fractal dimension is large. Then the initial hydraulic conductivity value at the beginning of the rainfall infiltration is much big, which results in the speed of rainfall infiltration is great and the saturated degree of soil inside of slope changes quickly leading unsaturated soil to become saturated. The passive pore water pressure generated by the increase of degree of soil saturation makes a negative impact on the instability of soil slope under rainfall.

2. EFFECT OF AIR-ENTRY VALUE

The effects of air-entry value on the stability of soil slope due to rainfall infiltration are shown in Figs. 45. The air-entry value, ψ_e is 1kPa, 10kPa, 20kPa and 50kPa respectively and the fractal dimension, D remains 2.35. The style of the slope safety curve changes a lot with different air-entry values in Figs. 45. In Fig. 45a, the air-entry value, ψ_e is 1kPa and the decrease of factor of safety occurs during 16~20h after rainfall began. As the air-entry value, ψ_e is 10kPa, shown in Fig. 45b, the decrease of slope safety factor occurs at 8h after rainfall began. At the beginning of rainfall infiltration, the decrease happens in Figs. 45c-d, in which the air-entry value is 20kPa and 50kPa respectively. This means that a delay time appears which is depended with the air-entry value. And it is obvious that small air-entry value decides a relative long delay time. Besides, the factor of slope safety remains minimum during another 24 hours after rainfall stops in Fig. 45a, which has a low air-entry value (ψ_e=1kPa). While ψ_e=50kPa, the recovery of safety factor curve in Fig. 45c occurs and dramatically. This is also due to the style of hydraulic conductivity function, as shown in Fig. 35. The initial hydraulic conductivity value becomes large when the air-entry value increases with fractal dimension, saturated hydraulic conductivity and saturation degree remain the same.

3. EFFECT OF SATURATED HYDRAULIC CONDUCTIVITY

The effects of saturated hydraulic conductivity on stability of a homogenous soil slope are reflected through the relationship between F_s and k_s, shown in Fig. 46. The saturated hydraulic conductivity, k_s, was varied with five different values of 10^{-8}m/s, 10^{-7}m/s, 10^{-6}m/s, 10^{-5}m/s and 10^{-4}m/s for a homogeneous soil slope of constant soil type S10, 2.35 (ψ_e=10kPa, D=2.35), H_w=2m, H_s=6m, α=33.7°, subjected to rainfall for 24h with six rainfall intensities of 3, 9, 15, 36, 100 and 360mm/h. All the plots in Fig. 47 shows that soil with k_s values of 10^{-8}m/s and 10^{-7}m/s, respectively, are less affected by rainfall. Contrarily, soil with k_s values of 10^{-5}m/s and 10^{-4}m/s are greatly affected by rainfall. It is suggested that soil slopes with a low saturated hydraulic conductivity are relatively safe under short-duration rainfall infiltration and for the soil slopes with a high saturated hydraulic conductivity the stability is affected by the short-duration rainfall greatly. This means that slopes with low saturated hydraulic conductivity need a long-duration rainfall to intrigue the instability.

The F_s(min)-k_s critical curve was presented in the broken line in Fig. 47 as follows:

$$F_s(\min) = \frac{a}{1 + be^{-ck_s}} \tag{67}$$

where F_s(min) is minimum factor of safety; k_s is saturated hydraulic conductivity; a, b and c are the fitting parameters and e is natural number (i.e., 2.71…). Here, a=1.293, b=-0.492, c=1.2×10^5, and the corresponding coefficient of correlation, r^2 is 0.9998.

8.2. Effect of rainfall intensity

The relationship between minimum factor of safety, F_s(min), versus logarithmic of rainfall intensity, I_r are plotted in Fig. 47. The semi log plot shows that generally the F_s(min) and I_r relationships follow a sigmoid shape in Fig. 47. There are two inflect points for the sigmoid shape line. The F_s(min) is almost constant at very low rainfall intensities for all soil types. And it starts to decrease rapidly when the first inflection point is reached. The trend in F_s(min) versus I_r relationship observed in Fig. 47 can be described by a sigmoid equation (MMF Line) as the form of

$$F_s(\min) = \frac{ab + cI_r^{\,d}}{b + I_r^{\,d}} \tag{68}$$

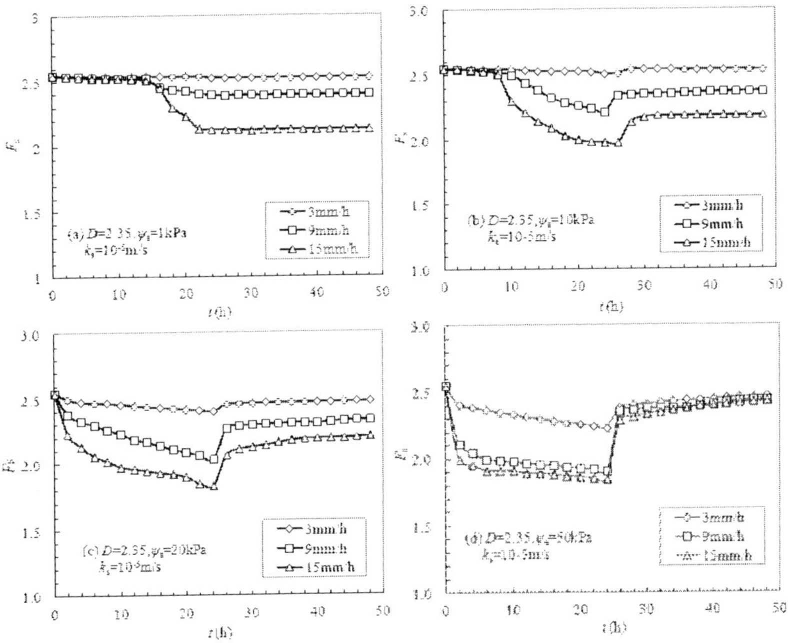

Figure 45. Effect of air-entry value on the slope stability

where F_s(min) is minimum factor of safety; I_r is rainfall intensity; and *a, b, c, d* are fitting parameters. The values for the fitting parameters and the corresponding coefficient of correlation, r^2, for the sigmoid line in Fig.47.

8.3. Effect of initial water table location

The effects of initial groundwater table location on stability of a homogenous soil slope are reflected through the relationship between F_s(ini) and F_s(min) with H_w, shown in Fig. 48. The initial depth of water table, H_w, was varied with five different values of 0.5, 1, 2, 2.5, 5, 7.5, 10 and15m for a homogeneous soil slope of constant soil type S10, 2.35, -5 (ψ_e=10kPa, D=2.35, k_s=10^{-5}m/s), H_s=6m, α=33.7°, subjected to rainfall for 24h with five rainfall intensities of 3, 9, 15, 36, and 100mm/h. The relationship between F_s(ini), and H_w shown in Fig. 48 appears to be linear

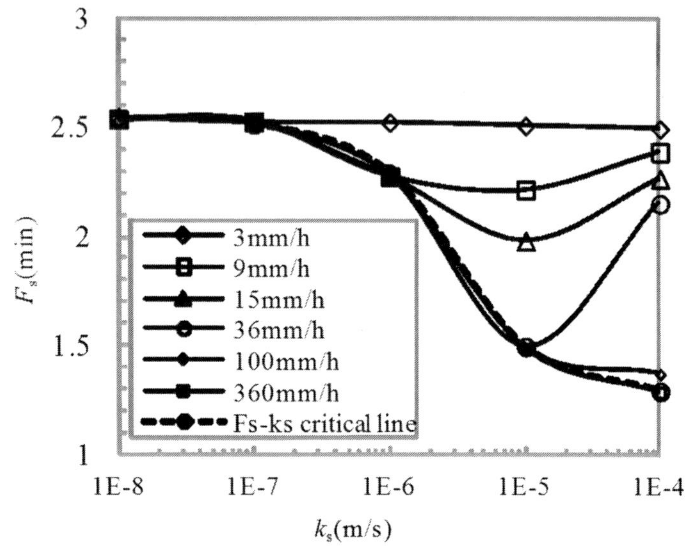

Figure 46. Relationship between K_s and minimum factor of safety

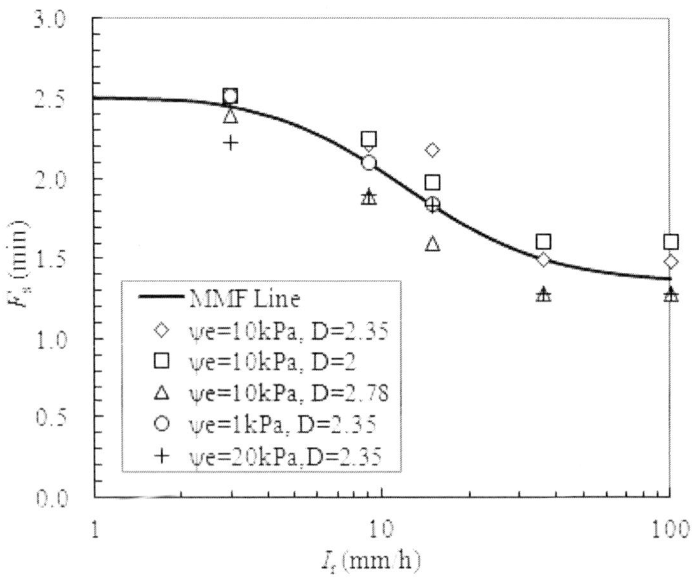

Figure 47. Effect of rainfall intensity on variation of minimum factor of safety

up to a depth of 7.5m beyond which F_s(ini) remains constant. This is because the initial pore-water pressure profiles generated for the slope at H_w=7.5m as same as those generated when H_w=7.5m. This is due to the limiting pore-water pressure of −75kPa adopted in the analyses. In Fig. 54 it also shows that the rainfall intensity is the dominated factor impacting on the reduction in factor of safety, *Fs*. And the initial depth of water table, H_w, mainly determines the value of initial factor of safety, F_s(ini). The F_s(ini) is smaller for slopes with a shallower H_w which means that the soil slope stability is much lower. Therefore, slopes with a shallow H_w are more likely to fail due to a rainfall compared with slopes which has a deep H_w.

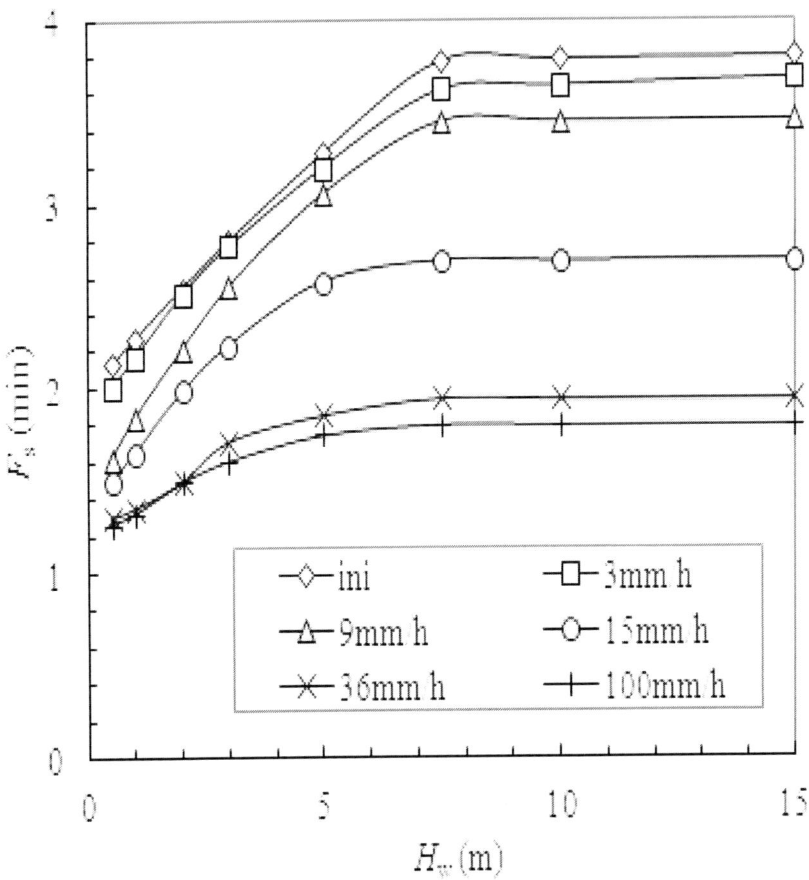

Figure 48. Effect of initial groundwater table depth on factor of safety

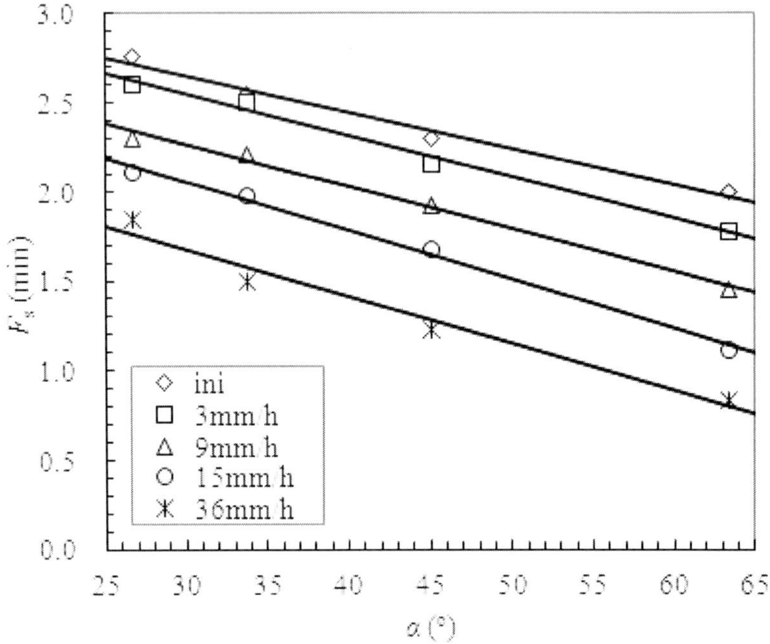

Figure 49. Effect of slope angle on factor of safety

8.4. Effect of slope angle

The effect of slope geometry on the stability of a homogenous soil slope is evaluated in terms of slope angle (α) and slope height (H_s). The effects of slope angel on the stability of a homogenous soil slope are reflected through the relationship between F_s(ini) and F_s(min) with α, shown in Fig. 49. The slope angel, α was varied with three different values of 26°, 33.7°, 45°and 63° for a homogeneous soil slope of constant soil type S10, 2.35, -5 (ψ_e=10kPa, D=2.35, k_s=10^{-5}m/s), H_s=6m, H_w=2m, subjected to rainfall for 24h with five rainfall intensities of 3, 9, 15 and 36mm/h. The relationship between F_s(min) and α shown in Fig. 49 appears to be negative linear. In general, the higher the slope angle, the lower the initial factor of safety and the minimum factor of safety. Because a steep slope will yield a lower factor of safety compared with a flat slope.

8.5. Effect of slope height

The effects of slope angel on the stability of a homogenous soil slope are reflected through the relationship between F_s(ini) and F_s(min) with H_s, shown in Fig. 50. The slope height, H_s, was varied with four different values of 3m, 4m, 5m and 6m for a homogeneous soil slope of constant soil type S10, 2.35, -5(ψ_e=10kPa, D=2.35, k_s=10^{-5}m/s), H_w=2m, α=33.7°, subjected to rainfall for 24h with four rainfall intensities of 3mm/h, 9mm/h, 15mm/h and 36mm/h. Fig. 50 shows that initial factor of safety decreases exponentially as the slope height increases. It also suggests that high slopes are generally easier to fail under rainfall due to the low initial factor of safety. The reduction in factor of safety for a high slope is smaller and occurs at a slower rate compared with a low slope.

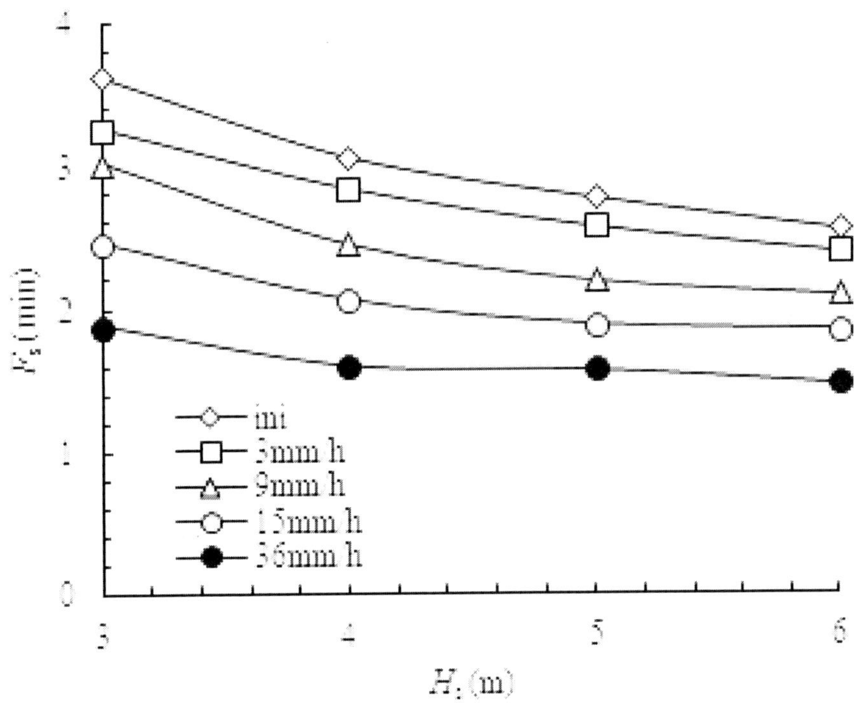

Figure 50. Effect of slope height on factor of safety

NAN. ACKNOWLEDGEMENTS

The National Natural Science Foundation of China (Grant No. 41272318) and State Key Laboratory of Ocean Engineering are sincerely acknowledged for their financial support.

REFERENCES

1. Abramowitz M, Stegun I. Handbook of Mathematical Functions. Dover Publications, New York, NY. 1970
2. Avnir D, Jaroniec M, An isotherm equation for adsorption on fractal surfaces of heterogeneous porous materials. Langmuir, 1989, 5, 1431–1433.
3. Brooks RH, Corey AT. Properties of porous media affecting fluid flow. ASCE J Irrig Drain Div 1966, 92:61–68.
4. Brutsaert W. Some methods of calculating unsaturated permeability. Trans Am Soc Agr Engrs 1967;10:400-404.
5. Brutsaert W. The permeability of a porous medium determined from certain probability laws for pore size distribution. Water Resour Res 1968; 4: 425-34.
6. Brutsaert W. A concise parameterization of the hydraulic conductivity of unsaturated soils. Advances in Water Resources, 2000, 23, 811-815.
7. Burdine NT. Relative permeability calculations from pore-size distribution data. Trans Am Inst Min Engrs 1953;198:71-7
8. Burdine NT. Relative permeability calculation from pore size distribution data. Transactions of the American Institute of Mining Engineers, 1953, 198, 71–78.
9. Carman PC. Flow of gases through porous media. New York: Academic Press, 1956.
10. Childs EC, Collis-George N. The permeability of porous materials. Proc Roy Soc A 1950; 201: 392-405.
11. Collins,B.D., Znidarcic. Stability analyses of rainfall induced landslides [J].Journal of geotechnical and geoenvironmental engineering, 2004,4,362-372.

12. Corey AT. Pore-size distribution. In: van Genuchten MTh, Leij FJ, Lund LJ, editors. Proceedings of the International Workshop, Indirect Methods for Estimating the Hydraulic Properties of Unsaturated Soils, University of California, Riverside, 1992. p. 37–44.

13. Crawford JW. The relationship between structure and the hydraulic conductivity of soil. Eur J Soil Sci 1994, 45:493–501.

14. Delage P, Audiguier M, Cui YJ, Howat MD. Microstructure of a compacted silt. Can Geotech J 1996;33:150-158.

15. Dirksen C. Unsaturated hydraulic conductivity. In: Smith KA, Mullins CE, editors. Soil analysis physical methods. New York: Dekker; 1991. p. 209–69.

16. Fatt I, Dykstra H. Relative permeability studies. Transactions of the American Institute of Mining Engineers, 1951, 192: 249-255.

17. Fuentes C, Vauclin M, Parlange J-I. A note on the soil–water conductivity of a fractal soil. Transp Porous Media 1996, 23:31–36.

18. Fuentes C, Haverkamp R, Parlange J-Y. Parameter constraints on closed-form soil water relationships. Journal of Hydrology, 1992, 134, 117-142.

19. Gates JI, Tempelaar-Lietz W. Relative permeabilities of California cores by the capillary pressure method. Drilling and Production Practice, Am Petrol Inst 1950:285-298.

20. GEO-SLOPE International Ltd. Seepage Modeling with SEEP/W2007 [M].Third Edition, Canada, 2008,3.

21. GEO-SLOPE International Ltd. Air Flow Modeling with AIR/W2007 [M].Third Edition, Canada, 2008,3.

22. GEO-SLOPE International Ltd. Stability Modeling with SLOPE/W2007 [M].Third Edition, Canada, 2008,3.

23. Gimenez D, Perfect E, Rawls WJ, Pacheoaky Ya. Fractal models for predicting soil hydraulic properties: a review. Eng Geo 1997;48:161-183.

24. Hoffmann-Riem H, van Genuchten M.Th, Flühler H. A general model of the hydraulic conductivity of unsaturated soils, in: van Genuchten M.Th, Leij FJ, Wu L. (Eds.), Proceedings of International Workshop, Characterization and Measurements of Hydraulic Properties of Unsaturated Porous Media. Riverside, CA. 22–24th Oct. 1997. University of California, Riverside, 1999, pp. 31–42.

25. Hunt AG, Gee GW. Application of critical path analysis to fractal porous media:comparison with examples from the Hanford site. Adv in Water Resources 2002;25:129–46.

26. Irmay S. On the hydraulic conductivity of unsaturated soils. Trans Am Geophys Un 1954;35:463-467.

27. Jarvis NJ, Messing I. Near-saturated hydraulic conductivity in soils of contrasting texture as measured by tension infiltrometers. Soil Sci Soc Am J 1995;59:27–34.

28. Kahr G., Kraehenbuehl F, Stoeckli HF, Muller-Vonmoos M. Study of the water –bentonite system by vapour adsorption, immersion colorimetry and X-ray technique: Part II. Heats of immersion, swelling pressures and thermodynamic properties. Clay Miner. 1990, 25, 499–506.

29. Khaleel R, Relyea JF. Evaluation of van Genuchten–Mualem relationships to estimate unsaturated hydraulic conductivity at low water contents. Water Resour Res 1995; 31: 2659–1668.

30. Kozeny J. Ueber kapillare leitung des wassers im boden, sitzungsberichte, akad der wissensch. Wien, Math-Naturw Klass Abt IIa 1927;136:271-306.

31. Mandelbrot BB, The Fractal Geometry of Nature, W.H. Freeman, New York, 1982.

32. Matsuoka H, Soil Mechanics. Morikita Shuppan Co., Ltd., 1999, p. 61.

33. Mualem Y. A new model for predicting the hydraulic conductivity of unsaturated porous media. Water Resources Res 1976, 12:513–22.

34. Polubarinova-Kochina PYa. Theory of ground water movement. Princeton, NJ: Princeton University Press, 1952. p. 613 (translated from the Russian by DeWiest, JMR 1962).

35. Purcell WR. Capillary pressures-their measurement using mercury and the calculation of permeability therefrom. Trans Am Inst Min Met Engrs Petrol Devel Technol 1949;186:39-46.

36. Rahardjo H, Ong TH, Rezaur RB. Factors controlling instability of homogeneous soil slopes under rainfall [J]. Journal of geotechnical and geoenvironmental engineering, 2007,12,1532-1543.

37. Ravichandran N, Krishnapillai S. A Statistical Model for the Relative Hydraulic Conductivity of Water Phase in Unsaturated Soils. International Journal of Geosciences, 2011, 2, 484-492.

38. Rieu M, Sposito G. Fractal fragmentation, soil porosity and soil water properties. Soil Sci Soc Am J 1991; 55: 1483–1489.

39. Schaap MG, Leij FJ, van Genuchten MT. Rosetta: A computer program for estimating soil hydraulic parameters with hierarchical pedotransfer functions, J. Hydrol., 2001, 251, 163-176.

40. Simunek T, Wendroth O, van Genuchten MT, et al., Water Resources Res. 35 (1999) 2965.

41. Smettem KRJ, Kirkby C. Measuring the hydraulic properties of a stable aggregate soil. J Hydro 1990, 117: 1-13.

42. Stingaciu LR, Weihermüller L, Haber-Pohlmeier S et al. Determination of pore size distribution and hydraulic properties using nuclear magnetic resonance relaxometry: A comparative study of laboratory methods. Water Resources Research, 2009, 46: W11510.

43. Sugii T, Uno T, Hayashi T, Proceedings of the 31st Japan National Conference on Geotechnical Engineering, Kitami, Japan, 1996, pp. 2075

44. Toledo PG, Novy RA, Davis HT, Scriven LE. Hydraulic conductivity of porous media at low water content. Soil Sci Soc Am J 1990;54:673–9.

45. Touma J, Vauclin M. Experimental and numerical analysis of two-phase infiltration in a partially saturated soil. Transport in Porous Media, 1986, 1, 27-55.

46. Touma J. Comparison of the soil hydraulic conductivity predicted from its water retention expressed by the equation of van Genuchten and different capillary models. European Journal of Soil Science, 2009, 60: 671-680

47. Tyler SW, Wheatcraft SW. Fractal process in soil water retention. Water Res Res 1990, 26:1047–54.

48. Uno T, Kamiya K, Tanaka K. The distribution of sand void diameter by air intrusion method and moisture characteristic curve method. Proc Japan Soc Civil Engrg, 1998, 603(III-44): 35-44.

49. van Genuchten MTh, Leij FJ. On estimating the hydraulic properties of unsaturated soils. In: van Genuchten M.Th, Leij FJ, Lund LJ, editors. Proceedings of the International Workshop, Indirect Methods for Estimating the Hydraulic Properties of Unsaturated Soils, University of California, Riverside, 1992. p.1-14.

50. van Damme H. Scale invariance and hydric behaviour of soils and clays. CR Acad Sci Paris 1995, 320: 665–81.

51. van Genuchten MTh. A close form equation for predicting the hydraulic conductivity of unsaturated soils. Soil Sci Soc Am J 1980;44:892–7.

52. Vogel HJ, Roth K. A new approach for determination effective soil hydraulic function. Eur J Soil Sci 1998;49:547–56.

53. Watabe Y, Leroueil S, Le Bihan J-P, Influence of compacted conditions on pore-size distribution and saturated hydraulic conductivity of a glacial till, Can Geotech J, 2000, 37:1184-1192.

54. Wheeler SJ, Karbue D, Constitutive modeling, In: Alonso EE, Delage P, Proc. 1st Int Conf Unsat Soils Rotterdam: AA Balkema, 1996.

55. Wyllie MRJ, Spangler MB. Application of electrical resistivity measurements to problem of fluid flow in porous media. Bull Am Ass Petrol Geologists 1952, 36:359-403.

56. Xu YF, Sun DA. A fractal model for soil pores and its application to determination of water permeability. Physica A 2002;316(1-4): 56–64.

57. Xu YF, Dong P. Fractal approach to hydraulic properties in unsaturated porous media. Chaos, Solitons & Fractals, 2004, 19(2): 327-337

58. Xu YF, Sun DA, Yao YP. Surface fractal dimension of bentonite and its application to determination of swelling properties. Chaos, Solitons & Fractals, 2004, 19(2): 347-356.

59. Xu YF, Calculation of unsaturated hydraulic conductivity using a fractal model for the pore-size distribution, Computer and Geotechnics, 2004, 31(7):549-557.

60. Xu YF. Fractal approach to unsaturated shear strength. Journal of Geotechnical and Geoenvironmental Engineering, 2004, 130(3): 264-273

INDEX

P

S

U

W